THE **Intel Trinity**

HOW ROBERT NOYCE,

GORDON MOORE,

AND ANDY GROVE

BUILT THE WORLD'S

MOST IMPORTANT COMPANY

Michael S. Malone

HARPER
BUSINESS

An Imprint of HarperCollins*Publishers*
www.harpercollins.com

HarperCollins books may be purchased for educational, business, or sales promotional use. For information, please e-mail the Special Markets Department at SPsales@harpercollins.com.

All photographs courtesy of Intel unless otherwise noted.

FIRST EDITION

Designed by Fritz Metsch

Library of Congress Cataloging-in-Publication Data has been applied for.

ISBN 978-0-06-222676-1

14 15 16 17 18   OV/RRD   10 9 8 7 6 5 4 3 2 1

Contents

Part IV: The Most Important Company in the World (1988–1999)

Part V: The Price of Success (2000–2014)

Part VI: Aftermath

Introduction: Artifacts

It was a night for contradictions. After an unusual January day in which the temperature had soared to nearly 70 degrees, the evening had just begun to cool to its proper temperature when the first of the cars—Mercedes, BMWs, chauffeured Lincoln Town Cars—began to pull up to the anxious platoon of waiting valets.

The well-dressed men and women slowly climbed out of the cars with the careful and unsure movements of the elderly—an unlikely sight in Silicon Valley, home of the young, quick, and infinitely confident. And as these venerable figures made their way into the entry hall, took their wineglasses from the passing trays, peered at one another's name tags, and grinned and embraced in recognition, their sincere happiness seemed more like that of a Midwestern high-school reunion than the nervous and wary bonhomie taking place around them in the restaurants and bars of the Valley.

The venue for this party—the private screening of a new PBS *American Experience* documentary titled *Silicon Valley*—was itself something of a contradiction. Built on what had once been San Francisco Bay marshland inhabited only by burrowing owls, hard by the long-forgotten site (symbolic for what was to come) of a roller-coaster manufacturer, the curved glass-and-steel building—looking like the offspring of an airport terminal and a multiplex cinema—had originally been built as the corporate headquarters of high-flying graphic computer maker Silicon Graphics Inc.

But in a vivid reminder of the old Valley rule that whenever a company builds a fancy new headquarters you should short the stock, SGI had cratered—and the building, not yet filled with desks and bodies, was abandoned. And there it sat, a white elephant in the middle of a real-estate slump, until the least likely of tenants appeared: the Computer History Museum.

This odd tenant had not been especially welcome upon its arrival in the Valley in 1996. For the previous twenty years, it had been the Computer Museum of Boston—a celebration of just the kind of mainframe-computing Big Iron that Silicon Valley had been founded to defeat. But the vanquished computer industry, like generations of other, human, outsiders, had come to Silicon Valley to make a fresh start—and the Valley, as always, had allowed it to earn its place. The big old computers were still there on exhibit, but largely pushed to the back, supplanted by the genius of the local companies—Intel, Apple, Hewlett-Packard, Cisco, Google—that now ruled the digital world.

More important, though, than even the artifacts on display (because in high tech even the hottest new inventions are quickly forgotten), the Computer History Museum had found a new role in the Valley for speeches, reunions, and honors celebrating the pioneers and inventors of the digital age. And now that those founding figures, like the men and women gathered on this night, had grown stooped and gray— some of them gone forever—the museum had found its true purpose in preserving the collective memories and acquired wisdom of the industry's pioneers . . . preserving them until future generations needed them once again.

Once, that date had seemed far off. Silicon Valley had always been about the future, not the past, about what could be rather than what had been, no matter how glorious. Nobody ever got rich and famous in high tech by looking over his shoulder; rather, you built better telescopes to spot the Next Big Thing as it appeared on the horizon. You had only to walk the exhibit area at the Computer History Museum to see a middle-aged engineer excitedly describing some machine he'd worked on three decades before to a young engineer, looking bored and unimpressed. Out in the Valley, on any given Saturday, you could see lines at local junior college recycling lots dumping off tons of "obsolete" computers, printers, and other devices often little more than a couple years old.

Moore's Law, the biennial doubling of computer chip performance that had accelerated the pace of innovation and become the metronome of the modern world, was named after the grandfatherly-looking man who at this moment was making his way into the museum. Moore's Law guaranteed that change would be so central to modern life that there would be precious little time left for nostalgia. When you are being chased by demons, your only chance of survival is to keep racing forward as fast as you can; looking back can only scare you. Worse, as Moore's Law had been warning for a half century now, it wasn't even enough just to go fast. Rather, you had to go faster and *faster,* progressing at a pace humanity had never before known, *just to keep up.*

That unimaginable pace, from living with numbers—subscribers, transistors, bandwidth, processing speeds, mass storage—that had no precedent in human experience, was now daily life in Silicon Valley and indeed in the entire digital world. And nowhere was this pace more torturous and unforgiving than just seven miles down the road at Intel Corporation, the home of the microprocessor and Moore's Law, the company pledged to the death to maintaining the flame of the Law. The company upon which every enterprise in Silicon Valley—and almost every human institution on the planet—now depended.

And yet despite all of these forces pushing Valley life forward, this evening event represented something brand-new in the story of this restless community. Suddenly, after more than fifty years of incessantly, unrelentingly being news, Silicon Valley had finally and unexpectedly become *history.* It would now be studied by schoolchildren and be the subject of college courses and endless doctoral dissertations. All of that was occurring already around the world, but as anecdotes and sidebars. Now it would join the main narrative of twentieth- and twenty-first-century history.

And that was the biggest contradiction of all—because that official history, certified by the most watched history program on America's public television network, had built its tale of Silicon Valley around the story of Intel—and in particular, its charismatic cofounder, Robert Noyce—the company that was the very embodiment of living in the future and indeed had chained itself to that future. And now it would be the poster child of Silicon Valley's past.

How would Intel, one of the most valuable companies in the world, builder of the microprocessor engines that powered the global Internet economy, reconcile its endless need to drive chip technology forward at the exponential pace of Moore's Law with this growing retrograde pull of its glorious past? There was no obvious answer, but somehow it seemed that the search for a solution to this conundrum of contradictions would begin at this overdue event on this unlikely evening.

Perhaps tellingly, though almost every one of the two hundred people at the premiere were in some way connected to the story of Intel Corporation, only the handful of PR executives who were managing the event for PBS were current Intel employees, people who worked every day in the world of social networks, smartphones, and embedded controllers. Given that it was still early evening, it could only be assumed that the rest of Intel's 6,000 employees in the Bay Area (of the 107,000 around the world) were still at the office, battling everything from the encroachments of competitors to the endless Sisyphean challenge of Moore's Law. Even Intel's CEO, Paul Otellini, who had over his forty years with the company worked with just about everyone in the hall—and owed his career to many of them—was missing, presumably in action.

Instead, almost everyone else in attendance belonged to an older, now officially legendary, world of minicomputers, calculators, digital watches, and most of all, personal computers. For them, the Internet Age and the dot-com bubble represented the end of their careers, not the beginning—and Facebook and Twitter were phenomena they read about in their retirement, the tools and toys of their grandchildren.

But even if they were no longer in the game, they had the consolation of knowing that, as the documentary was about to show, they had changed the world. They were the business heroes of the second half of the twentieth century, and their legacy was secure, unlike that of the current generation of Intel employees, who labored for a company that had dangerously and uncharacteristically misstepped with the arrival of mobile computing and was now struggling to regain ground lost both to giant unforgiving competitors like Samsung and to hot up-and-comers like ARM. For the modern Intel, unlike the historic Intel being celebrated on the screen, the jury was still out.

It was these aging men and women who not only had built the

most important company of the age, but arguably had shaped the form of the modern world. It had made many of them very rich—a quick mental audit of the gathering suggested a total net worth of nearly $50 billion—and many sensed that the process had now begun to make them immortal.

Two of the men now being escorted by their families into the museum were already living legends. Both looked much older than anyone remembered them, and that recognition drew muffled gasps from even their old friends and workmates.

Gordon Moore, the giver of the great law, now spent most of his time in Hawaii, making only occasional visits to his foundation in Palo Alto. He looked comparatively healthy—but those who knew him well also knew that this was only a recent positive turn. Scores of people from his past—from founding investor Art Rock to old lieutenants like Ed Gelbach and Ted Hoff—made their way over to shake Gordon's hand, and with the gracious humility, like an old Sunday school teacher's, that had always characterized him, Moore greeted each in turn.

Beside Moore stood a slightly smaller man, his face showing the effects of Parkinson's disease but radiating intensity out of proportion not just to his size and health, but to anyone else in the hall. Andy Grove, the greatest and most ferocious businessman of his generation, met each handshake—from old compatriots like Les Vadász to old competitors like Federico Faggin—with the same fiery look.

Together these two men, with the help or threat of everyone else in the hall, had built Intel Corporation into the most innovative company the world had ever seen. And then through an impossible gauntlet of challenges—including on occasion their own misguided stubbornness—they had built Intel into the most valuable manufacturing company in the world. And through Intel's products and commitment to Moore's Law, they had made possible the consumer electronics revolution that now defined the lives of three billion people, with millions more joining every day. Humanity was now richer, healthier, smarter, and more interconnected than ever before because of what they achieved. And now humanity was beginning to recognize that fact.

In this story, they were the heroes—and everyone else had been a player, and happy to have been a part of that story. And if there was pride in what they had accomplished, it was bittersweet, because everyone knew that this was not only a celebration, but perhaps a last gathering as well. Anyone who doubted that had only look at the two elder statesmen at the center of this gathering. And so as guests in turn shook hands with Moore and Grove, each understood that it might be for the very last time. Tonight would be a celebration, but it would also be a last gathering of the tribes who had built Silicon Valley. And so each fixed that handshake in mind as a final memory.

Regis McKenna, the marketing guru who had led Intel into the big world and helped to devise the most influential high-tech branding program of all time, had been muttering about the flaws in the documentary, which he had already seen. "It's not bad," he said, "but it suffers a little from being too . . . *East Coast*, if you know what I mean. There's a bit too much credit given to the government for making all of this possible. There's too many clips of rockets."

Then, seeing Grove and Moore enter, he slipped away—to return a few minutes later, grinning but with tears in his eyes. "When I got up to Andy, he bowed and said, 'Teacher.' The last person who called me that was Steve Jobs."[1]

Despite the bonhomie and nostalgia, there was also the unsettling sense at the gathering that this reunion was haunted by the one irreplaceable person who wasn't there, the man who had made all of this—Intel, Silicon Valley, the digital revolution—possible. The man who was the subject of the new documentary and whose life had defined that of everyone else in the hall—including the reporters, waiters, and museum staff.

Bob Noyce.

Everywhere one looked, there was a reminder of Noyce, the third member of the troika, or as they were called, the Trinity: Noyce the charismatic father, Grove the truculent son, and Moore, the holy spirit of high tech. It wasn't just the sight of his widow, former Apple executive Ann Bowers, or the endlessly repeated comment among the crowd that "I wish Bob was here." It wasn't even the little glass case full of artifacts—Noyce's employee badge from Intel, samples of some of

Fairchild's earliest products, the ur-notebook of Noyce's original notes on the development of the integrated circuit—that underscored his genius as a scientist, not just as an entrepreneur.

Rather, it was everywhere, a ghostly presence. When Grove and Moore slowly made their way to the display case to look at its contents, everyone else stood back, all registering the third figure who should have been there among them. Bob Noyce, the patriarch, their old partner, looked up at them from his employee badge with the burrowing focus for which everyone remembered him. He had always been older than the rest, first at Fairchild among the twenty-year-olds of his own generation, and then later, with his graying temples, at Intel among his Baby Boomer employees. But now they had grown ancient, while he, frozen in time, had remained in middle age, at the height of his fame and power.

Thanks to the amnesia of Silicon Valley and the digital world, he had almost been forgotten by each subsequent generation of techies as they elbowed their way to their own fame at a thousand dot-com companies, and then at Google, Facebook, and Twitter. Noyce: "Saint Bob," the man who founded their community and created the device upon which all of their empires rested, the scientist who had been behind two of the greatest inventions of the last century, the man once known as the Mayor of Silicon Valley, had been reduced to a small exhibit in the Intel Museum to be glanced at by thousands of schoolchildren passing by on tour. There were no statues to Bob Noyce, no streets named after him, no industry prizes, no Nobel Prize, because he had died five years too soon, not even the global fame and notoriety that accrued to the young man he had mentored and to whom he had served as a surrogate father: Steve Jobs.

But now, as Silicon Valley at last became history, as other once high-profile figures died or faded into obscurity, and as the tally of real achievements was finally being made, the spectral figure of Robert Noyce began to once again emerge into full relief. Bill Shockley's favorite among the young geniuses he gathered around him in the Valley's first modern company—and the leader of the so-called Traitorous Eight, who mutinied under his harsh management and started Fairchild. The head of Fairchild Semiconductor and the irreplaceable figure at the top of perhaps the greatest collection of talent ever assembled in a single company. The man who invented the true integrated circuit, then a decade later presided over Intel, the company he cofounded, when it invented the microprocessor—and thus the critical figure of the modern digital world. The victorious commanding general of Silicon Valley's fight against the Japanese business onslaught of the 1980s. The man who first bridged the abyss between Silicon Valley and Washington, DC.

The rediscovery of Bob Noyce had begun at the turn of the century, a dozen years after his death, when Jack Kilby of Texas Instruments was awarded the Nobel Prize for the invention of the integrated circuit. Newspapers around the world carried the news, inevitably with the added comment—graciously echoed by Kilby himself—that had he lived, Robert Noyce would have inevitably shared in the award.

It was enough to keep Noyce's story alive, and as the second decade of the new century began—as Steve Jobs's life ended after the most amazing run of innovation in American business history, as politicians made their pilgrimage to the Valley for money and validation, and as the struggling economy turned to the Valley for rescue—the attention slowly began to shift to what made Silicon Valley so *different* from everywhere else—and where that difference had come from and who had led that change.

So storytellers from everywhere began to look to the Valley for answers. Soon, besides major books on Jobs and on Hewlett and Packard, a celebrated movie was made about the controversial birth of Facebook, and a documentary about Valley venture capitalists—even a short-lived reality show about young high-tech entrepreneurs. Apple cofounder Steve Wozniak appeared on *Dancing with the Stars* and became a cameo icon on shows like *Big Bang Theory*. And in Boston, documentary producer Randall MacLowry approached PBS giant WGBH and proposed a major documentary about the founding of Silicon Valley—centering on Dr. Robert Noyce and Intel.

And so on this night, the story had at last come full circle. The once most famous figure in Silicon Valley, who had then been almost forgotten, was now to be restored to his rightful place at the center of the story. The Trinity of Intel, reduced to just a duo for a generation, was now being restored into the firmament. And as Andy Grove and Gordon Moore stood beside each other and stared into the display case at the sacred artifacts—the early notebooks containing Noyce's original drawings of the semiconductor integrated circuit, some of the resulting first samples of those chips from Fairchild, and the Intel employee badge, with the once older man staring up at them with a much younger face than their own—once again it was three of them again, Noyce and Moore and Grove. The friendships and feuds and the victories and losses were behind them now. They stood together again in the cold, clear light of history.

The crowd now made its way up the stairs into the museum's main auditorium, an austere room with a curved ceiling like a giant Quonset hut, the pipes and ductwork exposed above, the cement floor hard below. The dressed tables with their platters of elegant snacks did little to mitigate the harshness of the venue. This too was a legacy of Silicon Valley, and Intel most of all. Reaction to East Coast social and organizational hierarchies, from which many of them had come to California to escape, had led to the egalitarian—and often spartan—work culture of the Valley.

In many ways, no company had taken that further than Intel. To walk into the company, even in the early 1970s, when the corner office was still the dream of every salaryman in the world, was to encounter a sea of cubicles. It wasn't even apparent which Intel building was corporate headquarters, because Andy, Gordon, and Bob were often scattered to different buildings. A secretary would have to lead you through this rat's maze of carpeted half-walls until suddenly you found yourself in, say, Noyce's "office"—which was just a larger cubicle, the Great Man sitting at the same white vinyl pull-out desktop as everybody else—except with the National Medal of Invention stuck on the wall where everybody else kept a picture of their kids or their last vacation.

No one had ever again taken this egalitarianism so far. Hewlett and Packard may have joined their fellow workers in the lunchroom. And Steve Jobs may have temporarily replaced the title *secretary* with *associate*. And Yahoo may have turned its lobby into a playground, while Google held its meetings on Ping-Pong tables, but the men and women who ran those companies still reserved private offices to themselves. Only Intel's founders backed their avowed philosophy with real actions—most of the time—and made subsequent generations of Valley leaders feel forever guilty for falling short of this ideal. And so if these new entrepreneurs and executives couldn't live up to this leveling attitude, they could at least look the part by stripping away all artifice and elegance from their work environments.

Thus the standard Silicon Valley office building—concrete walls, grass berm, glass and steel, and white stucco walls inside. And so thirty years on, when Silicon Graphics designed its new headquarters, it created only a more sophisticated version of the same old High Valley Austerity—and the museum hall in which the guests now found themselves could have been just as easily repurposed as a Porsche showroom or a corporate R&D lab . . . or frankly, anything but a museum.

There was a last brief burst of table hopping, and then the general manager of KQED, the San Francisco PBS station that owed much of its existence to Valley money (and no doubt hoped to nail down more tonight), took the stage and called the gathering to order. He introduced a local congressman, who had insisted upon being on the program no doubt with the same motives as the station manager. Then MacLowry came onstage to provide a brief description of the documentary—a needless summary of the story written in real life by the people in front of him.

And then the lights began to dim. Everyone looked up at the screen in rapt expectation—except one figure. Andy Grove, his face now less ferocious than appraising, craned his head around to look back at the crowd, recording one last memory.

Then the screen lit up—and the children of Silicon Valley, the founders of Intel, watched as their story unspooled.

PART I | The Fairchildren (1957–1968)

CHAPTER 1

The Traitorous Eight

To understand Intel and the three men who led it, you must first understand Silicon Valley and its beginnings. To do that, you need to know the stories of Shockley Transistor, the Traitorous Eight, and Fairchild Semiconductor. Without that understanding, Intel Corporation will remain—as it does to most people—an enigma.

Silicon Valley began on a warm September morning in 1957 when seven key employees of Shockley Transistor of Mountain View, California, decided to quit their jobs and strike out on their own.

Whatever their fears, they were sure they were doing the right thing. Their boss, William Shockley, was one of the world's greatest scientists; they had felt honored when he recruited them, and they certainly had been proud when he was awarded the Nobel Prize for Physics soon after they joined the company. But Shockley had proven a nightmare boss: mercurial, paranoid, arrogant, and dismissive. If he thought so little of them, if he distrusted them so much, why had he hired them? It was time to go. Now.

But the seven men weren't sure about the disposition of the eighth and most important member of the team: Bob Noyce, their natural leader, a charismatic athlete and scientist who had quickly proven to also be a born businessman. He was the first among equals. Without him they were still quitting, but they weren't sure they could succeed. Even as they pulled into the driveway of Noyce's Los Altos home, the seven still weren't sure he'd join them, and so they were greatly relieved when they saw Bob striding down his front walk to join them. He was in. Shockley Transistor was doomed. And in its place stood Fairchild Semiconductor, the home for many of them for the next decade.

The eight individuals, forever known by the epithet that Shockley gave them when they resigned—the Traitorous Eight—included Robert Noyce, Gordon Moore, Jay Last, Jean Hoerni, Victor Grinich, Eugene Kleiner, Sheldon Roberts, and Julius Blank. Between them, they represented what was probably the finest accumulation of young talent in solid-state physics anywhere in the world, including even the research groups of IBM and Motorola. Shockley, with his brutal hiring process, had made sure of that. Indeed, that hiring stands as his most valuable contribution to Silicon Valley. But none of these men knew anything about running a business. To their credit, they were smart enough to realize that fact.

It is sometimes forgotten that Fairchild was far from the first electronics company in the San Francisco Bay Area. In fact, even as the Eight walked out of Shockley's company, that history was already more than a half century old. For much of the first half of the twentieth century, beginning with local boys experimenting with wireless radio in the teens to vacuum tube makers in the twenties to the brilliant students in the thirties, such as Bill Hewlett, Dave Packard, and Russ Varian, who lingered in town after finishing Fred Terman's celebrated electronics program at Stanford University, the Valley had been a hotbed of electronics innovation and entrepreneurship. It awaited only a spark to ignite it into a full-fledged technology business community.

That spark was the Second World War. Suddenly, the little companies that had been struggling on commercial contracts found themselves buried in huge and lucrative government orders. At HP, Hewlett went to war, while Packard slept on a cot in his office and managed three shifts of women working around the clock. In the process, out of necessity, he learned to manage by establishing objectives and then entrusting his employees to meet those targets. When they did so, he gave them more responsibility . . . and to his amazement, the company not only ran itself, but did a better job than when he was directing it from the top down. He also found that these women workers were more productive when they were treated like members of a larger family—and that included giving them enough flexibility in their hours to let them deal with sick kids and other personal matters.

Other companies in the area made the same discoveries, and though none went so far as David Packard, most implemented personnel policies that were far more progressive than their East Coast counterparts.

The war also brought another, more sweeping, effect upon life in the Santa Clara Valley. More than a million young men passed through the Golden Gate on their way to fight in the Pacific. For many, their brief stay in San Francisco remained a treasured memory of good times and good weather before the long, often brutal years that followed. Moreover, during their tour of duty, many of these farm boys and store clerks were trained to deal with state-of-the-art aircraft and electronic instruments. They saw the future and wanted to be part of it. And as they made their way home after VJ Day, now armed with their GI Bill, many decided that their old civilian life was no longer enough. Instead, they would finish college quickly, get married, have babies, and head to California to take part in the next big Gold Rush—and not necessarily in that order.

By the end of the forties, driven by demand from a shattered Europe, a consumer explosion from the postwar wedding and baby booms (not to mention television), and renewed defense spending for the Cold War—the US economy was once again on fire and racing toward the greatest period of expansion in the nation's history. The postwar migration to California had begun. Many of the migrants went in search of the thousands of new aerospace jobs in Southern California. But almost as many Midwesterners and Easterners headed instead for the San Francisco Bay Area—especially once the Lockheed brothers (from Los Gatos, but having made their riches in Burbank) identified the future of their industry in space and decided to tap into the area's high level of scientific education. Soon Sunnyvale's Lockheed Missile and Space division was the Valley's largest employer. Other Eastern firms began to arrive: Sylvania, Philco, Ford Aeronutronics, and most important, IBM. Big Blue set up shop in San Jose intent on using local talent to develop a new form of magnetic memory storage: the disk drive.

The technology these companies and their scientist/engineer employees were building had evolved as well. Over the previous twenty years, since the founding of Terman's lab, electronics had evolved from simple instruments designed to control and manage the flow of electricity in wires and vacuum tubes. The war had brought radar, microwave, and the first computers. Now a new revolution was about to hit—one that would not only lead them to redesign all of their existing products, but set them on the path to even greater inventions and riches.

This revolution began at Bell Labs in New York City. Just before the war, two scientists, John Bardeen and Walter Brattain, watched a lunchtime lecture about a singular new material. It looked like a small slab of silicon glass—a perfect insulator—so the attendees weren't surprised when, after wires were attached to each end and current was introduced, . . . nothing happened. But then the demonstrator aimed a flashlight at the center of the slab—and their jaws dropped as the current suddenly passed through the wire. As the demonstrator explained, the silicon had been "doped" with an impurity, typically boron or phosphorus, that gave the material a unique property: when a second current was introduced at right angles to the original, a kind of chemical "gate" opened that allowed the original current through.

Profoundly impressed, Bardeen and Brattain made plans to investigate this new *semiconductor* as soon as their current research was done. But then history got in the way. It wasn't until 1946 that the two men were again free to investigate semiconductors. They progressed quickly—until they ran into several technical snags. The two scientists were brilliant men (Bardeen would eventually become the only person to win two Nobel Prizes in physics, the second for explaining superconductivity), but now at an impasse, they decided to go down the hall to get some help from an even more brilliant physicist, William Shockley.

No doubt they went with considerable trepidation, because Shockley was notoriously arrogant and difficult. In the end, their hopes *and* fears were realized. Shockley did indeed solve their problems, but now the reputations of Bardeen and Brattain would be yoked with Shockley's forever.

In light of its eventual fate, the device that the two scientists eventually fabricated—the transistor—was remarkably crude, almost Neolithic-looking in its first incarnation. It appeared to be a little metal arrowhead—with an unbent paper clip stuck on its back—plunged into the flat surface of a tiny irregular slab of burned glasslike germanium. But it worked, brilliantly. Even in this most primitive form, the transistor was faster, smaller, and consumed less power (and gave off much less heat) that the vacuum tubes it was designed to replace.

The transistor could be used in almost every application where vacuum tubes were the current standard—and in a whole lot of new applications, like portable radios and avionics. With all of that demand, a lot of people and companies quickly got very rich. And like many scientists before and since, Dr. William Shockley looked out from Bell Labs on all of this entrepreneurial and corporate fervor and asked: why are *those* people getting rich on *my* invention? As is always the case, there were other factors at work as well, including Shockley's resentment that (rightly) Bardeen and Brattain were being given more credit by Bell Labs—and the world—for the transistor and his abrasive style, which, legend has it, had left him with nothing but enemies at the laboratory.

But it wasn't all about Shockley's personality. His genius was in play as well: having now studied solid-state technology for years, Shockley was convinced that germanium was a dead end, mostly because the requisite crystals could not be grown pure enough for higher levels of performance. Silicon, he had concluded, was the future of the transistor: not only could it be made purer but it was, after all, among the most common substances on earth.

So already scheming how to get rich and famous off his discovery, in 1953 Shockley took a leave of absence from Bell Labs and headed to California and a teaching position at his alma mater, Caltech. Within the year, Texas Instruments began making silicon transistors, both validating Shockley's theory and further spurring him to go out on his own. Arnold Beckman of Beckman Instruments offered to back Shockley if he would work inside his company, but when Shockley's mother got sick, he used the opportunity to convince Beckman to let him move north and join her in Palo Alto. There he set up shop as Shockley Transistor Laboratories and tried to recruit his old workmates at Bell Labs. When that failed—apparently no one in New Jersey ever wanted to work with him again—Shockley put out word that he was going to build the industry's most advanced transistors . . . and that he was looking for the nation's best and brightest young scientists to come help him change the world.

The tragedy of William Shockley is that, when he arrived in the

Bay Area, he held everything in his hands to do just that. He had a mammoth reputation—one that was growing even greater as rumors spread that he might share the Nobel Prize. Thanks to that reputation, his call for top young talent was answered with a blizzard of job applications from which he chose—as history would show—eight young men of extraordinary talent, including two of world-historic importance. He had a technological vision (beginning with a revolutionary new "four-layer diode" transistor) that would eventually define a trillion-dollar industry—and the acumen to get there before anyone else. And he had picked a location to start his company that, once again, history would prove to be the most fertile for high-tech company creation on the planet.

And yet Shockley failed. And he failed so completely that, other than the residual notoriety of his outrageous views on race and intelligence, all that is really remembered of the man—once lauded as the greatest applied scientist since Newton—is his failure at Shockley Semiconductor.

What happened? The simple answer is that Shockley proved to be such a terrible boss—paranoid, contemptuous of his subordinates, and arrogant—that he drove away that same brilliant young talent that he had so successfully recruited just a few months before. All true. But there were a lot of bad and tyrannical bosses in 1950s America, and few ever faced a widespread mutiny in which the entire middle management walked out with no real job prospects. Bad as he was, it is hard to believe that Bill Shockley was the worst boss in America in 1956. So how did he become the bête noire of Silicon Valley history?

There are several answers to that.

The first is context. After HP survived the wrenching postwar layoff, Hewlett and Packard (spurred in no little part by their wives) set out to find a new kind of management style that was more congruent with the casual, nonhierarchical style that characterized Northern California. Through the 1950s, they built upon the policies that the companies had first implemented during the war. Soon HP was famous for flexible hours, Friday beer busts, continuing-education programs with Stanford, twice daily coffee and doughnut breaks—and most important, employee profit sharing and stock options.

Even the physical nature of the company reflected this new kind of enlightenment. In his last great innovative contribution, Fred Terman, now provost at Stanford, set aside hundreds of acres of rolling pastureland adjoining the university to be leased by his old students and their companies. The result, the Stanford Industrial Park, was and still is the most beautiful and elegant of industrial parks—one of the wonders of the industrial world. Even Khrushchev and de Gaulle asked to see it on their visits to the United States. There, in their great, hill-hugging glass buildings, surrounded by this utopian vision of commercial work as paradise, and enjoying a work culture that had no equal in business history, HP-ers registered the highest levels of loyalty, morale, and creativity ever seen in the business world. And on weekends those employees could even camp and play in an entire valley, Little Basin, which HP had purchased for them in the mountains above Palo Alto.

Hewlett-Packard took this enlightened management further than any company in the Valley (indeed the world), but it wasn't alone. A tour of the peninsula in the mid-1950s would have found one company after another—Varian, Litton, Sylvania, Philco, Lockheed—offering flattened organizational charts, greater trust in employees, recreation programs, and (at least for that buttoned-down era) more relaxed work environments. Just a mile from where Shockley would establish his company, on any spring afternoon, one could watch a Little League game at Mountain View's Monte Loma School—where little Steve Jobs would soon play—featuring teams like Sylvania Electric competing against Ferry-Morse Seeds. There were vast picnics and other social events at Lockheed. And at NASA Ames, an early computer terminal was set up in the lobby of one of the buildings for the children of employees and their friends to use—like little Steve Wozniak. Here, in the thick of the Baby Boom, the best Valley companies understood the importance of family.

It was into this world that Shockley came—a man considered dysfunctional as a manager even in Murray Hill, New Jersey. And the young men he hired, most of them moving to California with families in tow, had only to look around and realize they had made a bad decision. Shockley would have been a miserable boss even in the Northeast, but on the San Francisco Peninsula, compared to what was going on around him, he seemed like the Bad Boss incarnate. As if to show how little he had in common with the enlightened business executives around him, he even eschewed the Stanford Industrial Park for a little cinderblock building between the railroad tracks and a shopping center on the Valley floor. If HP and its counterparts were reaching for the stars, Shockley's grubby little storefront suggested that he was going to slug it out in the dirt.

Still, it took a little while for the frustration and anger of the new hires to reach a boiling point. Shockley (to Bardeen and Brattain's dismay) did get to share the Nobel Prize for the transistor. He celebrated by taking his new team to a nine a.m. champagne breakfast at Dinah's Shack, a popular local restaurant. In retrospect, this was not only the greatest moment in Bill Shockley's life, but also the high point of Shockley Labs. His new employees were thrilled—most had dreamed of working with a world-class scientist, and now here they were with just such a great man.

Unfortunately, the excitement didn't last long, and the memories of that moment quickly faded in the harsh light of daily work with Shockley. It quickly became obvious to them that their boss knew little about running a business and even less about motivating employees. On the contrary, he seemed unable to settle on a business model or a realistic product strategy for the company. As for his subordinates, he treated them—even Bob Noyce, who became his favorite—as fools, and worse, potential traitors. New ideas that were brought to him were either dismissed by Shockley or shown to outsiders for their opinions, Shockley being unable to make a realistic appraisal of his own. He demanded that his employees take lie-detector tests. And he even darkly suggested that his employees were trying to undermine him.

It all became too much for those eight brilliant young men who had been Shockley's original recruits. They understood that here, in Santa Clara Valley, some of their neighbors had started great companies in their garages. They knew there were companies just blocks away where employees were *happy*, where they looked forward to going to work; places where their bosses actually *trusted* them to make the right decisions. And they wanted to be those entrepreneurs, and work in those kinds of places—even if they had to build them themselves.

As is often the case with paranoids, Shockley's worst fears became self-fulfilling prophecies. Now those employees really did begin plotting against him—or more precisely, plotting to escape him.

Recalled Jay Last, one of the Eight: "One evening we met at the house of Vic Grinich [another member of the Traitorous Eight] to talk about our next move. We were all downhearted, sitting in this dark-paneled room. We could get jobs easily, but we liked working together. That night, we made the decision to find some way that we could work as a group. But we were asking, 'How can we get a company to hire a group of eight people?' "[1]

That was in March 1957. It took six months but, inevitably, there came that fateful morning in September. As a car full of four of them pulled up in front of Bob Noyce's Los Altos home, they feared that their mutiny couldn't survive without him. But Noyce did join them, in what Shockley would consider the greatest betrayal of all. That day, the Traitorous Eight served Shockley their notice and walked out of the lab and into history.

CHAPTER 2

The Greatest Company That Never Was

The least celebrated part of the story of the Traitorous Eight is that of their wondrous adaptability. They were young; they were smart; they were emerging from a nightmare . . . and nothing was going to stop them.

In retrospect, what is particularly striking about the founding of their new company was how quickly and easily the Eight scattered to tackle the duties that best fit their skills: Bob Noyce and Jay Last took on photolithography, the basic step in transistor manufacturing. Gordon Moore and Jean Hoerni, two of the best young solid-state physicists in the country, teamed up to improve diffusion, the process by which the gas impurities turned silicon into a semiconductor. Sheldon Roberts grew silicon crystals. Victor Grinich, aided by the company's first employee, Murray Siegel, devised the specifications for the company's first product: the 2N696 "double-diffused-base" transistor, a project originally begun at Shockley Transistor but back-burnered there.

The speed at which this young team (the oldest was just twenty-nine) organized itself and set to work—moving far more quickly than they ever had with Shockley—suggests that a real leader was now in charge. It was the first glimpse of the charisma and command of Robert Noyce.

As impressive as all of this activity might have seemed, had anyone been watching, it was the eighth member of the team whose work mattered most. Courtly and with a soft Austrian accent, Eugene Kleiner had taken on the task of finding a company willing to hire the team intact. Kleiner had actually begun the process even before the Eight left Shockley by contacting a friend of his father's who worked at the investment firm of Hayden, Stone & Company in New York City. His letter, which tried to explain the complex technology that the Eight proposed to manufacture, would normally have been casually tossed away had it not been spotted by a new member of the firm ambitious to make his mark: Arthur Rock.

Rock in turn convinced his boss, Bud Coyle, to fly with him out west and take a look at the fledgling start-up. Today, when hundreds of new tech firms each year tap into tens of billions of dollars of investment money held by established venture-capital firms, it's hard to imagine the thought processes of these two men, especially given what they saw when they arrived in the Valley.

Having left Shockley, the Eight were now based in two locations. One was Victor Grinich's garage in Palo Alto, where he and Murray Siegel designed test systems, and the rest of the team in a rented building a couple miles away on Charleston Road in Mountain View. In Grinich's garage, the two men quickly realized that there were no test instruments in existence that could do the work they needed—so they invented their own. Indeed, they invented almost everything, setting precedents that stand today.

For example, as Siegel would recall, the team needed a workbench. "We had no idea how high it should be. So one day in my motel room—I still hadn't bought a house yet—Vic and I took telephone books and stacked them on a table while we stood next to it. When the telephone books hit our midsections—we're both about the same height—we decided that was the height we'd want. That ridiculous bench is an industry standard today."[1]

Over at Charleston Road, the work environment was even more primitive. That facility, which was soon expected to produce some of the most advanced electronics in the world, at first didn't even have electricity.

Recalled Siegel: "We would work until dark. As the days got shorter, so did our work hours. Outside, however, there was a construction line pole with power that we attached wires to so we could at least do sawing and such. I remember Vic Grinich out there that fall with gloves on, a muffler, a hat and his pipe, with a heater nearby plugged into the line."[2]

It was this crude little operation that Rock and Coyle first visited. Amazingly, they didn't run for their lives. Instead, they saw an opportunity to try out a new investment model they had devised, one in which Hayden, Stone would act as intermediary between a corporate investor and this new team. As Jay Last recalled, "Art said, 'The way you do this is you start your own company.' We were blown away. There was no concept of funding a group back then. Hayden, Stone agreed to find us a backer. . . . It was really the start of venture capital."[3]

In the weeks that followed, Kleiner and Rock huddled and came up with a list of thirty potential investors in the new company. Today that list, which is still in Rock's filing cabinet, is risible: it included companies as unlikely as United Shoe, North American Van Lines, and General Mills.[4] But once again, they were cutting a new trail, working with few precedents. And while it seems obvious now, the pair was stunned and disappointed when they were turned down by all thirty.

But Rock had one more company to try. It was Fairchild Camera & Instrument Company, a forty-year-old business that had made its fortune in aerial photography and innovative airplanes. Founder Sherman Fairchild was himself a legendary risk taker—indeed, a prototype for today's high-tech entrepreneurs—and it's likely that he saw himself in the eight kindred spirits and decided to back them.

Even more, he saw the potential of Robert Noyce. He would later say that Noyce's passionate and visionary presentation of his vision for the future of the transistor was ultimately what persuaded him to invest. What Noyce explained and Sherman Fairchild eventually believed was that by using silicon as the substrate, the base for its transistors, the new company was tapping into the most elemental of substances. Fire, earth, water, and air had, analogously, been seen as the elements of the universe by the pre-Socratic Greek philosophers.

Noyce told Fairchild that these basic substances—essentially sand and metal wire—would make the material cost of the next generation of transistors essentially zero, that the race would shift to fabrication, and that Fairchild could win that race. Moreover, Noyce explained, these new cheap but powerful transistors would make consumer products and appliances so inexpensive that it would soon be cheaper to toss out and replace them with a more powerful version than to repair them.

This was Noyce the visionary, and in that single presentation, made in 1957, Noyce not only accurately predicted the future of the electronics revolution for the next half century, but he also gave a preview of what would in time be known as Moore's Law. It's no wonder that Fairchild Camera and Instrument invested $1.5 million into the new company. The Traitorous Eight, plus Hayden, Stone, officially owned the new company, but Sherman had the right to buy them out after five years—and the little firm would be called Fairchild Semiconductor, a wholly owned subsidiary of FC&I.

Said Jay Last, "We didn't realize at the time the legacy we'd leave. . . . Thank God Shockley was so paranoid or we'd still be sitting there."[5]

With the Fairchild money in place, the team now got to work in earnest. Initially Kleiner, as the financial guy and the man who had made the connection with Fairchild, was officially in charge of the start-up. But Kleiner, one of the kindest figures in Silicon Valley history, was not a natural CEO. Bob Noyce was. And as the head of operations, he was the acknowledged leader. Everyone at the new Fairchild Semiconductor knew—and the parent company would soon learn—that Noyce was the real leader of the operation, and on him its fortunes would rise or fall.

To his credit, Bob Noyce knew two things: first, that if Fairchild Semiconductor was going to survive, it needed to get its first transistors built and brought to market quickly. And second, if the new company was going to be *profitable*, it would have to find a revolutionary and low-cost new way to fabricate those transistors, or Fairchild would get steamrollered by big transistors makers with their factories and economies of scale.

It was a credit to Noyce's leadership—and his courage—that within three months, Fairchild had a prototype of its first transistor and was in Oswego, New York, pitching it to IBM for use in that company's XB-70 bomber avionics contract. Even more amazing, Big Blue agreed to buy from the tiny company, which had just moved to its new offices not far from Shockley's lab. As Fairchild's future HR director Jack Yelverton would tell PBS a half century later, "Bob had the ability to charm anybody. He had a great smile, he had a quick wit. And when he walked into a room, people would sit up and pay attention."

There was the legendary Noyce confidence, too, and that would play a key role in making Fairchild, and later Intel, look far bigger than it really was—and buy enough time for the company to grow into its ambitions.

But now, fledgling Fairchild had to deliver on that order—one hundred transistors, all in silicon rather than the traditional but more fragile (this was for a supersonic bomber, after all) germanium. Payment would be $150 per unit: thirty times the standard industry price. If the company could do it, it would not only be on a strong financial footing, but having IBM as a customer would put the company among the elite among transistor companies. Other big orders would be destined to follow. But if the company failed, the industrywide humiliation—and the resulting lost orders—might shutter Fairchild almost before it began.

Noyce, in his first important display of decisive leadership, split Fairchild into two technical teams, under Moore and Hoerni, to pursue alternative designs for the IBM transistor. Five months later, ahead of deadline, Fairchild delivered the devices. So inexperienced was the company with the basic operations of business that Last had to go to a local store to buy Brillo pad boxes to ship the transistors in.

IBM was happy, and almost overnight Fairchild was invited to bid on a number of prestigious contracts. By early 1958 it had won—over huge competitors, including Texas Instruments—the US government contract to supply transistors for the guidance systems on Minuteman nuclear ballistic missiles. The company quickly had visions of growing rich as the leading supplier of transistors to the military and NASA. Fairchild was flying high, but it was soon brought back to earth.

As was customary, Fairchild was required to deliver samples of its new chips to government inspectors to see if they met "mil-spec" for performance under the stress of temperature, pressure, and g-forces. And the results were catastrophic. The government testers found that with many of the Fairchild transistors, they merely had to tap the device with the eraser end of a pencil to cause it to fail.

Last: "All of a sudden, we didn't have a reliable product. We realized that when we were sealing these up, little specks of metal would be loose inside the can and short out the device sometimes. We were really running scared. It would've been the end of the company. We needed to solve the problem."[6]

Noyce, as always, stayed cool. Once again, his solution was to divide his technical staff into two teams under Moore and Hoerni and set them in competition with each other to come up with a new fabrication process. Hoerni's team not only won, but in the process Hoerni himself came up with one of the great technological leaps of the century.

He called it the planar process, and he already had been quietly working on the concept for almost a year. It had its roots in the photo-lithography techniques used in the printing of pamphlets, posters, and low-cost art prints (and today in mass publishing). Essentially, the design of the transistor was drawn out by hand in a large, sometimes wall-size, format, then photographed and reduced into a tiny transparency. Typically, in those early days, two or three of those transparencies were created, representing each layer of the circuit.

A silicon wafer—a tubular silicon crystal sliced like a salami—was then coated, as with a photograph, with a photo-sensitive chemical. Then when a strong light (later UV and laser) was fired through the transparency onto that surface, the dark areas and lines on the transparency would leave areas on the wafer unexposed. These unexposed areas were then etched away with acid and then either dosed with the semiconductor impurities ("diffusion") or plated with a metal conductor or insulator. This process was repeated for each transparency.

Hoerni's prototype, which he kept simple, resembled a tiny bull's-eye distorted outward on one side into a teardrop shape. It is one of the iconic images of the digital age. Hoerni had set out to find a more efficient way to fabricate transistors; what he accomplished was infinitely more remarkable: he transformed the way human beings looked at reality.

Until that moment, manufactured objects were almost always three-dimensional: automobiles, telephones, Coke bottles, and even transistors. With the planar process, Hoerni made the world flat and two-dimensional. He had bridged the very different worlds of printing and electronics and in doing so had created a manufacturing process that not only could make transistors in ever smaller sizes (you just reduced the transparencies and used a higher wavelength of light), but also could manufacture them in almost infinite volumes just as you would by printing more copies of a book page. Not only that, you could multiply the volume of transistors produced in each print run merely by multiplying the number of transistors you drew on each transparency—like large sheets of postage stamps.

As much as anyone, Hoerni, with his planar process, defined the look of the modern world, in which trillions of transistors—as we will see, embedded in larger chips—are manufactured *each day*. Unfortunately, because of a restless personality that made him one of Silicon Valley's first serial entrepreneurs—often leaving companies just before their big success, as with Fairchild—he remains the least celebrated of high tech's pioneers.

With the planar process, Fairchild Semiconductor had the breakout technology it needed to become not just a major player in the transistor industry, but possibly the industry leader. The new Fairchild planar transistor not only solved the reliability problem with the Minuteman project, but it instantly rendered almost every other transistor in the industry obsolete.

Every other semiconductor company faced annihilation, and prudently most went hat in hand to Fairchild to license Hoerni's invention. The rest tried to copy it—leading to years of litigation.

Now Fairchild was making revenues not just from its transistors (IBM, for example, quickly came back with a big order for the new devices), but also from licenses. It was a recipe to print money. And the company did just that—sending most of it back east to the coffers of the mother company. If Sherman Fairchild had any doubts about his investment, they now evaporated, and he made plans to exercise his buyout as soon as he was able to do so.

Needless to say, Fairchild's management and sales team were feeling pretty cocky when they walked into a March 1959 trade show in New York City, the biggest industry show of the year. They owned the seminal product in the industry—as well as the hottest new product, the 2N697 planar transistor.

But the grins were soon wiped off their faces when they walked by the Texas Instruments booth and saw that company's announcement of its pending patent for a brand-new type of multiple-transistor device—a complete circuit on a single chip—that threatened to make their transistor as obsolete as Fairchild had just made the competition.

As the Fairchilders would eventually learn, TI's patent application was based on work done nine months earlier by a junior engineer at the company named Jack Kilby. Kilby, being a new hire, had not yet earned a summer vacation—and so was required to work those months in the broiling TI headquarters while most senior staff was gone. With little to do, he spent his time doodling new design ideas in his notebook. One of those designs, which used tiny wires to link together multiple transistors on a single semiconductor substrate, instantly struck Kilby as being supremely important, and that September he showed it to management. Kilby's bosses not only recognized the importance of Kilby's design, but just as important, that it was an idea whose time had come. It appeared to solve the growing problem in the transistor business of making transistors both smaller *and* more reliable.

It was TI's first public announcement of Kilby's design that rocked the Fairchild team on its heels at the trade show. With the 2N697 planar transistor, Noyce and his crew had thought they were going to be the stars of the show. Now Texas Instruments was trumpeting a new technology that could crush them the way they had just crushed others, such as Philco. The stunned team returned to Mountain View intent on catching up with this new threat.

The story of Fairchild in this era is one of brilliant inventions consistently trumping anything that its bigger competitors threw at it. And Fairchild's response on this occasion was the biggest one of all— ultimately, it not only defeated TI and others, but changed the direction of human history.

Like Kilby, Noyce had been quietly working on a solution of his own to the transistor miniaturization problem. The Minuteman mess had driven him forward in pursuit of a more reliable design, and Hoerni's planar process had made his idea infinitely more feasible. Now, with the TI announcement, Noyce knew that the time had come. He polished up his notes—the same ones that, fifty-three years later, would be enshrined at the Computer History Museum—and presented them to his technical team.

Reinforcing the notion that great inventions are almost never created in a vacuum but are the result of multiple researchers pursuing an idea whose time has come, Noyce's designs solved the same problem that Kilby's did—that of putting multiple linked transistors on a single slice of silicon. But Noyce's design had three advantages. First, he developed it months later than Kilby, and thus had seen that much more development of transistor technology. Second, unlike Kilby, who had worked with traditional "mesa" transistor technology, Noyce had the planar process—which would prove the best solution to the multitransistor paradox. And most of all, because of the first two, plus the fact that Bob Noyce had an idea of genius, his design had the ultimate advantage over Kilby's by actually *working*—that is, it could be manufactured in huge volume at a low price and function reliably in the real world.

"It was as if a door had been flung open. The scientists at Fairchild suddenly looked down into a bottomless abyss microscoping from the visible world into that of atoms—an abyss that promised blinding speed and power, the ultimate machine. When they let their minds wander they realized that not just one transistor could be put on a chip, but ten, maybe a hundred . . . for Christ's sake, *millions*. It was dazzling. And it was as thrilling as hell."[7]

Fairchild called Noyce's design an "integrated circuit"—the IC, the chip—and it would put the silicon in Silicon Valley, as the region would be dubbed a decade in the future. It is a good candidate, along with the Farnsworth television, the Bardeen/Brattain/Shockley transistor, and its future stepchild, the microprocessor, for the title of greatest invention of the twentieth century. Like the transistor before it, the integrated circuit would eventually win its inventors the Nobel Prize. But unlike the transistor team, only one of these inventors would be alive to receive the award.

The integrated circuit changed everything. Up to this point, Fairchild Semiconductor was the hottest new company in the transistor industry, a scrappy up-and-comer unequaled in its innovation and risk taking. Now, with the invention of the integrated circuit, *there was no transistor industry*. More precisely, this entire multibillion-dollar industry filled with huge, mature, and wealthy companies was now essentially obsolete. There would still be demand for transistors in the years to come, but they would never again dominate the tech world. Now their trajectory was clear: a long slide to commoditization and oblivion. As for Fairchild Semiconductor, it was now king of the electronics world; the owner of a technology to which the rest of the industry, including the biggest companies, like General Electric, IBM, and Hewlett-Packard, had to conform.

It wasn't long before Fairchild became one of the fastest-growing commercial companies in history, eventually jumping from the original Traitorous Eight to twelve thousand employees in less than a decade. Ironically, but in a glimpse of the Valley culture to come, by the time the company reached this size most of the original Traitorous Eight were long gone.

In later Valley companies, this break typically occurs at the initial public offering, when founders chafe at the new rules being imposed by the new corporate bureaucrats. But in 1961 at Fairchild Semiconductor, the confrontation was between the relatively conservative scientists who had founded the company and the small army of wildcatting business professionals who swarmed to Mountain View to be part of the Next Big Thing. Many of these newcomers would be business titans forever associated with the chip industry—men like Charlie Sporck and Jerry Sanders. But for now, they were just rookies, trying to find a place, make their mark, and transition Fairchild Semiconductor from a start-up into a real company.

Fairchild was not only growing rapidly, but its corporate culture was changing. Founders like Last—the "old" lab guys—were being supplanted a new breed of hip, confident, and whip-smart employees who cared less about some old feud with that guy Shockley they saw drinking at Main Street Bar & Grill, and more about beating the hell out of TI and Motorola and ruling the electronics world.

At the beginning of 1959, Fairchild Semiconductor had $500,000 in sales and one hundred employees. It also had a brand-new general manager, Ed Baldwin. Baldwin seemed a loyal and competent employee . . . right up until the day he unexpectedly walked out with his team (and other walkouts from Bell Labs, General Transistor, and other firms) to start his own company, Rheem Semiconductor Inc.

It was only after Baldwin was well gone that Fairchild discovered he hadn't left empty-handed. Rather, he had taken with him a copy of Fairchild's "recipe book" on how to manufacture its transistors—in other words, the planar process.

Recalled Noyce, "We found out later that somebody had been hired, not from Fairchild, but a high school teacher or something like that, who was asked to sit down and study the book. And we got that testimony. So it was a pretty flagrant case."[8]

The Baldwin episode had two effects upon Fairchild. On the one hand, it made the company much more vigilant about protecting its intellectual property, wording its employee contracts, and managing employee departures. But at the same time, the precedent had been set; and in the back of the mind of more than one Fairchild employee was a memory of just how easy Baldwin's departure had been—and how quickly he had found the capital needed to start a company of his own.

Baldwin was replaced by Charles Sporck, a taxi driver's son from New York, who had proven his worth as a manufacturing expert at General Electric. Sporck was tall, slightly menacing even when he was happy, and much more capable than eloquent. More than anyone, it was Sporck who turned Fairchild into a real company.

Less visible to outsiders, but at least as important, Sporck often made decisions and adjudicated disputes when Noyce refused. Many years later, Sporck would remember that "Bob's biggest problem is he had great difficulty saying no. If two department heads had different opinions as to what they wanted to do, it was whoever was there last got the right decision, 'cause he always gave you a yes."

This weakness in Noyce as a manager was more than compensated for by a kind of genius—part of his legendary charisma—for attracting the most talented people to follow him with almost boundless loyalty. So even as Sporck was straightening out manufacturing, Moore was driving product development in the labs, and Tom Bay was revolutionizing high-tech marketing, a fourth, less celebrated figure, Don Rogers, ran Fairchild's sales operations and made several crucial hires. One of them was Don Valentine, who initially took on Fairchild's Southern California sales office. Valentine, who would go on to become perhaps the most successful venture capitalist of them all (Apple, Atari, Oracle, Cisco, and Google, among others), in turn hired a number of young salesmen and marketers who would also go on to become icons of the Valley, including the flamboyant Jerry Sanders (with whom Valentine regularly feuded), A. C. "Mike" Markkula (who would be the third founder of Apple Computer), and Jack Gifford (Intersil and Advanced Micro Devices).

It was the talent of these young men—none of them was yet thirty—along with the hundreds of others who arrived at Fairchild in the early sixties that enabled the company to achieve a level of innovation and success that few companies have ever known before or since. Indeed, from 1961 to 1963, Fairchild was one of the most remarkable companies in business history, and very likely the greatest concentration of raw scientific and business talent ever in such a small, young company. It is for those years that the "Fairchildren," as they made their way out into the business world, were most nostalgic—about which they forever asked themselves what might have happened had they managed to stay together and rule the semiconductor world for one or two generations.

What makes the Fairchild of this era such an enduring legend isn't just the history-making technology or even the mind-boggling concentration of future industry leaders and tycoons, but the sheer insanity—the high-living, hard-partying, making-up-the-rules-as-they-went culture of Fairchild during its golden years.

Just to repeat the anecdotes of the era is insufficient. There were the endless after-work drunken gatherings at the nearby Wagon Wheel saloon, where women employees were hustled, marriages were broken, feuds were fed, and in time, employees were stolen by competitors. There's the image of design genius Bob Widlar, the craziest Valleyite of them all, frustrated over some new invention, taking an ax outside and chopping down trees planted as part of the landscaping. Perpetually drunk, Widlar got into a bloody fistfight with a competitor on the floor of a trade show and once wandered a deserted Fifth Avenue in Manhattan in a snowstorm trying to walk to a sales call in New Jersey. And then there was the time Widlar's compatriot Norman Doyle wandered the halls of Fairchild dressed in serape and sombrero as a Mexican bandito, complete with machete, rifle, and twin bandoliers of bullets.

But those are just the legendary moments. To truly understand life at Fairchild in the early 1960s under the laissez-faire leadership of Bob Noyce, in a company filled with well-paid Young Turks convinced (with good reason) that they were part of the hottest company on the planet, one needs to appreciate the dynamics of everyday life at the company. Only then can one appreciate the daily madness of working at Fairchild.

One of the best descriptions of that time comes from Marshall Cox, who joined Fairchild in 1962. He, too, would go on to become a major Silicon Valley entrepreneur, but on the day of his job interview, when he forgot his résumé and his car broke down on the way to the meeting at the SoCal sales office, he assumed that he had no future either with Fairchild or the Valley.

"I showed up at the interview about three thirty, and I was just livid. I figured that it was a total waste of time at that point, so I barged in the door and said to Don Valentine, who was the regional manager, "I'm the three-o'clock appointment. I'm thirty minutes late, I don't have a résumé, and I'm not a double-E [i.e., he didn't have an electrical engineering degree]. If you want to end it right now, I don't give a shit, but enough's enough.""

Valentine hired an astonished Cox on the spot.

A week later, Marshall Cox flew up to the Bay Area for the first phase of Fairchild's annual sales meeting. He barely understood a single presentation during the course of the day. Happily, "by five o'clock all the presentations were over and we went upstairs for refreshments. Now, the second floor wasn't finished at this time, so it was like a big attic upstairs filled with about fifty or sixty guys—and the company served us brownies and scotch. That was it: brownies and scotch. I thought, boy this is some weird company. And my kind of guys!"[9]

Within a year, Cox had been so assimilated into the Fairchild culture that he was going along on sales calls pretending to be one of the company's chief engineers in order to impress the prospect and close the deal.

Jerry Sanders, the brilliant superstar salesman who was constantly the subject of gossip and envy (he was once falsely accused of wearing pink trousers to a sales call at buttoned-down IBM), loved his time at Fairchild despite—or perhaps because of—being the subject of so much controversy. He explained the experience this way:

"I think everybody was drunk most of the time in the semiconductor industry. But I think it was only because we worked so hard. . . . And the guys who didn't know they had to work hard to drink hard didn't make it. So a lot of these guys just fell out. But, no, I think we were a hard-charging gashouse gang. We were fantastic. But we worked our asses off."[10]

Presiding over this big frat house, like the coolest dad in the neighborhood, was Bob Noyce. He was running the hottest company on the planet—and increasingly making both himself and his crew comparatively wealthy for young engineers of that era. And if he was often indecisive, that's what Charlie Sporck was for, especially now that the latter had been named general manager and finally had the power to make those decisions.

Bob Noyce brought three things to Fairchild. First, as much as any chief executive in electronics, he was as good as if not greater than any scientist who worked for him—and that gave him not only the ability to understand the most advanced research being done in Fairchild's labs, but also the vision to see where it might lead. Second, he was indifferent to corporate hierarchies and the trappings of power. That created a leveled work environment at Fairchild that had little precedent in American business—and would prove the germ of the Silicon Valley culture that has now spread around the world. Third, with his natural skills as a leader, he probably held this volatile group together longer than a lesser leader would have. Even as things began to turn bad for Fairchild, the company's top talent still resisted leaving Noyce because he was a father figure, a mentor, and in many ways, the person each of them wanted to be.

At one extreme, perhaps no one at Fairchild worshipped Bob Noyce more than Jerry Sanders. Working out of Fairchild's Southern California sales office, having joined the company after his dream of being a movie star had been dashed, living a Hollywood lifestyle perpetually beyond his means, Sanders saw in Noyce the wise, understanding father he had never had as a tough kid on the rough streets of Chicago. Noyce, in return, perhaps because he saw his own brasher side in the young man, always treated Sanders like a baby brother, forgiving mistakes that he might not in others, always cutting Jerry special breaks.

Another important Fairchild hire, an intense and brilliant young Hungarian refugee named Andrew Grove, had a different experience. He arrived at Fairchild in 1964 fresh from earning his PhD in chemical engineering from UC Berkeley and was anxious to make up for lost time. Ironically, given their future together, Noyce's charisma was largely lost on Grove. As he would say many years later, "I had nothing but unpleasant, discouraging dealings with him as I watched Bob manage a troubled company. . . . If two people argued and we all looked to him for a decision, he would put a pained look on his face and maybe said something like, 'Maybe you should work that out.' More often he didn't say that, he just changed the subject."[11]

The hyperdecisive Grove may have been frustrated with Noyce's indecisiveness, but much of that had to do with Grove's job working in R&D with Gordon Moore. Grove deeply admired Moore's brilliance, but unfortunately for him, his eventual job as assistant director of process development essentially made him the guy assigned to figure out how to standardize the fabrication of new designs in the lab so that they could be mass produced in the factory. Because the two camps were perpetually feuding, it was the ultimate thankless job at Fairchild. Marshall Cox would recall:

"Andy had kind of a funny rap at Fairchild in those days because as an operations guy, . . . he was always developing these processes that when you put them into a real production environment, they wouldn't work. Now, R&D's position on all this was: 'Hey, if you guys weren't such douche bags, you could absolutely take this process and put it into production.' Well, Grove, because he was a process-development guy, got a lot of nailing for being a guy who didn't develop practical kinds of things; they said he wasn't an operations guy. But he, let me tell you, I think the sonofabitch proved his point at Intel. He [was] a great operations guy."[12]

As was his personality even then, Grove raged at the unfairness of his position, but also waited patiently for his main chance to break out of this trap. As it turned out, it was already heading his way.

By 1965, thanks to both these structural problems and growing industry competition, Fairchild was starting to lose its industry leadership. In particular, at Motorola a new star had emerged: Lester Hogan. In many ways—they were both famous scientists—Noyce and Hogan were a lot alike. But Hogan didn't hesitate to make decisions . . . or fire an employee. And now whenever Fairchild looked in the rearview mirror, it saw Motorola coming up fast.

Beyond the internal skirmishing and the resurgent competition, there were two other factors now challenging Fairchild. Ultimately they would work hand-in-hand to destroy the original company.

The first of these was Fairchild Camera & Instrument. Sherm Fairchild and his executive team realized pretty quickly what they had in

Fairchild Semiconductor. But what they never did understand was that this exciting little division out in California wasn't just creating innovative new products, but a whole new corporate culture—one with very different definitions of success and equally different expectations about how to get there.

Never was this stark difference made more apparent than in the spring of 1965, when, on his own initiative, Noyce got up before a major industry conference and in one fell swoop destroyed the entire pricing structure of the electronics industry. Noyce may have had trouble deciding between conflicting claims of his own subordinates, but when it came to technology and competitors, he was one of the most ferocious risk takers in high-tech history. And this was one of his first great moves. The audience at the conference audibly gasped when Noyce announced that Fairchild would henceforth price all of its major integrated circuit products at *one dollar*. This was not only a fraction of the standard industry price for these chips, but it was also less than it cost Fairchild to *make* them.

A decade later, the Boston Consulting Group would look at Noyce's move and use it to devise a new pricing strategy based upon the learning curve. As Noyce reasoned at the time, why not set the price from the start where it will be a year or two hence—and capture customers and market share dominance before the competition could respond? Noyce also had the courage (and the power) to actually implement it—without any real evidence that it would succeed beyond his own faith in its logic. Figuratively and literally, Bob Noyce had pushed all of the chips to the center of the table and bet the fate of Fairchild Semiconductor. And because he was Bob Noyce, as lucky as he was brilliant, he won.

As the industry reeled, Fairchild regained lost market share. And the profits soon returned. The Noyce legend grew. According to Noyce's biographer, Leslie Berlin:

"Less than a year after the dramatic price cuts, the market had so expanded that Fairchild received a single order (for half-a-million circuits) that was equivalent to 20 percent of the entire industry's output of circuits for the previous year. One year later, in 1966, computer manufacturer Burroughs placed an order with Fairchild for 20 million integrated circuits."[13]

Those orders had a powerful impact on Fairchild's market value. Berlin: "In the first ten months of 1965, [Fairchild] Camera and Instrument's share price ballooned 447 percent, shooting from 27 to 144, with a 50-point growth in the month of October alone. This was the fastest rise of any stock listed on the New York Stock Exchange at the time. Sales and profits hit another record high."[14] The capstone came when mighty IBM, its captive semiconductor operations still the largest in the world, decided to license the Fairchild planar process.

Fairchild Parent rewarded Fairchild Child's success the way all East Coast companies of the era did: it kept a sizable chunk of the profits to fund other company operations, and it promoted the people at the top of the division to a fancier position and a better salary for a job well done. Back in New Jersey, it didn't cross anyone's mind that this was exactly the *wrong* response to an egalitarian company that shared both risk and reward among all of its employees, whose executives had moved to California precisely to get away from the Old World of business, and which needed to plow most of its profits back into product development to stay ahead of the competition in a fast-moving take-no-prisoners industry.

In perhaps the most celebrated piece of journalism about Silicon Valley, *Esquire*'s profile of Bob Noyce, Tom Wolfe found the perfect image to capture the unbridgeable chasm between East and West: a chauffeur waiting for John Carter, the president, outside the Mountain View headquarters of Fairchild Semiconductor:

"Nobody had ever seen a limousine and a chauffeur out there before. But that wasn't what fixed the day in everybody's memory. It was the fact that the driver stayed out there for almost eight hours, *doing nothing.* . . . Here was a serf who *did nothing all day* but wait outside a door to be at the service of the haunches of his master instantly, whenever those haunches and the paunch and the jowls might decide to reappear. It wasn't merely that this little peek at the New York–style corporate high life was unusual out here in the brown hills of the Santa Clara Valley. It was that it seemed *terribly wrong.*"[15]

In keeping with this old-school business style, FC&I promoted Noyce to the title of group vice president and shoved under him not just semiconductors, but instrumentation, graphics products, and several other divisions that apparently seemed to corporate management to fit within the omnibus notion of "electronics." That semiconductors was essentially a chemical industry that only produced electronic devices as an end result apparently never crossed their minds.

For both professional and personal reasons, promotion to corporate was just about the last thing that Bob Noyce wanted. Noyce's gut told him, once again accurately, that his presence was crucial to keeping Fairchild Semiconductor together.

But FC&I gave him little choice but to accept. And so, even as he began to make his regular trips to New Jersey, the void he left in his wake began to be filled by management figures moving up one place in turn. The biggest consequence of Noyce's departure was that he named Charlie Sporck as the new general manager of Fairchild Semiconductor. With the decisive Sporck in charge and a senior management beneath him that would soon lead the semiconductor era of Silicon Valley, this should have been the moment when Fairchild turned everything around and regained its leadership.

Instead it all went to hell.

Exactly what happened is still the subject of debate among old Fairchilders and modern industry historians. What is obvious is that several things happened at once.

For one thing, the entire electronics industry slumped. The idea that high tech has an inherent, structural four-year boom-bust cycle is well known today, now that the Valley has passed through a couple dozen such cycles of different severity. But in 1966–1967, it was a brand-new concept, probably driven by overproduction in anticipation of orders from the new UHF television market, the military ramp-up to the Vietnam War, and the NASA space program.

A second factor was that a new fabrication method had emerged in the semiconductor industry—metal-oxide semiconductor (MOS) technology. Because it was faster, cheaper, and more power efficient, it was already being recognized as the future of the industry. Yet while other companies raced to convert from traditional bipolar to MOS, Fairchild ignored this revolution. Said one Fairchilder, "Everybody knew that MOS was the wave of the future. It was obviously the technology for goddamn calculators. And yet we couldn't get anybody to work on it."[16]

However, the growing success of former Fairchilders in this new market wasn't lost on the current Fairchilders. After all, they told themselves in words that would be repeated in thousands of engineers' heads in the years to come, *those guys used to work here. They were no different than us—certainly no smarter. And now look at them. They're rich. Why aren't we?*

CHAPTER 3

Digital Diaspora

A door had opened. And the first person to walk through it was Fairchild's resident renegade genius, Bob Widlar. Looking past his drinking, fighting, and erratic behavior, Widlar had one of the most creative minds in tech history. His forte was linear devices—nondigital circuits such as amplifiers—that were the last great canvases for solitary creative artists of electronics. The only thing that mattered more to Widlar than his linear devices was his freedom. From the tree chopping at Fairchild to driving up with a goat in his Mercedes to nibble the uncut lawn at National Semiconductor to, in his last years, emerging every few months from his compound in Mexico to deliver his latest design to Linear Technologies, Widlar's comparatively short (he died at fifty-three) life was devoted to ripping away every constraint—including lack of money—on that freedom. And so one morning in 1966, Robert Widlar announced that he was quitting Fairchild Semiconductor to join a struggling little semiconductor start-up down the road in Santa Clara called National Semiconductor.

As with all departing employees, at his exit interview, Widlar was asked by Fairchild to fill out a six-page questionnaire explaining the reason for his departure. Widlar instead wrote, in big block letters, one word on each page: I . . . WANT . . . TO . . . GET . . . RICH. Then, because he never wrote his name, he signed the last page with an X. It was Fairchild's first warning of what was to come.

That's when the one company operation that no one worried about—manufacturing—began to fall apart. This was the last place anyone expected problems; after all, Charlie Sporck was a manufacturing god in semiconductors. And with a mountain of back orders,

Fairchild's manufacturing lines were hardly sitting empty—on the contrary, the company was adding shifts. And yet out of the blue, yield rates collapsed, delivery dates slipped, and Fairchild could no longer produce enough working chips to meet demand at the very moment it had a chance to maximize revenues.

By the end of 1966, Fairchild Semiconductor could deliver chips on only one third of its pending orders. It couldn't even bring new products to market because they got stuck in a limbo between development and manufacturing.

It was then that a third reason for Fairchild's collapse surfaced. The real reason for the company's fall—or so believed many future Fairchildren as they looked back on their old employer—was that the parent company had used Fairchild Semiconductor as a cash cow, for too long taking home the precious profits the division needed for working and investment capital. Now Fairchild Semiconductor was beginning to physically break down under the onslaught of so many new orders. Underscoring this argument was the dizzying pace that Fairchild Semiconductor had maintained over the previous eight years. In that time it had grown from the original Eight to twelve thousand employees, with annual sales of $130 million (in 1966 dollars)—just about as fast as any manufacturing company has ever grown. A lot of tech companies have grown at a fraction of that pace and been crippled by growing pains. Short of money and long on odds, Fairchild's weaknesses had finally caught up with it.

But a fourth explanation, this one given by Sporck himself, may be the best one. He was coming unglued with frustration dealing with Fairchild Camera & Instrument. Not a man to bend his will to anyone, Sporck hated almost everything he saw at corporate . . . and realized to his fury that he could do almost nothing about it—as he told Jerry Sanders, "I don't want to feel like a kid pissing in his pants. I want to run my own show."

Other forces were at work as well. The semiconductor industry boom had created a huge gravitational field for talent, and by the mid-1960s, there weren't enough electronics engineers, solid-state physicists, and computer scientists in the country to meet the demand. So chip companies turned to the only solution they had left: they raided their competitors.

Fairchild Semiconductor, bursting with talent and the locus of the electronics industry's attention, was the ultimate target for headhunters and corporate recruiters. There was only one solution, Sporck realized, and it was what Fairchild Semiconductor had always wanted: freedom. And that in turn meant two things: let the semiconductor division keep more of its profits for reinvestment, and give employees *stock options*.

Noyce had been begging corporate for just such an employee incentive/reward program for years. and his pleas had fallen on deaf ears. Noyce quickly joined Sporck in his demands.

They were wasting their breath. In what ranks alongside Shockley's treatment of the Traitorous Eight as one of the biggest mistakes in Valley history, Fairchild Camera & Instruments turned the two men down. At Fairchild Semiconductor, it would be salary or nothing at all.

Shockingly but understandably, the next Fairchilder to go was Charlie Sporck. Sporck: "It started to dawn on us that, hey, you know, . . . maybe there is something to this starting up a new company."[1]

Widlar's departure had shown Sporck where to look: National Semiconductor—which, now that it had the best linear designer on earth in Bob Widlar, could be turned from an also-ran into a superstar. Sporck schemed for months, all while quietly assembling a team of Fairchild stars, from top scientist Pierre Lamond to young Floyd Kvamme. And then in March 1967, he pulled the trigger. Over drinks at another favorite local watering hole, Chez Yvonne, Sporck told Noyce that he was leaving to join National Semi as its new president and CEO.

Noyce didn't even try to talk Sporck out of it. He would later say, "I suppose I essentially cried when he left. I just felt like things were falling apart, and I just felt a great personal loss, frankly. You know, working with people that you're fond of, then having them break apart, was I would almost say devastating."[2]

Now it was chaos at Fairchild Semiconductor. Said one Fairchilder, "I was somewhere between stunned, scared and brokenhearted."[3] In an effort to restore some kind of normalcy, Bob Noyce took temporary control of the operation in order to buy time until he could find a replacement for Sporck. But now the talent pool was greatly diminished. It is no wonder that Noyce looked back and said, "I just felt that things were falling apart."[4]

Only now did Fairchild Camera & Instrument act to stem the bleeding. Its response was predictable: it called Noyce back to New Jersey for endless rounds of meetings to go over every detail of the division's operations, from advertising expenses to the organizational chart. Noyce was reduced to doodling "Try to get East Coast out!" in his notebook.

When Noyce returned to Mountain View, he quietly took Gordon Moore aside. "Bob came to me and said, 'How 'bout starting a new company?' Well, my first reaction was, 'Nah, I like it here.'"[5]

It finally dawned on Fairchild corporate (with the help of stock analysts, the media, and shareholders—the stock had lost almost half of its value) that it had made a terrible mistake. It agreed to create three hundred thousand new shares of stock for an option plan for about a hundred Fairchild employees, mostly middle managers. For about half of those recipients, it was the first stake they had ever had in the company.

But it was too little, too late. Chip companies were offering thousand-share bonuses to Fairchilders just to jump to their firms. By now, Fairchild was bleeding talent from every door . . . and every booth at the Wagon Wheel. Marshall Cox: "It had a real funky effect on everybody at Fairch. We used to go to the Wagon Wheel on Friday night, and slowly [it] turned into 'Well, it's Friday, who did National grab this week? Son of a bitch.' They always got some good engineering [guy], some process guy or something. It really was negative. It was like, 'Jesus Christ, Tom [Bay], sue their fucking asses, what the hell's wrong?' But he wouldn't."[6]

Noyce knew that the only way to restore Fairchild Semiconductor to its old glory was to make the division the heart of a new mother Fairchild, one that stripped away most of its other operations and focused upon recapturing dominance of the semiconductor industry. But to make such a wholesale change in such a giant and venerable company, Noyce realized that he would have to become the CEO of Fairchild Camera & Instrument. And in his heart of hearts, he didn't want that job.

Noyce: "I guess I simply didn't want to move up into that sort of environment. I wanted to stay in California. The corporate headquarters were in New York. Besides, Fairchild Camera had a number of businesses in it that I wasn't particularly interested in—like the reconnaissance business, the printing-press business, things like that. And also, just looking at my personal life, I had gotten a whole lot more enjoyment out of starting things from scratch than running a large company—and in a sense I still do. The other personal motivation was simply finding out by myself whether the success of Fairchild had been a lucky fluke or whether it was something I had done well."[7]

The company then announced that it was searching for a new CEO. Noyce—and the rest of the chip industry—assumed he would be offered the job. Instead, Fairchild passed over Bob Noyce completely as being insufficiently experienced. It would go down as the third great stupid judgment by the old corporate world regarding the new world of tech. From now on, the youngsters would be in charge—first of Silicon Valley and eventually of the global economy.

Gordon Moore, in his diplomatic way, would later tell an interviewer that "Bob was kind of ticked off" at the board's decision. That probably doesn't quite cover the fury that Noyce felt. He had saved corporate from a slow oblivion, made them rich, handed them control over the hottest new industry in the world, . . . and their response was that he still needed to grow up.

Once again, as he had a month before, Noyce walked over to Gordon Moore's laboratory and asked the same question. This time Gordon knew that Bob was deadly serious. Noyce had already talked with Art Rock, now a big-time venture capitalist with the Rockefeller family fund, Venrock. When Noyce told his old friend that he was quitting Fairchild, Rock teased him by saying, "It's about time!" But then he told Noyce that if he could convince Moore to leave as well, and if the two of them would throw in some of their own money to show their commitment, Rock would find them the money they needed to start a memory company.

This time, when Bob proposed to Gordon that they quit and start a new company, Moore agreed.

Noyce was now ready. He flew back to Syosset and gave his resignation directly to Sherman Fairchild. Sherm Fairchild, demolished—like Shockley, he had begun to look upon Bob as his Golden Child—begged Noyce to stay on a little longer and find his replacement.

Noyce agreed. He immediately recommended that the new head of Fairchild Semiconductor should be Dr. Les Hogan of Motorola. Sherm Fairchild quickly dispatched board member Walter Burke to Phoenix to make the offer.

Hogan received the offer with more amusement than interest. Why would he want to leave Moto, at the moment of his greatest glory, to take over a troubled company full of prima donnas, many of whom were already walking out the door? He refused Burke's offer.

Next to call was Sherman Fairchild. Hogan thanked him for the offer but said no. But then arrived the one man in the entire industry whom Hogan considered his equal: Bob Noyce. Hogan remembered being amazed: everyone knew that Fairchild Semiconductor was Noyce's creation; why would he want to leave?

"I said, 'You're nuts, you should be the president of the company.' And Bob said, 'I'm quitting. I'm going to start another company.' I said, 'Why? What you've done is good, you may have some problems, but they can be fixed.' But Noyce was adamant."

In the end, Sherman Fairchild found the one reason why Les Hogan would jump to Fairchild: money. Piles of money. So much money ($120,000 per year, ten thousand shares of Fairchild stock, and a $5 million loan to buy ninety thousand more shares—all in 1968 dollars) that it became a standard unit of monetary measurement in high tech: other CEOs saw themselves as being paid in fractional Hogans. Hogan even demanded—and got—what was considered impossible: the corporate headquarters of Fairchild moved to Mountain View.

Still, as Hogan would say later, "I wouldn't have gone if Bob Noyce [had not encouraged me]. I had great respect for Bob Noyce and he's a great salesman."[8]

In his departure letter to the Fairchild board, Noyce opened his heart: "The reason for my resignation is more basic. As Fairchild has grown larger and larger, I have enjoyed my daily work less and less. . . . Perhaps it is partly because I grew up in a small town, enjoying all the personal relationships of a small town. Now we employ twice the total population of my largest 'home town.' More and more I have looked with longing to the earlier days of Fairchild Semiconductor, when there was less administrative work and more personal creative work in building a new product, a new technology, and a new organization. . . .

"My plans are indefinite, but after a vacation I hope to join a small company in some area of high technology and to get close to advanced technology again in this manner, if I have not been away too long. The limited resources of any small company will be a handicap, but I have no large scale ideas. I do not expect to join any company which is simply a manufacturer of semiconductors. I would rather try to find some small company which is trying to develop some product or technology which no one has yet done. To stay independent (and small) I might try to form a new company, after a vacation."

It was the end of the Fairchild Semiconductor of legend, the Greatest Company That Never Was. The more than one hundred companies that spun off from the mother firm would fill the Valley with entrepreneurial fire and competitive fervor—and would make the region the heartland of the Digital Age, of which it would be the greatest contributor. And yet despite this enormous subsequent success, the aging Fairchildren would always have one great regret: What if the old Fairchild had managed to keep together? Wrote a local journalist a few years later:

Silicon Valley's most famous firm has never really existed. Rather it is an illusion, a magic-lantern show that shines memories through a small company now gone for fifteen years and projects a distorted and gigantic shadow upon the present. Sometimes, when this mythic firm is spoken of, one imagines it still alive, possessing minds of unmatched brilliance, unbeatable products and sales the sum of those of [the Valley's semiconductor companies] combined. It is the largest, most innovative and most exciting semiconductor company in the world.[9]

It was also the start of another legend, this time of a very real company, Intel Corporation. Intel would start with just the reputations of two men and within a generation sit at the center of the world's economy, the multibillion-dollar corporation on which the entire electronics revolution would rest. In the process, it would invent, and then build by the billions, the most complex products ever mass-produced. And it would accomplish this with an almost pathological adherence to a pace of change, established by one of its founders, that had almost no precedent in human history.

This relentless and unflagging pursuit and propulsion of the rapidly evolving world of digital technology at the most basic component level—and preservation of the law that the technology embodies—has been the story of Intel for more than forty years. And that pursuit has taken the company on a merry chase to great riches and irrevocably changed the company in profound ways on several occasions. These different incarnations of the company can be seen as different eras of the Intel story. These eras, as you will see, track with the sections of this book:

- The *Start-Up Era* ("Start-Up") begins with Intel's founding in the late 1960s and continues until the company's initial public offering in 1971. During that era, Intel was a memory-chip company—one of many, but with a particularly distinguished pedigree. The company's survival was far from assured, especially in the first couple years, when it struggled to get its fabrication yield rates under control. In the end, the company survived and even thrived in the memory business, but both outside the company (runaway demand attracting new competitors) and inside (the birth of the microprocessor), the seeds were being planted for the end of Intel's memory-chip business.
- The *Microprocessor Wars* ("The Spirit of the Age") lasts until the end of the 1980s. During these years, Intel and its highly innovative and brilliant competitors perfected the microprocessor and fought nearly to the death to own this hugely lucrative new market. This era ended with Intel's victory in winning the era's defining personal computer contract with IBM.
- The *Industry Titan Era* ("The Most Important Company in the World") at Intel extends through much of the 1990s. In it, the company, now led by Andy Grove, almost competes only with itself—keeping up with the demands of Moore's Law and the endlessly rising expectations of its customers.
- The *Global Giant Era* ("The Price of Success") extends through the end of Grove's and Moore's involvement with Intel. In the period, Intel

becomes the most valuable manufacturing company in the world—but also faces a series of scandals and controversies that mar this historic success.

- The *Post-Founders Era* ("Aftermath") continues to the present, and begins with Craig Barrett assuming the CEO position and continues through the leadership of his replacement, Paul Otellini, and his early retirement and replacement by Brian Krzanich. These three CEOs have faced not only a series of devastating global recessions, but also a semiconductor marketplace undergoing radical shifts with which Intel has struggled (not always successfully) to keep up.

But this is more than just a story of technology or of business. And while the natural human tendency is to focus on tech, the real story of Silicon Valley has always been as much the people as the products. And it is around those people, in particular the three men who founded and led Intel during its first four decades—Robert Noyce, Gordon Moore, and Andrew Grove—that this narrative will revolve.

PART II

Start-Up (1968–1971)

CHAPTER 4

The Ambivalent Recruit

The announcement of Bob Noyce's departure shook Fairchild to its core. Charles Sporck's leaving had been a painful wound that the company had survived and from which it would eventually recover. But Sporck had been an employee, albeit a very high-ranking and widely respected one. Noyce, on the other hand, was a *founder*, and beloved. Seeing Sporck leave was like having your older brother join the army; Noyce leaving was like having your father go out for cigarettes . . . and never return.

Everyone at Fairchild ached with the loss, and dread of what would come next could be found in every office at Fairchild Semiconductor.

The media—what there was—covered the story pretty much as Fairchild corporate presented it. Noyce wisely played along, though Gordon Moore felt no obligation to do so. Under the headline RESIGNA-TIONS SHAKE UP FAIRCHILD, the *San Jose Mercury-News* carried the lead: "More turbulence has shaken Fairchild Semi-Conductor Division, the rags to riches Mountain View Electronics giant." The story went on to note: "It was the latest in an inexplicable wave of executive resignations at the firm during the past year."

"They were among the eight men who built Fairchild from a tiny laboratory into a worldwide firm employing more than 14,000 persons with yearly sales approaching $200 million. Their departure leaves only two of the original founders, Julius Blank and Dr. Victor Grinich.

"Contacted Wednesday, Noyce said he wanted some time to free himself of administrative responsibilities and return to his first love—engineering work in a small, growing company.

"'I had a lot more fun in 1958 than I'm having in 1968,' he said in explaining why he decided to give up the post. He faced a future of 'pushing more and more paper,' and decided to stop.

"Noyce said he has had discussions about joining firms which he declined to identify and said he will mull over the possibility of forming a new company. Moore has indicated he plans to form a new semiconductor company."[1]

While most readers of this story (and others) appear to have believed it, some of Noyce's old employees and new competitors were more skeptical of both his statements and his motives.

Don Valentine, now one of the patriarchs of high-tech venture capital, told a Stanford University interviewer in 1984, "I remember when I was leaving in 1967, Bob Noyce in the interview said, 'You know, it's really too late to start a semiconductor company, Don. Why don't you just stay here? We're doing exceptionally well. We'll make it better.' And I said, 'No. My—my destiny is to move on.' [It's an] anecdote that I remembered in 1969 when Bob was leaving to start Intel. I called Bob and I said, 'Bob, two years ago you told me it was too late. How come you're leaving in 1969 to start the semiconductor company?' "[2]

Noyce's response was to offer Valentine a job at his new company. Ironically, given his VC future, Valentine declined, choosing to stay with an established company like National Semiconductor rather than join an unproven start-up.

Though few at Fairchild knew, Noyce and Moore had already been meeting secretly in Noyce's home study to sketch out their new company. By the time they finally submitted their resignations, they had already concluded that they really would get into the memory-chip business (where Moore's Law told them that the company that kept up with the breakneck pace could sweep the market) in MOS technology (again unexpected, given Fairchild's long resistance to this process). They'd also convinced—it likely wasn't difficult—Art Rock to serve as chairman of the board.

The two founders had even begun recruiting. Among the first on board was one of Fairchild's best marketing executives, Bob Graham. It is a glimpse into the chaotic times in which all of this was occurring—the violent, revolutionary year of 1968—that Graham was having a meeting at Noyce's house when the television in another room announced the assassination of Robert Kennedy. The two men were as shocked as the rest of the world—and Noyce, representing Fairchild in an industry tribute to RFK in *Electronic News*, would write that "instead of drawing the people closer together for the mutual good, our society seems to be polarizing itself into antagonistic groups, each with little regard for the rights of the other."[3]

That was Noyce's public response. In a telling glimpse into how the normally opaque Noyce saw himself as a historic figure, Graham would later tell Charlie Sporck that Noyce took the assassination personally, as a warning that life was short and he needed to find his destiny quickly.

A second and even more important member of the new team essentially hired himself. Andy Grove: "Gordon didn't say no. I don't remember what he said. I mean, he didn't exactly hug me. But then he hasn't exactly hugged me or anyone else in my presence, then or any other time. So we started feverishly talking about what was to be Intel."[4]

Andy Grove was more than just Gordon Moore's number two in Fairchild's R&D department; he was also Moore's most ardent disciple. In the older man he saw the plainspokenness, scientific rigor, and intellectual honesty he most admired. This admiration led Grove to have an even greater allegiance to Gordon Moore than to Fairchild.

Early that summer, both Grove and Moore had been scheduled to attend a technical conference in Colorado. Grove arrived a day earlier—and so, when Gordon finally appeared, Andy quickly set about his usual task of briefing his boss on what had happened at the conference to date.

It was a point of pride to Grove to play this role as consigliere to his hero. So he was crestfallen when Moore, seemingly distracted, barely listened to the update. Frustrated, Grove finally asked what was wrong. As he would recount many years later to Charlie Sporck, Moore replied, "I've decided to leave Fairchild."

Without any hesitation, Grove announced, "I want to go with you." There was no way that, now that he had found his mentor, Andy Grove was going to lose him.

To his relief, Moore presented no objection to his announcement. On the contrary, Moore continued the conversation, revealing his future plans, notably the creation of a new memory chip company. It was only after a few minutes, during which Grove had already become emotionally fully committed to this new venture, that Moore added as an aside, "By the way, Bob Noyce is involved with this."

Grove: "I was not happy about that. My first reaction to hearing that was 'Oh, shit' "

Even Charlie Sporck, who had known both men for a half century and who was a great admirer of Bob Noyce, was taken aback when, many years later, Grove told him this. Why, he asked, such a negative reaction to Noyce?

Grove replied, "I found Bob aloof, indecisive—watching staff meetings where people were devouring each other—and Bob would look detached." As already noted, Grove would make a similar comment a few years later when PBS interviewed him for its documentary on Fairchild, Intel, and the birth of Silicon Valley. To Leslie Berlin, he would say that at the staff meetings he was allowed to attend, Noyce let "people bite into each other like rabid dogs" while Noyce sat there wearing "a pained expression and a slight, somewhat inappropriate smile. His look was either 'Children, would you please behave,' or 'I want to be anywhere but here'—or some combination of the two." It was an attitude Grove dismissed as that of a man who "did not argue, [but] just suffered."[5]

The last straw for Grove, according to his biographer Richard Tedlow, was the ugly—and for Andy, illuminating—case of a certain individual named Wilson, who was one of the two division managers at Fairchild. As Grove would tell it, Wilson had become a serious alcoholic, and his erratic behavior was beginning to damage the company just at the moment it was facing both heated competition on the outside and manufacturing problems on the inside.

It is a glimpse into Grove's mind that he would come to blame the Wilson situation, or more accurately, Noyce's lack of response to it—even more than corporate indifference, lack of capital, stock options, or even Sporck's defection—as the real cause of Fairchild's collapse:

"The reason Fairchild disintegrated was because neither Bob nor Gordon was capable of removing a Wilson, who was staggering in at eleven o'clock for a nine o'clock meeting breathing alcohol. I mean it wasn't subtle. Nor is it hearsay. I cooled my heels in Wilson's office for two hours, with a huge technical problem involving [the] recall of integrated circuits on a major scale, before he showed up.

"I had witnessed Bob Noyce being absolutely inactive and paralyzed as the Wilson/Fairchild [situation] was disintegrating. Because I subbed for Gordon at Noyce's staff meetings a number of times and it was awful. . . . I had very little appreciation for Bob."[6]

Even Tedlow, Grove's official biographer, feels obliged to note after this quote that, while Grove was unable to accept Noyce's behavior in this matter, he was perfectly willing to forgive his hero, Gordon Moore, for exactly the same failure of character.

So why, given his antipathy toward Bob Noyce, did Andy Grove decide to risk everything and throw in his lot with his idol, but with a boss for whom he had little respect as a manager and whom he privately saw as an overrated prima donna?

There are a number of possible answers. One is that Grove appreciated, almost before anyone else in Silicon Valley, that in any given electronics industry, there are only a handful of entrepreneurs, managers, and scientists who give that industry its innovation and dynamics. Success follows those individuals wherever they go. It was true then and it's true now. So it's possible that even at this early date in the Valley's history, if only unconsciously, Grove sensed that history was making a turn—that the center of gravity in the Valley's chip industry was shifting out of Fairchild.

It's also possible that Grove had a presentiment (and an accurate one at that) that the Hogan era at Fairchild would not be a welcome one for the number two R&D executive of the old order.

Grove also might have thought that the new company would resemble the current one: that is, that he would work with his idol, Gordon Moore, and continue to create a whole new generation of breakthrough integrated circuits in the laboratory, while Noyce would remain Mr. Outside, talking with investors, customers, and the press. In that scenario, they would rarely cross paths—and Grove would be spared the frustration of watching Noyce's leadership (or lack thereof) in action.

A more likely possibility is that Grove talked his way into the new company because at this point in his career, he would have followed Gordon Moore to hell if that had been the only choice.

Finally, there is one last but highly improbable possibility. Andy Grove would one day be America's greatest living businessman, but it is probably too much to expect even from him—then just a lab scientist—to have concluded that neither Noyce nor Moore was capable of running the company they dreamed of building . . . and that soon enough the door would open for him to assume the job of chief executive. Still, it *was* Andy Grove, and he was more than capable of thinking several steps ahead of his current predicament and moving decisively in response. Still, if that was his strategy in 1968, it was the wrong one: it would be a decade before he took the reins of the company.

Meanwhile, almost from the first day, Andy Grove had good reason to worry about the two men to whom he was now reporting. So casual and unrigorous were the two men about the process of company building that in all of those months of secret meetings, they never got around to giving the company a name. For a while they used N. M. Electronics, but that sounded like one of those old-line instrument companies that could still be found in aged Quonset huts and shabby tilt-up industrial parks around the Valley. Next they imitated Hewlett-Packard and tried Moore-Noyce—which looked good until the phrase was said aloud . . . and sounded disturbingly like "more noise," that is, the background static of leaking electrons, the biggest fear of chip-designing engineers.

Only after those dead ends did Gordon and Bob settle on Intel, an amalgam of Integrated Electronics, which also neatly produced a portmanteau word that sounded like *intelligence*. It is telling about Noyce and Moore's partnership that this sloppy and apparently undirected process resulted in one of the most compelling names in high-tech history. The process no doubt drove Grove half mad, but it's also doubtful that he could have come up with an equally successful result.

It is also important to remember that, unlike Moore, Andy Grove was not a founder of Intel. He earned a salary, but no stock—Chairman Art Rock had set up the initial distribution of shares that way. Also unlike the two founders, Grove could be fired—a horrifying prospect for a man with two young children at home—which helps explain why he kept his own counsel about his opinions regarding Bob Noyce. Thus Noyce and Moore could afford to play around and make decisions along a pathway defined by their own eccentricities. Grove, by comparison, was merely an employee. And even that status was tainted by the founders' sloppiness with details. Grove recruited Leslie Vadász, who became

Fairchild's other "mad Hungarian." Vadász, whose life had followed a trajectory similar to Grove's and who would ultimately become Intel's longest-serving employee, through some error in paperwork was given the employee number 3; Grove was given number 4, even though he was hired first.

Noyce and Moore shrugged this off; after all, they had split their ownership of Intel fifty-fifty and had all but flipped a coin to see who would be CEO and COO. But it rankled Grove. He was still trying to make his reputation in the world, and so even the littlest slights mattered—especially in the years to come when the two founders would playfully tease the apparently humorless Grove with jokes about the "first hire" or "employee number four." Grove could only steam.

But this resentment went even deeper: only in his retirement did Andy finally admit that during these early years he was almost continuously terrified: "I was scared to death. I left a very secure job where I knew what I was doing and started running R&D for a brand new venture in untried territory. It was terrifying. I literally had nightmares. I was supposed to be the director of engineering, but there were so few of us that they made me director of operations. My first assignment was to get a post office box so we could get literature describing the equipment we couldn't afford to buy."[7]

But once again, whatever inequities Grove felt had been placed upon him had come from *both* founders. So why target his ire almost solely at Noyce?

In the end, Grove's biographer Tedlow concluded that there were two reasons. The first was that "Grove hates to be manipulated. People who have worked closely with him have marveled at the tenacity with which he pursues reality. Mere charm will not get you very far with him.

"Second, and perhaps more important, Grove and Noyce had more in common than Grove himself might appreciate. They both aspired to positions of leadership. They both enjoyed the special aura that surrounds special people. They both have been show-offs. They have loved the limelight. They both were performers. Noyce loved to act, and he enjoyed singing madrigals with groups he organized. Grove has a deep, sonorous voice, believed in 'strategy by speech making,' and possessed a superb and carefully crafted public presence. Grove loved opera from his youth and is more than a little knowledgeable about it."[8]

To this can be added the fact that the two men also shared the trait of toughness. Grove was famous for his table-pounding and shouting, but anyone who ever faced Bob Noyce's ire remembered it as even more frightening. Andy would flare and forgive. But when you crossed Noyce, that affable demeanor would drain away and his dark eyes would go black—leaving the poor recipient feeling pinned to the back of his chair. It was a devastating and often unredeemable experience.

Ultimately, Tedlow decided, the greatest similarity between Noyce and Grove was that they were "magnified" men, "operatic"—which made Grove uniquely positioned to see the prima donna in Bob Noyce even as he struggled to control the same penchant in himself. Unlike the infinitely more self-actualized Gordon Moore, the two men deeply craved attention. "There was probably some jealousy in this relationship," Tedlow concluded, though he failed to note that it mostly went in only one direction. "Grove had to be the star of the show, but so did Noyce. Two stars can't [both] get top billing."[9]

This competitiveness and jealousy, often presented as ruthless honesty, never really left Grove—as if Noyce's legacy in death irked him just as much as had Noyce's leadership in life. He was still making those honest (and ungracious) remarks about Noyce two decades after the latter's death.

And yet, even if Andy could never admit his similarities to Bob, as the years passed and he became just as celebrated as his onetime antagonist, as he too found himself a high-tech icon, and as he dealt with the challenges of being chief executive and public spokesman upon whose words the fate of entire industries rose and fell, he increasingly *became* the man he resented. Andy Grove in the 1990s, at the peak of his power and fame, was unexpectedly witty—if not playful, then at least an accomplished game player, ready to take big risks and prepared to trust his hunches and intuitions. He forgave his enemies and even let attacks roll off him. And as disease led his body to slowly betray him, Andy Grove for the first time even accepted the possibility of failure. How much of this he found in himself and how much he unwittingly learned from Bob Noyce, is something we will probably never know.

As if all of this isn't complicated enough, consider one more unlikely fact. Although both Noyce and Grove loved and admired Gordon Moore, neither of them had much to do with him outside of work. Part of the legend of Moore is that he was a man of simple interests. For Gordon and his wife, Betty, an ideal vacation was a drive into the mountains to go rock hunting—even after the Moores had become billionaires and could easily have bought their own mountains. There are many anecdotes of Gordon at elegant dinners and gatherings, happily engaged in conversations about minerals. It can be said that the Moores' private life was so comfortable and self-contained that it was almost impenetrable.

On one emblematic occasion, a camera crew was eating lunch in a studio control room awaiting word from the lobby that Dr. Moore's limousine had arrived . . . and scrambled when they discovered to their shock that Moore was wandering around the darkened studio looking at the electronics in the cameras. It seemed that Moore had driven himself up from Intel headquarters, parked in the back, found an unlocked door, and walked in. And rather than drawing any attention to himself, he had kept himself busy looking at the equipment.

Given all of the glowing tributes the other two—especially Grove—made over the years about Gordon Moore, it probably comes as a surprise how little time he spent with either of them once he drove out of the Intel parking lot each evening. But what is even more surprising, especially once you appreciate Grove's resentment of Noyce, is to discover how much social time—even weekends—the two men (and their families) spent together. Noyce even taught the Grove family to ski over the course of several trips to Lake Tahoe.

All of this doesn't quite fit with the contemporary image of Noyce and Moore being close buddies, Grove as worshipful disciple of Moore, and Grove perpetually challenging Noyce for control of Intel. But neither does it fit with the earlier image of the three men as a brilliant troika harmoniously working together in a common cause. Frankly, the evidence argues that Intel may have been the most successful technology company ever founded by a dysfunctional start-up team—certainly by such a team that stayed together.

The more one looks at the three founders—or two founders plus one—who started and ran Intel for the next twenty years, the less they look like partners and the more like family: argumentative, conspiring, bickering, resentful over past slights, occasionally jealous and blaming one another, and yet at the same time tied together by far deeper binds than commerce, proud of each other's victories, filling in for each other's weaknesses, willing to drop their own animosities to find a common enemy, and more successful as a team than they ever were as individuals. And ultimately, even after death, each of their fates is tied inextricably to those of the others.

There is indeed something almost biblical about the relationship of Noyce, Moore, and Grove, made historic by the extraordinary success the three achieved together. Noyce was the beloved and charismatic but often indolent and unreliable Father. Grove was the brilliant but truculent Son in a perpetual Oedipal battle with Noyce, while always protecting Moore, a Son for whom reaching even the highest pinnacles of success and fame was not quite enough. And Moore, the embodiment of kindness and decency, was the very Holy Spirit of the digital age, but also ethereal and unwilling to make the tough choices. It was an unholy Trinity, but also a perfect fit.

Most business partnerships—and this is as true in Silicon Valley as in the rest of the world—start out happy. Otherwise they wouldn't start at all. True, a sizable fraction end acrimoniously, but that acrimony is typically thrown into relief by the harmony of the start. Intel is that rare and probably impossible-to-reproduce example of a great company in which the founders began badly and grew in their mutual respect over time. And one of the reasons they succeeded was that they never let their personal antagonisms get in the way of their avowed duty as the protectors of Moore's Law and their commitment to making Intel the most successful electronics company in the world. It was that commitment that led them to mostly stay out of each other's way and respect each other's turf.

And it worked. Most people, even senior Intel executives, had little idea of any friction at all at the top of the organization chart. And if they did, it probably mattered little—if that's what it took to make Intel successful, they could live with it.

In fact, Intel was *very* successful, literally from day one. And that pushed any reckoning among the three men down the road.

CHAPTER 5

Intelligent Electronics

N M Electronics Inc. was founded on July 18, 1968. Art Rock had asked Noyce and Moore to each put in $500,000 of their own money as a show of commitment to other investors. The two men refused—having never gotten Fairchild stock, neither had the money—offering to put up half that. Rock merely shrugged, put in $10,000 of his own money, and went out in search of the rest.

In keeping with the plain and informal style of its founders, there was no official incorporation ceremony. The three men simply signed the requisite California state incorporation documents and returned to their work. Art Rock was designated chairman. Noyce and Moore became president and executive vice president, respectively, though both said later that the titles could have been easily reversed. "We don't care who has what title," Noyce told his son. "[Titles] are mainly useful for helping people outside the company figure out what you do." Years later, Moore would add, "He and I had worked together so long, we were very comfortable doing that."

NM Electronics in the those first few weeks was little more than a rented building, a few desks and chairs, and boxes of newly purchased lab equipment being unloaded by employees numbering fewer than a dozen. Everyone knew one another by first name, paychecks were made out by hand, and every person—including the two founders— was expected to pitch in as needed on anything from janitorial work to receptionist to installing the sophisticated fabrication equipment.

The fabrication area itself, once assembled, was hardly inspiring. Recalled Grove, "The fab area looked like Willy Wonka's factory, with hoses and wires and contraptions chugging along—the semiconductor equivalent of the Wright Brothers' jury-rigged airplane. It was state-of-the-art manufacturing at the time, but by today's standards, it was unbelievably crude."[1]

At this early stage, the company still wasn't sure what it would build: multichip memory modules? Schottky bipolar-type memory? Silicon-gate metal-oxide semiconductor (MOS) memory? The only way to find out was to actually try to build each type of product.

It proved to be what Gordon Moore would later call a "Goldilocks semiconductor strategy": "Multichip memory modules proved too hard and the technology was abandoned without a successful product. Schottky bipolar worked just fine but was so easy that other companies copied it immediately and we lost our advantage. But the silicon gate metal-oxide semiconductor process proved to be just right."[2]

But the company still had to build these MOS memory chips in volume at a competitive price. That proved to be a much greater challenge, one that would require some very creative thinking.

One such innovation was to round the rough edges of each chip's metal layer—a process called reflow glass processing—to prevent it from cracking, a serious problem. It was Gordon Moore's idea, and he was so proud of that company-saving patent that the certificate hung on his office wall for the rest of his career.

Les Vadász found himself in the thick of the MOS development project: "We worked around the clock. Joel Karp and I redesigned a good portion of the product while the first moon landing was going on. We listened to 'One small step for man' on the radio while we struggled to rework the chip."[3]

But even getting the design and the manufacturing right proved to still not be enough for success. As Vadász recalled, "When we got into [memory] production on the MOS process, we were shocked by the market response. For our first order we only got about one-third the price we had expected. Bringing down costs was as significant an achievement as the technical work."[4]

Meanwhile, there was still the matter of finding the money to underwrite this work. With $300,000 of that initial half million already committed to capital improvements and R&D and with a burn rate that would likely soon reach $20,000 per month (Noyce's annual salary alone was $30,000, one third of his final Fairchild salary but still a siz-able financial draw on an undercapitalized start-up), it was obvious to Rock that NM Electronics would need some kind of bridge loan to get under way while he went out in search of investors.

Thanks largely to Noyce's reputation and Rock's banking connections, that money, as a line of credit, was secured quite quickly. That bought time for Rock, often accompanied by Noyce, to find investors.

The new-company incubation system that today makes Silicon Valley unique already existed in 1969, but it was much less sophisticated (and less heavily funded) than it would be just a decade later. Modern high-tech start-ups typically pass through a number of investment steps, from angel investors to seed-round investors, then the series A, B, and sometimes mezzanine investments by venture capitalists, and only then, if they are successful, an initial public offering of stock. In the early rounds, the standard investment form today is a convertible loan, offered at discount, that can be converted to stock options at the time of the series A investment (the discount refers to the reduced price at which these options can be purchased, based on the value of those options as priced to the VCs). This series A round, typically ten times the value of the angel round(s), normally sells off about 35 percent of the company and costs the company a couple of board seats (occupied by representatives from the biggest venture investors).

When Art Rock was out looking for investors in Intel, the process was simpler, but the net had to be cast far wider. Today, he could visit fifty venture-capital firms within a half mile on Sand Hill Road in Menlo Park in the hills above Stanford University. In 1969, Rock went out around the country with five hundred thousand shares valued at $5 apiece. The $2.5 million to be raised would be in the form of convertible debentures. These were essentially interest-bearing IOUs that could be converted into stock at the time of the initial public offering . . . if it ever happened.

In keeping with the improvisation of the whole process, Rock, after huddling with Noyce for suggestions, assembled a pretty eclectic bunch, including the other six members of the Traitorous Eight; Hayden, Stone (the original investors in the Eight, now led by Bud Coyle, Rock's boss on that earlier investment); two friends, Max Palevsky (legendary entrepreneur, Democrat Party financier, art collector, and soon-to-be owner of *Rolling Stone* magazine) and his college buddy Fayez Sarofim; Raychem CEO Paul Cook (probably both for his experience and in thanks for his recent offer to Noyce and Moore to start their company inside Raychem); Gerald Currie (cofounder of Fairchild customer and military-technology supplier Data Tech, probably invited because Noyce sat on Data Tech's board); and the Rockefeller Investment Group (in theory a Hayden, Stone competitor, but in practice high-tech company boards are often composed of multiple VC firms to spread the risk).

Noyce, ever the loyal alumnus, also asked that Grinnell College— more precisely, members of its board of trustees—be given the opportunity to invest. The idea was that the trustees would buy the stock and then donate those shares to the school. What makes this part of the deal of particular historic interest is that one of the newest members of the board was the "Wizard of Omaha," Warren Buffett. While this investment would seem to violate Buffett's well-known investing rules, as he told Leslie Berlin in 2002, "We were betting on the jockey, not the horse."[5]

Rock set the price for each seat at the table at $100,000. Then, to show his commitment to the company (and not coincidentally, to make himself the largest shareholder), Rock bumped his own investment to $300,000. Despite a fabled career that made him one of the world's wealthiest men, Rock would later say, "Intel is probably the only company I ever invested in that I was absolutely, 100 percent sure would be a success—because of Moore and Noyce."[6]

To prepare for any questions from potential investors, Rock asked Noyce to prepare what has been called a business plan but was more accurately an extended executive summary for such a plan. Even then, what Noyce prepared was something of a masterpiece of imprecision—it said little more than that the new company would build semiconductors not like those currently on the market, as well as at a higher level of integration. As Noyce admitted, "Frankly, we didn't want people to know what we were going to be doing. We thought it would attract too many competitors too soon."

As anyone who has ever started a high-tech company knows, this is an amazingly confident statement. Entrepreneurs fear that in committing a product or strategy to print and giving it to investors they are risking new competition or even outright theft of their ideas. But ultimately, they have to come down on the side of disclosure—if only because investors won't accept anything less before putting up money. It is a measure of just how great was Bob Noyce's reputation in 1968 that he didn't feel obliged to say much more than "Gordon and I are starting a company that will make a new kind of chip"—note that he didn't even feel the need to say "memory" or "MOS"—and yet still knew that investors would line up with their checkbooks.

And he was right. Incredibly, no one even asked to see the business plan.

Now it was time for Rock to formally open Intel to outside investment. He began to systematically call each of the designated investors. It took much of the day. As the founders feared, word of Intel's open round quickly spread across Silicon Valley and the rest of the electronics industry. The number of people who wanted in was hardly limited to those folks Art Rock called; Intel's office had been besieged by interested investors almost from the day it was occupied, but now anxious outside investors were resorting to desperate means. They even called the founders' homes. Betty Moore quickly learned to feign ignorance. As for Noyce and Moore, in the weeks and months that followed, when they were approached by friends and acquaintances disappointed at not being allowed into the round, they learned to shift responsibility to Art Rock.

All had reason to be upset: one of those $100,000 investments, if still held today, would be worth almost $3 *billion*.

Just forty-eight hours after Rock began his phone calls, the investors had fully subscribed the $2.5 million Intel private offering. Rock joked later that this mind-boggling achievement might have been even quicker had it taken place in the era of e-mails, texting, and mobile phones: "Back then, people had to return calls."[7]

It was a turning point in Silicon Valley history. The news that Intel had pulled off a fully subscribed, multimillion-dollar initial investment round in just two days electrified the electronics world. It served notice on Wall Street that a new generation of high-tech companies was about to burst on the scene. And inside the walls of Fairchild, it bolstered a lot of secret burning ambitions to follow the lead of Intel and National Semiconductor and establish other new semiconductor start-ups.

Meanwhile, Fairchild was making this decision easy. As emotionally devastating as the Noyce and Moore departures had been, there had still been the redemptive hope that Les Hogan, with his extraordinary record at Motorola, would streamline Fairchild into a disciplined, profitable enterprise, that he would restore the company to its historic rates of growth and perhaps even make it the industry leader once again.

In the end, Hogan accomplished almost all of these things, but not with the Fairchild employees who welcomed his arrival. In what Hogan would later admit, to his lifelong regret, was the single worst executive decision of his career, he invited (or otherwise left the door open to) his old team at Motorola—nicknamed Hogan's Heroes after the then-popular TV show—to join him in Mountain View. It was a move that quickly, using Hogan's word, "destroyed" the already fragile morale at Fairchild.

For many Fairchilders, especially in the middle-management ranks, this was the last straw. The Sporck, Noyce, Moore, and Grove departures had convinced them against all cultural pressure and precedent that quitting to found a start-up was possible. The success of National (in taking over a troubled chip company) and Intel (in starting a chip company from scratch) had shown them that such a departure could be wildly profitable. Now the imminent prospect of having an outsider come in and take your boss's job—or even *your* job—was the final catalyst. The Fairchilders were about to become, famously and forever, the Fairchildren, a diaspora of hundreds of former Fairchild employees, high-ranking and low, who would abandon the company over the next few years to set up their own companies across the Valley floor from Redwood City to South San Jose. Many started or took over the leadership of new semiconductor companies: Advanced Micro Devices, Intersil, American Microsystems, Signetics, Computer Micro-technology, Precision Monolithics, and perhaps a hundred more. Others moved into the systems business with companies like Data General and Four-Phase. And hundreds more set up shop around the Valley providing the goods (laboratory equipment, office supplies, clean rooms, fabrication equipment) and services (public relations, marketing, legal, training, commercial real estate, banking) that the fast-growing chip companies needed to survive and thrive.

The few became dozens, then scores and hundreds, of former Fairchilders looking to stake a claim in what was quickly becoming a silicon gold rush. And in the process, between 1967 and 1973, they created not only a vast and prosperous business-technology community, but also the greatest start-up company incubator the world had ever seen. By the mid-1970s, it was possible to start with an idea on a single sheet of paper and then assemble enough home-grown talent and support—everything from people who watered the office plants to wealthy venture capitalists writing large checks—to create from scratch a billion-dollar enterprise. Nothing like this had ever happened before. And as this community began to develop its own internal culture—one that within a generation would be adopted around the world—it also began to lionize its own unique heroes: entrepreneurs.

Those Fairchildren who joined existing companies in the Valley quickly discovered that they were no longer surrounded by the best and brightest in tech as they had been at Fairchild. Neither were they often in an enterprise with a major tycoon backing it, like Peter Sprague at National Semiconductor, nor in one exhibiting the technological innovation to stay among the industry leaders. Frankly, for many it was a relief: a nice little niche company in a safe side-market with respectable growth, after the madness and breakneck pace of Fairchild Semiconductor. But for others, it was the worst of Fairchild: a suffocating and distant corporate owner that had little understanding (nor any desire to improve that understanding) about its minor division out in California. It's not surprising then, as the years passed and the bad memories faded, that Fairchild began to look more and more like a magical place.

For those Fairchildren starting their own companies, the disappointment was often even greater. The spectacular launch of Intel Corp. only threw this frustration into greater relief.

Jerry Sanders was the ultimate case. Several companies and a bunch of start-up teams just out of Fairchild approached him about becoming their CEO. That only shocked him: "What did I know about being president? I didn't even know the difference between a balance sheet and an income statement."[8]

Eventually it dawned on Jerry that it *might* be possible to create a chip company that played to his strengths, which were marketing and sales. In particular, such an enterprise might opt out of the innovation arms race that Intel and National Semi were about to ignite and instead get into the business of "second sourcing," licensing other companies' products and building them as a backup source for customers. With this idea, which he pounded out with newly fired Fairchilder John Carey, Sanders quickly found six more interested Fairchildren, and together in 1969 they founded Advanced Micro Devices. The rest of the team expected Sanders to become not only CEO, but also sales director—a job he wisely refused. His task, as he quickly discovered, was to go out and find enough investment money to keep AMD alive.

He very nearly failed. Sanders worked with one VC who turned out to have no idea how to raise money from his peers. Dropping him, Sanders went out on his own. Not knowing what else to do, he followed the path of his hero, Bob Noyce, and made an appointment to see Art Rock:

"I got nowhere with him. Art Rock said to me, 'It's no time to go into the semiconductor business, it's too late for that, there's no hope in semiconductors, plus just about the only investments I ever lost money on were ones where a marketing guy ran them.' So I said, 'Thank you very much,' and so much for Art Rock."[9]

Whether he was being sincere or just frightening off a potential future competitor to Intel, Rock was at least gracious enough to suggest some other investors whom Sanders might contact. But when Jerry called them, they only showed their ignorance of the industry by asking how AMD could possibly compete with General Electric and RCA—two electronics giants that had all but abandoned the semiconductor business.

In the end, it took a group of diverse investors combined with a group of old Fairchildren, Fairchild vendors, and even some Intel employees to finally—and barely—push AMD over the top of its investment goal of $1.5 million. Sanders: "Bob Noyce always said it took [Intel] five minutes to raise five million dollars—well, it took me five million minutes to raise five dollars. It was just grim. But I was dogged about it. I knew I had a story. I knew we could make money."

By the end of 1969, the first great wave of new semiconductor companies, almost all of them populated by Fairchildren, had appeared—transforming Santa Clara Valley. Over the next few months, these many companies targeted their markets, settled down to work—and began to capture the world's attention.

It wasn't long before the electronics trade press noticed. At the beginning of 1971, *Electronics News*, a tabloid that fashioned itself as the weekly news source for the industry, sent one of its reporters to the Valley to investigate this new phenomenon. The reporter, Don Hoefler, produced—beginning on January 11, 1971—a groundbreaking series of articles on this new semiconductor community. But Hoefler's greatest contribution was the dateline on the story—in later years, he alternately credited it to his buddy Ralph Vaerst and to an overheard conversation between two salesmen at the Marriott Hotel—which he thought best captured the explosion of new chip companies sprouting throughout the region.

It was "Silicon Valley, USA." And for the next decade, it was the silicon semiconductor companies of this Valley—most of all Intel, National, Fairchild, and AMD—along with Motorola and Texas Instruments, that would dominate the rest of the tech world and increasingly the US economy, as they fought to the death. They would not only remake the world, but also set it on a new and unrelenting pace.

CHAPTER 6

Robert Noyce: The Preacher's Son

Who were these three men, this trinity, who created Intel Corporation? We'll look first at the two founders, Noyce and Moore, and later, Grove.

Robert Norton Noyce was born on December 12, 1927, in Burlington, Iowa, a small city on the Mississippi River in the extreme southeastern corner of the state, about equidistant from Chicago, Des Moines, and Saint Louis. The family home was actually in nearby Denmark, a tiny village dominated by a Congregationalist church.

This was appropriate, as Congregationalism shadowed all of Bobby Noyce's youth. His father was a Congregationalist minister, as was his father's father; his mother, the daughter and granddaughter of Congregationalist ministers. Ralph Noyce was a small, shy, and intelligent man who had gone off to World War I from the family dairy farm in Nebraska, then returned to study Greek and Latin at Doane College before going on to graduate study in the School of Theology of Oberlin College. A born academic (and unfortunately, not a natural preacher), Ralph did very well at Oberlin, even getting nominated for a Rhodes Scholarship. At twenty-eight, preparing to begin his career in the ministry, he was introduced to a friend's sister, Harriet Norton, a sociology undergraduate at Oberlin who dreamed of becoming a missionary.

After meeting Ralph Noyce, Harriet set aside those plans and devoted herself to supporting her husband and then her four sons. Unlike her mild-mannered and retiring husband, Harriet Noyce was raised a city girl, in Chicago—and she kept an outspoken (apparently she never stopped talking), opinionated urban attitude for the rest of her life, even as she moved the family from one small prairie town to another across Iowa as her husband changed ministries.

Bobby was the third son in the family, and his arrival was a bit of a disappointment, as his mother wanted a girl. Bob inherited his intelligence from both his mother and father—indeed, he was one of those rare cases where the child appears to have been much smarter than both of his parents. As for the rest, Bob appeared to have inherited from his mother a fearlessness—in Bob it would approach a dangerous recklessness as an adult—and a love for being in the public eye. From his father, he seemed to have inherited both a deep studiousness and an even deeper aversion to personal confrontation and conflict. In his father, this emotional pacifism flattened his career; prairie congregations wanted a preacher who spouted fire and brimstone, not an indecisive compromiser who preferred that his flock like him rather than fear his wrath.

Somehow, as an adult, this odd combination of traits made Robert Noyce a great man, but also a hugely complex, often contradictory, personality. He was the smartest man in almost any room *and*, even in middle age, the best athlete. He was, as everyone who ever met him (except Andy Grove) acknowledged, hugely charismatic, yet also distant, private, and largely unknowable. He dismissed celebrity but took every opportunity to seize the limelight. He dismissed hierarchies and titles but never forgot his position at the top of the pile. And oddest of all, he chose a career—entrepreneur/business titan—that demanded a certain pragmatic ruthlessness in dealing with the careers of subordinates, yet he so wanted to be loved and admired that he was all but incapable of firing even bad employees. Rather, he kept his hands clean but let first Charlie Sporck and then Andy Grove do the dirty work—and then wasn't above treating them as corporate enforcers. He was a great scientist—in fact, a strong case has been made that he should have won *two* Nobel Prizes, one for the integrated circuit, the other for the tunnel diode—but turned his back on research to become a businessman. And not least, though he always professed his yearning to work in a small boutique company where he could escape bureaucracies and focus on the work he wanted to pursue, he built two huge companies—one of them among the largest in the world—and never showed any inclination other than to grow them as big as they could get.

It may just be that, for the people who knew him, Bob Noyce's extraordinary appeal was due to these contradictions, not in spite of them. And there was more as well, none of it captured in the mere enumeration of his great achievements. In person, Noyce was a deeply physical presence. He had a low, gravelly voice that was often, because he was constantly amusing in conversation, punctuated by a chuckle that was like a shallow cough. Comparatively small, but unexpectedly fit for a man of his age and position, Noyce moved quickly, and he often seemed ready to spring from his chair. For a man whose reputation had been first made in the theoretical, Bob Noyce always seemed to stand firmly on solid ground.

Like many famous individuals, Noyce's concentration was almost superhumanly intense—and in conversation, the other party would often feel that he or she had entered into a private space owned by Bob Noyce and were the sole focus of his attention. For those few minutes, you were Bob's intimate friend—and you were thrilled that he agreed with everything you said. You were reluctant for it to end, even though you sensed that this was the way Noyce treated everyone he met and that he made them all think he was on their side. And you couldn't wait to talk with him again.

Comparisons have been made between the experience of being around Robert Noyce and the even more famous "reality distortion field" said to have surrounded his greatest acolyte, Steve Jobs. In truth, they were very different. Noyce's appeal was warm and visceral; Jobs's, cold and ethereal. Noyce made you feel that important, vital things could be accomplished if everybody could just steel their courage, ignore the risks, and move forward together toward a difficult but achievable goal—and have some laughs along the way. Noyce made you feel that, while he would be in the lead, it would only be because he was better suited for that job, as you were to yours.

Steve Jobs, by comparison, invited you to change the world, to accept his vision as your own, to join—if he deemed you worthy—the select few creating a perfect, cool new reality . . . with the knowledge that if you ever faltered, proved unworthy, or in some unknown way irritated Steve Jobs, that you would be jettisoned, shunned, and left behind.

The difference between the two is best captured in the photos of Bob Noyce, with his wide grin, and Steve Jobs, with his tiny, knowing smile.

Ironically, while Jobs's "zone" would come to encompass billions of people, it was Noyce's vision that was far more sweeping. Jobs merely wanted the world to own his company's computers; Noyce wanted to usher in the digital age. In a legendary television interview about his invention of the integrated circuit, which Noyce called "a challenge to the future," he unexpectedly turned to the camera, as if directly addressing millions of people at home throughout the nation, and said, "Now let's see if you can top that one," and smiled. From anyone else, it might have seemed arrogant, but coming from Bob Noyce, it was confidence, the friendly challenge of one competitor to another on the playing field to top what he's just done. Bob Noyce was happy to beat the hell out of you—he lived for competition, so much that it often put both him and others at risk—but after he did, he helped you up, slapped you on the back, and told you how well you had done.

Most of these traits were apparent in Bob Noyce even when he was a boy. There is a story from Bob's childhood that, at the age of five, he got into a spirited Ping-Pong game with his father—and actually beat him. He was exulting in his victory when his mother casually said, "Wasn't that nice of Daddy to let you win?"

Furious, little Bobby Noyce shouted at his mother in reply, "That's not the game! If you're going to play, play to win!"

That anecdote speaks volumes about a kindly but ineffectual father, an overpowering mother who was willing to be a little bit cruel to control the moment, and an astonishingly bright and athletic little boy with a competitive will greater than either of theirs.

In a glimpse of his peripatetic childhood to come, Bob was just six weeks old when the family packed up and moved to Atlantic, Iowa, a small town in the western part of the state, about halfway between Des Moines and Omaha, Nebraska. It was an established congregation of about two hundred, and the new minister and his family were given a furnished house to live in that even included a study for Bob's father to research and prepare his sermons.

It wasn't a prosperous life. Looking back, Bob Noyce would say,

"My earliest memory of that period was that it was Depression time. The church wouldn't pay Dad, so they paid him in produce."

The Noyces were a classic preacher's family. Ralph Noyce spent Saturday afternoons in his study preparing his sermons and all day Sunday at church delivering them. Harriet, in keeping with her personality, became that familiar type of preacher's wife who throws herself into every uplifting religious and self-improvement group in town, from the Ladies Auxiliary to vacation Bible camp to holiday pageants. And of course, she was a constant presence at the church throughout the week, but especially on Sundays.

The older Noyce boys were expected, as one would imagine, to be part of the church's social and liturgical life as well—not just helping around the church, but also participating in various Sunday school clubs and groups, as well as filling in any missing roles in church plays. As Bob grew older, these expectations were placed upon him as well.

Of course, being the preacher's sons, the Noyce boys were also expected, if not to be angels, then at least to stay out of trouble. This they largely did, at least until Bob got a little older. And it helped that Reverend Noyce was home most of the time and could be easily drawn out of his study to play.

That comfortable life lasted until 1935 and Bob's eighth birthday. But there had long been warning signs that tough times were ahead. The stock market crash had been only a distant disaster, but by 1932 its shock waves—the rapid deflation, massive unemployment, and grinding poverty of the Great Depression—had hit the Midwest hard. Farm produce prices collapsed, dragging down family incomes with them. This led to a wave of farm foreclosures. In town, the retailers and service providers who made their living from the surrounding farms watched helplessly as their revenues fell in concert. Then in the emblematic event of the Depression, the local banks—overextended on loans that wouldn't be paid, suffering runs on accounts by depositors, and lacking sufficient collateral to cover those lost loans—began to shut down. This last often put the townies in even more desperate straits than the farmers.

This was certainly true for the Noyces. With a relatively safe job and a subsidized home, they had largely been immune to the first few waves of the economic downturn, though Ralph did see his salary cut in half and his office closed. But then the local bank closed, taking with it the money Reverend Noyce had borrowed off his life insurance to pay for the birth of son number four, Ralph Harold. The Noyces were reduced to dipping into the reserves they had put aside for putting the boys through college.

Life was now very different from the comfortable world Bobby Noyce had known as a little boy. Now he would grow accustomed to hobos showing up at the door (no doubt the Noyce house had been chalk-marked with the encircled x as an easy handout) asking for food, a small job, or a place to sleep. As the minister's wife, Harriet Noyce often felt obliged to provide all three, the last in an abandoned henhouse. Bobby also got used to sharing his house with strangers, as the church gave work to impoverished members of the flock while the Noyces provided daycare for their children. Meanwhile, Ralph Noyce's salary continued to drop, then disappeared altogether. By the middle of 1935, the church was five months in arrears and was reduced to paying the preacher in kind. Bobby Noyce remembered the produce and the hams that were brought to the house; he was too young to notice the truck-loads of corncobs to be burned in the furnace during the winter.

In the end, even that belt-tightening wasn't enough. Finally, Harriet and Ralph gave up—and in late October 1935, the Noyces packed up their possessions, Ralph gave one last sermon, and the family departed for Decorah, a small town in the northeast corner of the state, just a few miles from the Minnesota border.

Economically, Decorah was much more welcoming to the Noyce family, but culturally it was not. The world had changed; thanks to the seemingly endless hard times, it was harder now, almost apocalyptic. Ralph's gentle questioning of his own faith, combined with his intellectual preaching style, rubbed the citizens of Decorah the wrong way. He gave them homilies when they wanted, needed, redemption and divine intervention ("a standard Scandinavian/Norwegian Iowa town," Noyce would recall). In short order, the congregation complained, then bled away. Within two years, the Noyces were gone again, this time to Webster City, a fairly large town just north of Des Moines.

With the Webster City job, Ralph had, strictly speaking, been kicked upstairs to an administrative post with the Iowa Congregational

Conference. His new job, associate superintendent, required not only bureaucratic work, but also filling in for absent ministers and administering the Congregational youth education program for the entire state of Iowa. This last put Ralph Noyce on the road almost every day for the next year. During those twelve months, he covered twenty-five thousand miles—most of them on dusty, pitted dirt roads—and spoke before more than a hundred congregations and church groups.

Ten-year-old Bobby Noyce would remember that his dad was lucky to make it home for one Sunday dinner every six weeks during this period. Yet children adapt, and Bob would also look back on this time as the beginning of one of the most magical times in his life.

The biggest reason for this was his mother. Freed from all of the responsibilities of being a preacher's wife, she focused her attention on her sons. The two oldest Noyce boys, Don and Gaylord, were now heading into their teens, and while Don was sickly, Gaylord was a budding scientist and typically enlisted Bob on his expeditions and experiments. Harriet, among that distinct species of the era of a highly educated woman feeling trapped as a homemaker, devoted her time to teaching the boys the social graces and supporting them in their after-school endeavors, from dead-cat taxidermy to the fabrication of explosives. In an unpublished memoir, she would write, "I felt the sense of belonging as a person of worth for myself, and not just the minister's wife."

It was Bob Noyce's first encounter with experimental science—but just as important, it was a lesson in just how much could be accomplished with skilled intellectual and financial support (i.e., Mom) behind you.

The job in Webster City was not just a way to escape the Decorahs of a minister's life. Ralph Noyce had also taken it as a springboard toward a very specific goal, and he was willing to put up with the seemingly endless sacrifice of lonely roads and guest bedrooms to get there. And eighteen months after took the position, he once again packed up the family to move to his dream location and the family's final home: Grinnell, Iowa.

For Ralph and Harriet, Grinnell represented what they had dreamed of since the first days of their marriage. It was a small, impeccable prairie metropolis with seemingly a church on every corner. For the intellectual couple, it was a college town, and much of the town's life and culture revolved around Grinnell College, an obscure but top-quality school.

For young Bob, weary from being uprooted for the third time in four years, it was just one more hassle of settling in, quickly making new friends (a talent he would also use well as an adult), and orienting himself to the local school's curriculum. He was twelve now, pudgy and more than a little intimidated by his scholarly, academically brilliant older brothers.

But over the next few years, Bob Noyce would come into his own, and the young man who emerged from high school five years later would be confident in his academic skills, his athletic abilities, his talent for leading others, and his attractiveness to the opposite sex. The Bobby Noyce who went to junior high school in Grinnell that first day feared that compared to his brothers and in the eyes of his new classmates, he might be a loser. By the time Bob Noyce (he had adopted the more adult nickname) walked across the stage at Grinnell High School to receive his diploma, he knew that he was one of life's winners, destined to do great things.

It was a remarkable transformation—and a first tantalizing glimpse of the extraordinary adult. For that reason, it deserves a closer look.

CHAPTER 7

The Demon of Grinnell

Moving to Grinnell gave Bob the first real stability he had known in his life—even if it was economically shaky. Probably because of the stress of the previous years, Ralph Noyce suffered a small cerebral hemorrhage that left him partially blind and with some damage to his short-term memory. That reduced his endless traveling but also made the family's finances more precarious. So although the Noyces stayed in Grinnell through Bob's high-school years, they never enjoyed the financial stability to purchase a house. Harriet moved the family to a new rental home almost yearly. Still, the family remained in town and the boys at the same schools for nearly a decade.

That stability seemed to unleash the Noyce boys, but especially Bob. Over the years, he and Gaylord had continued their adventures, and now they took them to a new level.

In their most famous achievement, and the one most remembered by neighbors, the two boys built a glider.

Before the Baby Boom of the 1950s, a similar boom took place in the 1920s after the end of World War I. The result was that the neighborhoods the Noyces lived in during the 1930s often looked like American suburbs in the early 1960s—that is, swarming with kids. At the time that Bob and Gaylord decided to build their glider, the neighborhood they lived in had eighteen kids, seventeen of them boys. Bob, though he was among the youngest, quickly learned to be a leader of this gang, in part because of his growing confidence, but even more because of his quick imagination, which came up with endless new projects and schemes.

As one neighbor would say, "Harriet had her hands full. Those boys, especially Bob, were into devilment."

The impetus for the glider came from an event, now long forgotten, that had a surprising influence on that generation of boys. A former barnstormer was touring the country offering brief rides in his Ford Tri-Motor for one dollar. Thousands of American boys, usually accompanied by a parent, jumped at the opportunity—while many thousands more, too poor to pay, looked on with longing.[1] Bob Noyce got his flight and came away inflamed with a desire to build an airplane, or at least a glider, of his own.

Tellingly, he enlisted first his brother and then all of the kids in the neighborhood to help—even the lone girl, who was recruited to sew the wing cloth. The wood for the wings came from spindles used in rolls of carpeting at the furniture store owned by one of the local fathers. Harriet, as usual, played a part: she made the paste. The design came from an illustration in the family's encyclopedia, and the whole thing cost $4.53.

The finished product, which resembled the Wright/Lilienthal gliders of a half century before, even down to the rear stabilizer, was an impressive achievement by any measure, but especially so given that it was designed by a twelve- and a fourteen-year-old. At four feet tall and eighteen feet wide, it was a bit small in scale to lift an adult, but *in theory* it might be able to keep a boy aloft. It also lacked wheels or skids, so Bob enlisted a group of neighbor boys to lift the glider and run with it.

To everyone's amazement (except perhaps Bob's), the glider actually flew—slightly—as the crew held it overhead and ran down the street. It did even better when they built a small dirt ramp, ran up it, and flung the glider into the air. But still, that wasn't enough for the budding aeronauts. One of the older neighbor boys not only had just gotten his driver's license, but also had access to the family car.

It wasn't long before that car was racing down the street, the glider roped to its bumper . . . and baby brother Ralph, age seven, aboard as the reluctant pilot. The glider flew. More important, little Ralph survived the flight—and no one spoke about it again.

But Bob wasn't done. Now that he had financing, proof of concept, and a working prototype, it was time to bet everything on a true field test: he would try to jump off the roof of a barn and live. The barn was just across a field behind the house, and once he was ready, Bob marched over with his team carrying the glider.

Word spread throughout Grinnell, and a small crowd gathered—to cheer on the young man or watch him break his neck. They were joined by a photographer from the *Grinnell Herald* (it's never been determined how he learned of the event—it wouldn't have been beyond Bob, or his mother, to have called the paper).

Bob climbed the three or four stories to the barn's roof and the rest of the team handed the fragile, twenty-five-pound glider up to him. For anyone who was there, it was an unforgettable sight: a twelve-year-old boy, precariously balanced on the roof of the barn, lifting the glider, taking a deep breath . . . and leaping off.

It is all there in that moment: the ambitious project, the technical acumen, the leadership, the self-promotion, and courage that bordered on recklessness. And being Robert Noyce, he had the skill—and the luck—to survive and not die young and unknown. It wasn't much of a flight; it was more of controlled fall, but the glider exhibited enough lift (and Bob enough skill on his debut flight) to bring both down to a hard but safe landing.

And so even before he reached his teenage years, Bob Noyce had already made himself something of a local legend. And he had just begun. Though he would never lose his love of flying—once he became wealthy after Fairchild, he owned, and was an enthusiastic pilot of, multiple airplanes—his interest shifted, as did all young American boys of the era, to automobiles.

Within a year after his inaugural flight, Bob and his buddies were building their first car. "We made our first 'kluge' [primitive] motor vehicle from an old gasoline engine off a washing machine. Back in the days when rural electrification was just coming in, there were a lot of gasoline-driven washing machines that were being dumped, and so they were cheap."[2]

Bob's father would presciently say that Bob thrived on "adrenaline and gasoline."

And that was just the beginning. As Leslie Berlin recounts:

Noyce would show up at neighbors' houses, his pockets full of wires and clips, and ask to borrow the 220-volt outlet for the kitchen range so he could try to build the electric arc [cutting torch] *Popular*

Science claimed was capable of burning a hole through steel. He started smoking cigarettes. He and his friends enjoyed tipping over outhouses on the nearby farms, though attacks of conscience often sent them back to the scene of the crime, swearing and sweating in the stench as they righted the wooden building. They shot firecrackers off the slides at Merrill Park and from the roof of Gates Hall on the college campus. And while his older brothers' commitments to Congregationalism deepened in high school, Bob began spending less and less time at the old stone church at the corner of Fourth and Broad.[3]

It wasn't all fun and mischief. Given the family's straitened finances, Bob was expected to make his own money, and it wasn't long before his days and vacations were filled with jobs that ranged from delivering newspapers and special-delivery letters to shoveling snow, hoeing beans, and detasseling corn.

In the brief interval of junior high school, Bob Noyce had changed the entire trajectory of his life. He was now settled, well known, and increasingly sure of his abilities. Now Grinnell High beckoned—and he was ready to make his mark.

On Monday morning, December 8, 1941, freshman Bob Noyce joined the student body in the Grinnell High auditorium to hear the official announcement of news they already knew: America was at war.

For young Bob, the war changed everything—and yet changed nothing. For the first few years at least, until Don and Gaylord reached enlistment age, life at home went on pretty much as before. Bob grew ever more confident in his abilities and accomplished in his schoolwork. He delivered papers, hung out at the local malt shop, built his beloved gasoline-powered model planes, and after he got his driver's license, became even more popular once he was given regular use of his mom's 1939 Plymouth. Its high-compression straight-6 made Bob a demon on wheels, and he quickly built a reputation drag-racing through downtown. Yet he still made it to oboe practice—and with almost no one noticing, maintained straight A's.

In another telling glimpse of the businessman to come, Bob cleverly contracted with his neighbors to shovel the snow from their driveways for a fixed fee. Thanks to these contracts, the neighbors didn't have to hire local kids with each snow but could rest assured that the work would be done—and Bob, praying for clear skies, often was remunerated for more than the sum of the individual jobs.

At the same time, the war was omnipresent. Senior boys graduated and joined up; their exploits (and occasional deaths) were announced at assemblies and in the school paper. Doing his part, Bob participated in the regular bond and scrap drives in town and served in the civil air patrol, helping with the regular blackout drills.

In his car, Bob Noyce was a demon—and not above sneaking out into the countryside to siphon rationed gas from farmers' unwatched tractors. Out of the car (and sometimes in its backseat) he was catnip to the girls: ". . . all the girls were crazy about [him]," one classmate remembered. "They thought he was the most handsome thing on the face of the earth."

Marianne Standing, the most beautiful girl on campus, took one look at Bob Noyce and was smitten: "He was probably the most physically graceful man I've ever met. Just walking across the lawn . . . on a horse, even driving a car."[4] Noyce, for his part, was drawn not only to Standing's stunning looks, but also to her wit, the fact that her mother was a divorcée (which in those days made mother and daughter "fast"), and the fact that, like Bob, she liked to smoke cigarettes. Before long, there were two figures racing around town in the '39 Plymouth. Bob's mother disapproved, believing that "Marianne had a gift for trouble"— which was likely exactly what Bob was looking for.

In an effort to corral her most troublesome child, Harriet Noyce set her mind to both reforming Marianne (dinners now ended with hymn singing) and keeping Bob distracted with even more activities. One of these, which would prove unexpectedly transformational in Bob's life, was a series of odd jobs he and his brothers did around the house of Grant Gale, a member of their congregation and a professor of physics at Grinnell College.

Never shy, Harriet soon parlayed this relationship into a request of Professor Gale to let Bob, now entering his senior year, take Gale's Introduction to Physics course.

High-school students taking college classes are a lot more common now than they were in late 1944. But Bob was a special case. The school yearbook called him "the Quiz Kid of our class, the guy who has answers to all the questions." Not only was he acing all of his classes, but also he was obviously bored to death in physics, the most difficult science class the high school had to offer. It had gotten so bad that Bob started bringing a watch to work on under his desk during class—to the amusement of his classmates when he quickly put a jeweler's loupe in his eye every time the teacher turned to the blackboard.

Even then, Gale might not have entertained such a proposal had not he known Bob so well. He checked for proof that there was some small precedent for local high-school students to attend classes at Grinnell, found it, and invited his neighbor to join the class. So beginning in January 1945, Bob began attending both Grinnell High and Grinnell College.

Noyce would say that "my last year I took college courses, specifically in physics, just simply because I was relatively bored with the stuff that was going on in high school."[5] But it was much more than that: once again his talent had put him in the perfect position for his luck to work. Grinnell may have been a small Midwestern liberal arts college, but Grant Gale's classes, made even smaller and more personalized because of the draft (Bob was the only male student), were anything but parochial.

Gale was, in fact, a superb teacher, capable of engaging even the nonscientific types who took his physics survey course—and a godsend to anyone interested in the subject. He eschewed traditional theoretical learning—there was no homework in his class, and textbooks were for reference—and instead taught his students using real-life questions and solutions. He'd march them outside and have them throw a snowball against the classroom building to measure the force of the impact or measure the change in moment of inertia by a spinning skater when she pulled in her arms. Gale was short on patience and long on homilies, and sitting in the classroom, surrounded by co-eds, puzzling his way through physics experiments, was like paradise to Bob. Gale's "interest was infectious," Bob would later say. "I caught the disease."

Meanwhile, he still had to graduate high school. And when he did, even his friends were astounded when he was named class valedictorian.

They knew how smart he was, but he had been so nonchalant, had goofed around so much in class—and ultimately made school look so easy—that none of them figured he could possibly have the highest grades among them.

A few years earlier, this achievement would have seemed impossible to young Bob. High academic honors were for his serious older brothers. But now he had a growing sense that he was destined for great things. Many years later, he would say, "I began to feel I had a bit more than average ability."

In truth, the award had made Bob a bit of a smart-ass. That summer, he followed Gaylord to Ohio and Miami University, where his big brother was undergoing officer training. To fill the time, Bob took some courses at the school—and at the finish of one math course told the instructor that he was "getting a nice bit of review out of her course, even though I didn't attend classes."

Other than the occasional classroom visit, Bob kept himself busy. He caught an opera, hitchhiked a couple hundred miles to see Marianne, and when he got particularly bored, took up swimming. In the end, more than college, swimming was the leitmotif of Bob Noyce's summer of '45. He swam almost every day, getting himself into ever better shape—until one day, after he watched three members of the Miami U diving team work out, he decided to try it himself. By the end of the day, "after landing flat on my back only twice, I perfected the technique. Before I went home, I did both a half and a full gainer off the ten-foot board—Whoopee!"

If that wasn't a confidence builder, then the invitation that came a few weeks later surely was. It came from the head of the Physics Department, offering Bob a job: if he would come to Miami U, the department would put him on the faculty payroll as a lab assistant—grading papers, even teaching a few classes—even as he took classes of his own. Even cocky, nineteen-year-old Bob Noyce, coming off the best year of his life, was astonished by this offer, which always went to graduate students, never incoming freshmen.

Life always presents turning points, and they often come just when things are going their best. The trajectory of Bob Noyce's life at this moment was toward Miami University, where he would likely have become a distinguished professor of physics and perhaps made a few important discoveries and inventions in a long and respectable career.

Instead, he made the right choice, for mostly the wrong reasons . . . and his luck worked for him once again. Fearful that he would be lost on such a big campus—his words were "another insignificant student"— Bob decided instead to return home to Grinnell and its college, newly awarded scholarship in hand, and start as a freshman.

The college was thrilled—though the personal acceptance letter from the president must have given Bob some second thoughts: "Your brothers before you have performed in a distinguished manner. You seem to have the ability to perform equally well. We expect great things from you." It seemed that Bob would never get out from under his brothers' shadows—Gaylord had just been named to Phi Beta Kappa—and, having had a chance to escape, he instead had put himself back in the same trap.

Almost as if he was trying to make the little world of Grinnell forget his older brothers, Bob Noyce quickly buried himself in a blizzard of activities, from the diving team to chorus, drama, yearbook, and of course, girls. It made even the frenzied pace of his high-school years seem sluggish. And yet he continued to get the highest grades, especially in his major—even as those classrooms filled with returning GIs in even more of a hurry to get on with their careers. His electronics professor had him write his own exam; his calculus professor asked him to teach one of his classes. And still that wasn't enough: insufficiently challenged, Bob decided to prove every formula he was being taught in physics, thereby recapitulating the entire history of the field. It was the showy, outsize move of a young man wanting to put the world on notice. At Grinnell, at least, it worked.

Yet even that wasn't enough. Halfway through his freshman year, Bob decided to earn a varsity letter as well. None of his brothers had done that, at least. And diving, which was increasingly his private obsession, would be his field of battle. Each day he would practice at Grinnell's bizarre diving pool—because of a low roof, divers had to jump up into an alcove cut into the ceiling before they descended. Each night he would practice a technique he called "envisioning myself at the next level," in which he would imagine doing his dives perfectly. In this he would presage the "mental rehearsal" training of Olympic and professional athletes decades hence, but also invented a mental habit for himself that he would continue to put to use for the rest of his life.

And it worked. In his second year on the Grinnell diving team, Bob won the 1946–1947 Midwest Conference Diving Championship.

Everything—the grades and honors, the socializing, the clubs, and the diving—now seemed to be converging on a storied senior year at Grinnell, a year of achievement that would be talked about with awe for a generation to come.

And then it all fell apart.

CHAPTER 8

The Pig Thief

I t was probably inevitable. The frenzied pace at which he had been running, taking on more and more tasks and responsibilities and expecting to be the best at every one, couldn't be sustained forever, not even by Robert Noyce.

The first crack came at the 1947–1948 Midwestern Conference Diving Championship. He was returning as the current champion, and so it was expected by everyone, including his parents, who had come back to town to attend the event, that Bob would have a convincing win. Instead, he lost by two points. Others saw it as a hard-fought second place; Bob saw it as a failure.

It got worse, as if the first crack, once made, spread along the thinnest spots in Bob Noyce's perfect veneer.

The boys who had left for war had now returned to Grinnell as men. Having seen the horrors of the Pacific or the European theater and having lived as if their lives would be short, they brought a whole new culture to quiet Grinnell: loud, hard-partying, unconstrained by mere rules, jamming every second with action, and rushing to get done and back into the world of adulthood. Bob had been the budding Big Man on Campus, but now with these worldly twenty-four-year-old decorated vets jamming the recently half-empty classrooms and crowding every campus party, he once again had to struggle to find a place.

It helped that Bob's natural charisma made him friends among all of these returning veterans, but he also had to devote a lot of precious spare time to keeping up his bona fides. Thus, when his four-year dorm, Clark Hall, decided to make a Hawaiian luau the theme of its spring house party, Bob casually accepted the job of finding—stealing—the pig for the roasting spit. It was a cocky move—and a foolish mistake.

Then it got even worse. Bob's girlfriend of the moment informed him that she was pregnant. In later years, he would never talk about this time, other than that the girl got an abortion. How Bob felt about that decision is unknown, but for a young man who seemed to own the world a few months before, it was a long way down to the bottom of this mess. The dream of perfection was turning into a nightmare.

And it got worse still. Still upset and depressed by the pregnancy and its end, Bob realized to his dismay that the house party was rapidly approaching and he was expected to find the pig. With the kind of fuck-it-all attitude of a young man who thinks he's got nothing left to lose, Bob got drunk with a buddy, and the two of them went out into the night, across the golf course behind the campus, to a well-tended farm, grabbed a twenty-five-pound piglet, and hauled it back to Clark Hall. Any thought the two young men had that this might be a secret mission ended when their housemates decided to butcher the animal in the hall's third-floor shower. The blood-curdling squeals could be heard all over campus. Still, the luau was deemed a great success, and one last time, Bob was the hero.

Reckoning came in the morning. Bob and his buddy, feeling guilty, went back to the farm to apologize and offer to pay for the animal. That's when they discovered they had made a terrible choice of a target for their thievery. Berlin: "The farmer was the mayor of Grinnell, a no-nonsense man given to motivating his constituents through mild intimidation. He wanted to press charges. The college's dean of personnel, a recently retired army colonel, was also inclined towards the harshest punishment possible; a few months later, he would expel another of [Professor] Gale's advisees for swearing at his house-mother. Since the farm was outside the city limits, the county sheriff was called in."[1]

Years later, in Silicon Valley, when the pig story first came out, it was an object of amusement, a welcome bit of humanity attached to the increasingly redoubtable Robert Noyce. In the San Francisco Bay Area in the 1980s, such an act would have gotten a fine and perhaps a required letter of apology. But it wasn't funny in 1948, especially in Iowa. As a letter, written by the dean of personnel to Bob's parents, said, "In the agricultural state of Iowa, stealing a domestic animal is a felony which carries a maximum penalty of a year in prison and a fine of one thousand dollars."[2] The first part was obviously a holdover from the days of cattle rustling and frontier justice, but the fine was very real, and huge: it equaled most of Ralph Noyce's annual salary, or three years of Grinnell tuition. And even if he did escape the court's punishment, it was pretty likely that Bob would face expulsion from the college. And that black mark would assure that he would never be able to attend grad school and likely would never be able to find a job worthy of his talents.

In the end, Bob Noyce's future was saved by his physics professor, Grant Gale, and college president Samuel Stevens. Neither man wanted to lose a once-in-a-generation student like Bob Noyce, and so, though the pig-farming mayor was adamant at first, Gale and Stevens finally convinced him to accept a deal in which young Noyce would cover the cost of the pig and no charges would be filed.

That got Bob off the hook with the law, but there was still the matter of the college. Though he should have been expelled, Bob was let off easy: he was allowed to finish the school year but then was to be suspended for the first semester of his senior year. He would not only be barred from campus, but from the town of Grinnell as well.

He had dodged the bullet—though remarkably and tellingly, his parents didn't think so. After Bob, humiliated and ashamed, left town and joined his folks at their new home in Sandwich, Illinois, he was astonished to find that not only weren't his parents angry at him, they were actually furious at the college. Ralph Noyce had even drafted an angry note to Stevens ("the rest of us will have to be the more ready to accept youth's offer of repentance and desire for forgiveness even if Iowa hog farmers do not see it that way") that Bob failed to prevent him from sending. Part of the tone of that note no doubt reflected some of Ralph's own bitterness at his lost fortunes in Grinnell, but President Stevens must have been appalled at the lack of appreciation for what he had done for Bob.

As for Bob, he had to find a way to fill the next six months with productive work before returning to Grinnell—if he had the courage to face the school and the townspeople. Luckily, out of pity, his old math teacher lined up a job for him in the actuarial department at Equitable

Life Insurance in New York City. As Gaylord was now working on a doctorate at Columbia, Bob quickly had both a job and a new place to live for the next six months.

Though one might think that the jump from Iowa to Manhattan would have been intimidating, for Bob it was already old hat. He and a buddy had spent parts of two previous summers working as bartenders and waiters at a country club just north of New York City. It beat pitching hay in Grinnell, the pay was good, and the two young men could occasionally sneak down to the big city to play.

So Noyce settled in quickly. Initially excited by the prospect of a career as an actuary, Bob quickly found the work unspeakably boring but the girls at the office pretty and abundant. In the evenings he caught Broadway shows and hung out with artists and playwrights and played bohemian. Being Bob Noyce, he did his work brilliantly, as much as he grew to hate it, and even learned a few things about the power of money as a motivator of human behavior—"Little facts like how people really do unconsciously react to financial incentives: if you pay them to die, they'll die, if you pay them to live, they'll live . . . at least statistically."[3] He would later say that he also learned enough to be suspicious thereafter of the data used in developing statistics.

It was a long, but nevertheless rewarding nine months: "I went into it with the idea that this was a secure and comfortable place to be. I came out of it with the feeling that it was a terribly boring place to be."[4] He would never choose the safe path again.

He returned to Grinnell in January 1949, head held high, and quickly reimmersed himself in the busy life he had left. On the surface, everything looked the same. Bob had even been sufficiently far ahead in credits that he was able to rejoin his classmates as an equal, with the prospect of graduating with them in June. But underneath, everything had changed. The Bob Noyce in frenzied pursuit of perfection was gone, replaced by a more serious, mature Robert Noyce.

Now, almost magically, his luck returned. By a remarkable coincidence, Grant Gale had gone to the University of Wisconsin with John Bardeen, and his wife was a childhood friend of the scientist. Moreover, Bardeen's boss, Bell Labs president (and father of the undersea telephone cable) Oliver Buckley was a graduate of Grinnell—his two sons were currently students at the college—and he often sent the physics department his obsolete equipment.

In 1948, ironically even as Bob Noyce was learning annuities at Equitable Life, a few blocks away Bell Labs held a public press conference (rather than merely published a technical article) to announce John Bardeen, Walter Brattain, and William Shockley's invention of the transistor. As a public relations event, it was something of a failure—only the New York Times gave it even a few paragraphs, on page 46. Noyce, who would have a greater impact on the fate of the transistor than anyone, apparently didn't even see the announcement.

But out in Iowa, Grant Gale did. He not only cut out the article and posted it in the hallway of the Physics Department, but also in his next equipment request letter to Oliver Buckley, asked if he might throw in "a couple of transistors."

Needless to say, Buckley didn't have any spare transistors lying around—the devices were barely out of prototype and preorders were backed up for months—but he was kind enough to gather up a bunch of newly published Bell Labs technical papers and monographs about the transistor and sent them to Grinnell.

Noyce's return to Iowa took the same path and occurred at about the same time. At Grinnell, he found Gale in a fervor over these documents—and dove in with him. Gale would recall, "It would be an overstatement to suggest that I taught Bob much. . . . We learned about them together." Noyce would say that "Grant had an infectious interest in transistors that he passed on to his students. . . . I began to look at [the transistor] as being one of the great phenomena of the time."

Tellingly, the entrepreneur was already emerging. While Gale saw the transistor as a new field of study, Noyce saw that "it would be something good to exploit—well, maybe 'exploit' is the wrong way to put it—but I saw it as something that would be fun to work with."[5]

Unable to actually hold transistors in their hands and test them in circuits, Noyce and Gale were largely restricted to understanding their underlying physics. In particular, semiconductor technology, and the doping of insulators with atomic impurities so that a current running through this *semiconductor* could be controlled by a small side current—the same demonstration that had obsessed Bardeen and Brattain a decade before.

The pioneering solid-state circuits in early transistors like those being studied by Noyce and Gale were used as amplifiers to replace much larger, hotter, and more fragile vacuum tubes. A decade hence, Bob Noyce's great contribution would be to turn these circuits into on-off silicon *switches* or *gates* to create the binary ones and zeros at the heart of all digital circuits—and then figure out how to make them flat and interconnected in arrays.

For now, though, the most he and his professor could do was understand as much as possible the principles behind the transistor. And as much as that was a handicap to their full understanding of the device, it also gave them a focus (without the distraction of playing with actual transistors) that left them, by the end of Bob's senior year, with nearly as much expertise as the scientists at Bell Labs. That Noyce, still an undergraduate, could even understand some of these technical papers was a wonder in itself—though Grant Gale, who knew the young man as well as anyone, wasn't surprised.

Robert Noyce graduated from Grinnell in June 1949 with degrees in mathematics and physics, Phi Beta Kappa, and from his classmates, the Brown Derby Award for "senior man who earned the best grades with the least amount of work." Characteristically, he took the last with humor and the appraising eye of a businessman, telling his parents that the award was for the "man who gets the best returns on the time spent studying."

Bob had also been accepted in the doctoral program at the Massachusetts Institute of Technology and given a partial scholarship. He told Gale that he planned to focus on the movement—particularly the transmission—of electrons through solids.

Bob's MIT scholarship covered his tuition but not his room and board, estimated at $750 for the year. His parents couldn't help with those expenses, so Bob joined them in Sandwich, Illinois, and spent a hot and miserable summer in construction—usually working with the corrosive and flesh-burning (and, we now know, carcinogenic) wood preservative creosote. From this literally scarring experience, Noyce managed to learn another life lesson: he swore to never be reduced to manual labor again, . . . and within months of his arrival at MIT, he had turned that partial scholarship into a full one.

In the stereotypical story of the country college boy going to the big-city university, the young man is overwhelmed by the sophistication of urban life and a world-class educational institution. For twenty-two-year-old Bob Noyce, it was just the opposite. While in Iowa, he had been exposed to the absolute cutting edge of applied solid-state physics. Bob now quickly discovered that at mighty MIT, "there were no professors around who knew anything about transistors." Ironically, it was at this moment that, back in Grinnell, Grant Gale was receiving his first shipment of gift transistors from Bell Labs.

It was not a happy discovery. Noyce was reduced to cobbling together courses, mostly in the field of physical electronics, that covered pieces of what knew he needed to study in semiconductor physics. "The major problems in the field at that time were electron emissions from cathode ray and vacuum tubes. But still they had many of the same physical properties [as transistors]; you had to learn the language, the quantum theory of matter and so on."[6] The rest of his education he picked up by attending every one of the few technical conferences around the country devoted to transistor technology. It was at these events that he first met some of the current and emerging superstars in the field—men who would play important roles in his future career—such as William Shockley and Lester Hogan.

He had a rough start. Though a superstar at Grinnell, Noyce was just one of hundreds of top students from around the world in MIT's physics program, which that year included the future Nobelist Murray Gell-Mann. Moreover, his basic education at Grinnell, though brilliant in some areas, was deemed inadequate in others—ironically, after he bombed his placement tests, he was ordered by the department to take a two-semester undergraduate course in theoretical physics. Gale was so concerned about how young Bob would do that he wrote to the MIT Physics Department asking for regular updates on the young scholar . . . and he was not thrilled by the first reports.

Meanwhile, Bob had little money on which to survive. Dorm life on campus was just too expensive, so Bob shared a cheap apartment in a rough neighborhood in Cambridge and lived off as many meals as possible with his friends. Luckily, he had a number of those—two old Grinnell buddies (and neighbors—they'd helped with the legendary glider), the Strong brothers, lived nearby. Brother Gaylord was now married and living in New Haven, Connecticut. As was his way, Bob also quickly made new friends with several of his classmates. They were a mixed group—rich, poor, blue-blooded, immigrant—but they found common cause in struggling to survive their first semesters at MIT. Not surprisingly, Bob also quickly found a girlfriend—whom he sometimes ordered to stay away while he crammed for exams.

It was a pretty miserable existence. Bob tried to put on a good face for his parents, writing, "My only observation for comfort is that everyone I talked to did as badly as I did" and "life looks unpleasant in spots." But eventually, he confessed to them his unhappiness and frustration: "The whole of [my visit with Gaylord] served to point out to me how misdirected I am. These people have some worthwhile goals in life. It doesn't seem to me that I have. I keep hoping that I will get wrapped up enough in physics to forget this." Still, he told them, dinner at his brother's had at least been a nice respite: "Anyway, my materialistic interests flew out the window until I got back here and started to wonder how I was to stay alive."[7]

Still, as the weeks passed and Bob didn't flunk out, he began to grow in confidence. Perhaps he really could make it at MIT. Two classes during this period played a crucial role in his future. The first, on the quantum theory of matter, was taught by a legendary physicist, John Slater. Slater, who would write a classic book in the field, terrified most of his students—he never looked at them, but just filled the blackboard with equations while shouting out questions. But Noyce, who actually answered most of those questions, discovered he had a particular aptitude in the field—itself the fundamental science behind semiconductors.

The second course, in electronics, was less important than the man who taught it, Wayne Nottingham. Nottingham, already well known for his talent for building cutting-edge laboratory vacuum equipment, also had built a national reputation for putting on an annual Seminar in

Physical Electronics, to which he invited the biggest names in this new field. And while Nottingham's Electronics course was largely a bust for Noyce—it didn't even discuss transistors that first year—the seminar was a milestone in his education, not least because one of the speakers that year was none other than John Bardeen.

By the end of that first quarter, Bob had not only caught up with his classmates and survived his exams, but from that point on, he would be one of the program's academic leaders, earning honors in every course he subsequently took at MIT. As at Grinnell, he seemed to come out of nowhere: the apparently average student who suddenly surpassed all of his classmates and sprinted ahead. He also seemed to learn and retain information faster than anyone else in the program—to the point that his fellow students nicknamed him Rapid Robert, after a fellow Iowan, fireballing baseball pitcher Bob Feller.

Classmate Bud Wheelon, who would later become the technical director of the CIA, was so impressed by Noyce's brilliance—and its depressing contrast to Bob's squalid living arrangements—that unbeknownst to Bob he personally went to Professor Slater and asked for a paid teaching-assistant position for his classmate, something Bob would have been too proud to do himself. Slater sent Wheelon away, saying it was none of his business, but a few weeks later, Noyce was stunned to be awarded a special $240 fellowship that got him through his first year.

Now once again, as he had in high school and college, Bob Noyce moved to the front of his class: graceful, affable, and whip-smart. His study group, easily identified by Noyce's glowing cigarette in the study room of the Graduate House, quickly became the center of attention. Now, with the new grant, he was able to move to the House (instead of driving home in the wee hours of the morning) and make his presence even better known. Soon he was going to parties and dances, and he even organized a clambake (no pigs in Cambridge) that became notorious because none of the physicists or chemists knew how to tap a keg, so they drilled it instead, with the resulting exploding geyser of beer.

Soon enough, with his studies now in hand, Bob began to look for new challenges. He auditioned for and won a place as a baritone at Boston's Chorus Pro Musica, one of the nation's leading choral groups. He quickly began serially dating almost every pretty girl in the chorus. And they happily responded: Bob had also gone back to swimming—and according to reports from ladies at the time, he looked like a body-builder. Through these years, as his buddies looked on in envy, he went through a progression of girlfriends, though none of them lasted long.

By the end of his first year of grad school, Bob was beginning to once again exhibit that superhuman energy that seemed impossible to maintain. But this time he didn't crash. Rather, he kept going. He went back to model-airplane building, then took up astronomy (or more accurately, the construction of reflector telescopes); then he tried his hand at painting, naturally enough convincing a young lady to model for him in her underwear. He applied for a Fulbright Fellowship to France, won it, then turned it down.

Perhaps most remarkable, given his heavy schedule, Noyce also took up acting, performing in several musicals—and being Bob Noyce, soon enough he was the lead. His brother Gaylord came to one of those performances and had an epiphany watching his younger brother. As he told Berlin, Bob's natural confidence and charisma, rather than any particular talent, enabled him to "pass himself off as an expert" performer. "His tone wasn't that great or accurate, but there he was, singing a lead."[8]

That confidence wasn't lost on either the cast or crew. And during one such performance, at Tufts University, it caught the eye of the costume designer, a mousy little blonde named Betty Bottomley, with as much energy as Bob Noyce and an even more devastating wit. Friends compared her to Dorothy Parker for both that wit and her East Coast sophistication; most folks were just intimidated by her cleverness. Having always preferred brains over beauty and uninterested for long in any woman unless she intellectually challenged him (like his mother), Bob was smitten.

But first, there was a doctorate to be earned. That summer, at the end of Bob's first year, Professor Grant Gale received an update letter from MIT. It said:

"Mr. Noyce has been an outstanding student in all respects. . . . We are sufficiently impressed with his potential that we have nominated him for a Shell Fellowship in physics for the next academic year, and he has received this fellowship.

"You are to be congratulated on the excellence of the training which he has had, and we look forward to an outstanding performance by Mr. Noyce."[9]

Gale cherished that letter and kept it for the rest of his life.

CHAPTER 9

A Man on the Move

Bob was now exactly where he wanted to be. The Shell money, $1,200, paid for his next year's tuition, and a likely research-assistant job would take care of school after that. He wrote his parents: "When I came here this fall, I was hoping something like this might work out. It seems that my optimism was somewhat justified."[1]

During the summer of 1950, Bob worked in Boston at Sylvania, a company that would one day be a neighbor in Silicon Valley. He completed his oral examinations the next May. After that, he took a consulting job at an optics company, audited a course at Harvard that MIT didn't have, and became a research assistant studying vacuum tubes at MIT's Physical Electronics lab. But mostly, his attention was focused on finding an adviser and getting to work on his doctoral dissertation.

He chose Nottingham, largely because he wanted a hands-on experimental scientist, not a theoretician, and set to work on what would be "A Photoelectric Investigation of Surface States on Insulators"—a topic central to the physics of the new Bell Labs transistors, now about to burst into the public eye. It would prove to be a major challenge, as the test materials had to be heated to 1,000 degrees Fahrenheit, resulting more often than not in an exploded test apparatus. But Noyce was not a man to fail; he persevered and ultimately achieved measurable and repeatable results.

It wasn't all work. Nottingham, an outdoorsman, spent much of his winters at his ski cabin in New Hampshire and typically invited his graduate students to meet him there rather than at the college. Noyce, the prairie boy, took advantage of this opportunity to learn how to ski. As with diving, he took to it so quickly that no one remembered him learning—he just seemed to be skiing soon after he put on skis for the first time. It quickly became a passion—to the point that he eventually instituted a "no thesis, no ski" rule on himself to stay focused on his dissertation.

Once again, as he had with the glider and with swimming, Bob quickly took to the air and started teaching himself how to ski jump.

This time the result wasn't so lucky: he landed hard on a jump and got a spiral fracture of his right arm. The break was so painful and dangerous—a pinched nerve, a potential blood clot—that he was rushed to Dartmouth, where doctors removed a bone splinter and put the arm in traction for two weeks. It would be almost two months before Bob recovered enough to go back to his experiments at MIT, and he would be troubled by his right arm for the rest of his life.

He completed his dissertation in mid-1953. As he would later admit, the experiment was, if not an outright failure, then certainly inconclusive. He never did find any surface effects, but whether that was because they didn't exist or because Bob had poor experimental technique wasn't clear. He did console himself with the knowledge that, along the way, he had made an important contribution to the general knowledge about magnesium oxide. But it wasn't the triumph he was hoping for.

Berlin has found some silver lining in the intangible results of Noyce's research: "The most important lesson that Noyce learned from his dissertation could not easily be translated onto paper. He had developed outstanding laboratory skills as a solid-state experimental physicist. His early false starts had taught him how to prepare materials and how to keep them from contamination. He also understood photoelectric emissions, electrons, holes, quantum states, and the physical properties of solids."

All true and all useful pieces of knowledge for a young scientist anxious to make his move into semiconductors. But the reality is that probably the best thing that came out of Bob Noyce's doctoral thesis was that, even as a failure, it was big and ambitious enough that it didn't hurt his reputation as one of the rising stars in the field. He had met all of the leading lights in the still tiny world of transistor technology, and to a man they wanted him on their team. Now it was time for Robert Noyce, PhD, to go out and pick his upward path among them.

But first: that August, he married Betty Bottomley in Sandwich, Illinois. The Reverend Ralph Noyce presided.

Now for a job. It seemed a given that Noyce, the top student in applied electronics at the most influential electronics school in the country, would take a job as a research scientist at one of the great electronics companies—RCA, AT&T, or General Electric—and spend the rest of his career building up a collection of patents and prizes. But Bob Noyce had other plans. Throughout his life, he had always shown a propensity to not take the predictable path but to choose arenas where he could control his fate and stand out among his counterparts.

So to the surprise and dismay of the entire industry, he picked Philco, a second-tier consumer electronics company based in Philadelphia. Philco had started out as a battery company, then moved into radio, and was now making its mark in television. Recognizing that the future was solid-state, Philco had opened its own transistor laboratory, which had not yet made much of an impact.

Why Philco? It would mean a lower salary than at the bigger companies. And Philco was hardly at the cutting edge of the technology, certainly not compared with Shockley, Bardeen, and Brattain at Bell Labs. Noyce would later laugh and say, "Because the way I put it to myself at the time was that they really needed me. At the other places they knew what was going on, they knew what they were doing."[2]

Implicit in that statement was Bob Noyce's confidence that, in his first professional job, he already knew more than the veteran scientists in the R&D department of one of the nation's largest electronics companies. In fact, that might well have been true a couple years before: Philco had spent the war years fulfilling one giant contract after another to supply electronic equipment (mostly radar and radios) and components (vacuum tubes and batteries) to the US military.

In 1950, like most companies in the industry—though a bit tardily—Philco decided to get into the transistor business, assigning one of its senior R&D executives, Bill Bradley, to establish the program and hire talent. Bradley had already hired or internally transferred about twenty-five scientists and engineers to man the project when he made a job pitch to the hottest new PhD out of MIT.

Noyce's luck held at that moment, because even before he arrived,

Philco's transistor group had already come up with an innovative and proprietary new type of germanium transistor called a surface-barrier transistor. It was one of the first such devices to get around the Bell Labs patents from Bardeen, Brattain, and Shockley's original point-contact design. Better yet, it outperformed the older design, operating at higher frequencies (that is, faster) with less power. The military, which had been lending money to Philco to keep it at the competitive edge, was hugely impressed and prepared to make major orders. And because this design was not the result, strictly speaking, of a defense contract, Philco was free to offer the device to the commercial world. In expectation of this, the company had also developed an improved germanium-growing process. This one-two punch of improved substrate with an innovative new design had convinced Philco that it was about to be a dominant player in transistors.

Noyce's timing in his arrival Philco was perfect. The new transistor was slated to be introduced to the world just three months after he joined the company. He would be the superstar rookie at a firm that was about to go from also-ran to the industry innovator. Even luckier, his seemingly dead-end dissertation on surface physics now proved to be ideally suited to this new surface-barrier transistor. Thus, rather than arriving on the job and going through the usual apprenticeship, Noyce found himself enlisted into a key role in the most important product team at Philco.

Noyce's assignment dealt with the process of etching and then electroplating the surface of the germanium—in particular, knowing when to stop the first and initiate the second, a feat of timing that had to take place at the microscopic level and that had to be done quickly enough to get the transistor out of laboratory fabrication and into mass production if Philco was ever going to introduce it to the world. Bob came up with a brilliant solution that used a light beam whose intensity would fall as the crystal was etched away until the right threshold was reached. Philco was so impressed that the company filed to make it Noyce's first patent.

Bob had joined Philco in order to be a major player right out of the gate, and now he became just that. Even as the new Philco transistor was being introduced to the world and receiving considerable attention in the trade press, Bob was hard at work after receiving the prestigious assignment of coauthoring the technical paper that described the physics of how the surface-barrier transistor actually worked. The paper took months to write, and though Philco would eventually decide that it was too valuable to the company to publish, it enhanced Noyce's reputation within the company and—as word got out of this newcomer's role in this world-class assignment—throughout the high-tech world.

In modern parlance, Bob Noyce was now a player—and perhaps not just professionally. Imitating his immediate boss, a flamboyant Italian former artist, Noyce took to dressing casually at work—a radical move in that buttoned-down era of the Organization Man. He also enjoyed the attention of the many pretty girls working at Philco. He would later joke that it was lucky that many of them were named Betty, so he didn't have to worry about saying their names in his sleep.

Ultimately, though, it was Bradley, the department head, who assumed a role not unlike Grant Gale at Grinnell as mentor in this next phase of Bob Noyce's life: management. From Bradley, Noyce learned the power of the boss as cheerleader: the boss who encourages his subordinates to join him on a quest, who is an endless font of new ideas (even if most of them won't work), who deprecates his authority even as he enforces it. As one employee of the era would later say, Noyce "was very easy to talk to, very helpful, and very different from a typical manager." If he had a weakness as a manager, it was that he didn't suffer slow or shallow people easily—a political skill at which he would become more proficient in the years to come, dealing with government officials, reporters, and shareholders.

Noyce's first eighteen months at Philco were almost everything he had hoped for, not just in the technology, but increasingly in the business itself. He had emerged as one of the leading young lights in the semiconductor industry. His name was attached to the most promising new transistor design on the market (the heart of the first practical hearing aids, among other devices), and best of all, he had once again caught the attention of the leading figures in the field.

But it all turned sour pretty quickly. Philco was a military contractor, after all, and that meant dealing with all of the frustrating features of government bureaucracy, contract negotiations, and paperwork.

Moreover, the transistor division was just one small and as yet largely unproven part of a large, sixty-year-old company. As long as the transistor department had been a relatively independent intrapreneurial venture at Philco, Noyce had been insulated from all of the sclerotic internal operations of the company. But now that he was a supervisor on an important company product dealing with both senior management and key customers, Bob quickly discovered just how stultifying corporate life could be.

To make matters worse, the company was bleeding money, to the point that it abandoned all R&D. "It seems that Philco is not yet really convinced that research pays for itself in the long run," he wrote to his parents and brothers. He was left to perform menial tasks that he dismissed as "bullshit, waste, make-work, and lack of incentive."[3] As a measure of his depression, Noyce even (against all evidence) managed to convince himself that he was also a lousy boss.

Looking back, and compared to the work experiences of millions of other employees in miserable jobs, this brief interval in Bob Noyce's life certainly doesn't seem Dickensian enough to have had such a profound effect upon the future work culture of Silicon Valley—and by extension the entire high-tech world. After all, he was young, celebrated by his fellow professionals and admired by his workmates. And there was no question that even at Philco his career had no upper bounds.

Still, Bob Noyce quickly decided that he never would work in an environment that didn't put innovation first, and that he would rather quit his job and risk wrecking his career than accept such a compromise. He never did again. Studies have shown that having control over their own fates, even if they fail, is central to the personalities of entrepreneurs. For all of his early success, Bob Noyce was now feeling trapped and at the mercy of forces beyond his control. It was this sense of helplessness that would make Philco the synecdoche of all that was wrong with traditional corporate existence for the rest of Noyce's life—and its echo could be heard in his intense reaction to Fairchild's corporate heavy-handedness fifteen years later.

The situation continued to deteriorate at Philco. Strikes shut down two company plants, and Noyce found himself in meetings with strikers shouting outside. At the same time, the federal government decided to sue the company for antitrust violations. All that Bob could see for the indefinite future was endless budget cuts and layoffs, little chance for advancement, and for him worst of all, less and less chance to do science. He could *feel* the rest of the solid-state world passing him by.

Bob wasn't the only unhappy member of the Noyce family. Betty may have been even more frustrated. She was stuck in the Philadelphia suburb of Elkins Park, Pennsylvania, as the wife of a company scientist. A new baby son (Billy) kept her at home and away from any intellectual challenges or her beloved cultural institutions. Her husband was often *not* home, and when he was, he often was distracted, singing in a local choir or building things in their two-room apartment. Seeing little Billy cry every time his father picked up a suitcase only made the situation more desperate for his mother.

Even as things grew worse at Philco, the friction increased in the Noyce household. Betty had never much participated in the social side of Bob's work life—the parties and dinners and other gatherings—but now she was almost invisible. And everyone noticed that he never seemed to talk about her or the things they did together. One of Noyce's workmates would tell Berlin that the one time he ever saw Betty was when he went over to Bob's house to help with an air conditioner: "Bob seemed to have kept her in the back." Betty herself admitted that much of her social reticence came from being "too snobbish," as an upper-class young woman lost in dreary suburbia.

What can't go on, won't. Bob realized that, more than anything, he wanted to "walk away" from Philco, "and start over again somewhere else."

That opportunity came in late 1955. Happily, the tech world hadn't forgotten Robert Noyce. Westinghouse approached him with an offer to join its Pittsburgh transistor facility—the city had been a center for electronics at least since the creation of the landmark ENIAC computer during World War II—with a 25 percent raise, a 10 percent wage increase guaranteed for each of the next two years, and most of all a chance to get back into the game. On hearing the news, Philco matched the offer, made Noyce's temporary management position permanent, and offered to move Bob's job to its Lansdale headquarters—quickly giving Betty visions of "a little house and a big yard."

Bob struggled with the two offers for a week. He had joined Philco, he later recalled, dreaming more of fame than wealth: "My only real ambition was to be able to buy two pairs of shoes simultaneously, particularly after having grown up in my brothers' hand-me-downs." But now he had a growing family and a wife who expected a more genteel existence—and he himself was beginning to enjoy the perquisites of a professional career. In the end, despite the fact that he would certainly have more job security at the booming Westinghouse, Bob opted to stay at Philco, where he would still be the big fish in that small, troubled, pond.

But the wheels of history were already turning. Not long before he fielded these job offers, Noyce delivered a scientific paper in Washington, DC. Down from New York to attend the conference was William Shockley, who was already plotting to leave Bell Labs and start up his own transistor company back home near Stanford University. He had known of young Bob Noyce, of course, now for nearly a decade, and the presentation on this day was both a reminder and a glimpse into how much Bob had grown as a scientist in the intervening years. Shockley added Noyce's name to his mental short list.

By the end of 1955, the brief interval of happiness brought by the job offers had faded. The young family had now added a baby girl, Penny, but otherwise their lives had changed little. Philco was still in trouble, and the move to Lansdale had evaporated. Now, in the golden hand-cuffs of a raise and new job title, Bob Noyce was feeling more desperate than ever.

Then on January 19, 1956, Bob answered a ringing phone. "Shockley here," said the voice on the other end of the line.

"It was like picking up the phone and talking to God," Noyce would say. "[He] said he was starting this thing out here on the West Coast, and that he'd like to talk to me about joining him. Well, Shockley of course was the 'daddy' of the transistor. And so that was very flattering. And I had the feeling that I'd done my stint in the minor leagues and now it was time to get into the majors."

Shockley had left Bell Labs two years before—with considerable acrimony. He had spent one of those years teaching at Caltech, the other heading a Pentagon weapons group in DC. Now he had moved to the Bay Area with the dream of starting a transistor company and getting rich.

On February 23, Bob and Betty Noyce flew out to San Francisco on a red-eye flight, touching down at six a.m. There had been snow on the ground in Pennsylvania when they left; the Bay Area, by comparison, was enjoying one of those spectacularly warm and sunny winter days that annually makes it the envy of the rest of the country. Bob, at least, who had dreamed of California his entire life, knew he was home. "I had a brother who was teaching at Berkeley, and, you know, his letters were stories of sunshine and lovely weather. . . ." Betty knew only that she was now three thousand miles from her friends and family.

The old cocky self-assurance came flooding back. Bob *knew* that he was going to get the job with Shockley, that there was no going back to Philco and the East Coast. The appointment with Shockley was at two o'clock. At nine a.m., the couple met with a real estate agent, and by noon they had picked out and put a deposit on a house in Los Altos, a rustic but elegant little burg adjoining Palo Alto, for $19,000. Bob made it to the interview with Shockley on time—and walked out a couple of hours later with a job offer.

He had made it to the top, and he had found his home. Robert Noyce was now ready to become a legend.

CHAPTER 10

Gordon Moore: Dr. Precision

Long after everyone else in the Silicon Valley story is forgotten—Terman and Hewlett and Packard, Noyce, Zuckerberg, Brin, and Page, even Steve Jobs—Gordon Moore will still be remembered. Not for his career in high tech, though it is all but unequaled, but for his law—which future historians will point to as defining the greatest period of human innovation and wealth creation in history. Centuries after Moore the man is gone, Moore the Law will survive, either as the turning point in the beginning of a new era for mankind or longingly as the metronome of a brief, golden interval now lost.

Moore the man's relationship with Moore the Law is a complex one. As he was the first to admit, when he sat down to write that *Electronics* magazine article, his ambitions were small: "What I could see from my position in the laboratory was that semiconductor devices were the way electronics were going to become cheap. That was the message I was trying to get across in circuits and saw that the complexity had been doubling about every year. . . .

"It turned out to be an amazingly precise prediction—a lot more precise than I ever imagined it would be. I was just trying to get the idea across that these complex circuits were going to make the cost of transistors and other components much cheaper."

Precise. In Gordon Moore's world, that is the highest compliment. Of all of the millionaires and billionaires and powerful men and women that Silicon Valley has produced, Gordon is easily the most beloved. At one point the richest man in California, he seems to have no enemies. He is affable, kind, soft-spoken, and self-effacing. When his peers were buying the world's largest sailboats and most exotic cars, or pretending to be scruffy and bohemian, Gordon and his wife, Betty, still owned the pickup in which they spent holidays rock hunting.

But be inaccurate, be sloppy in your technique or your math, be *imprecise*, and a different Gordon Moore appears—not angry, but distant, cold, appalled. Henceforth, he will never treat you any less than politely, but he will also never take you quite seriously again. Gordon Moore may be kind, but he is also, always, correct.

Over the years, Gordon has learned to accept the fame of his law and the unique celebrity that has come with it. "I was embarrassed to have it called Moore's Law for a long time, but I've gotten used to it. I'm willing to take credit for all of it," he chuckles, "but all I really did was predict the increase in the complexity of integrated circuits and therefore the decrease in the cost."[1]

But one senses it goes even further. Gordon Moore has come to be proud of his law, not in *spite* of the fact that its reach has grown over the last fifty years to encompass memory chips, the semiconductor business, the electronics industry, and even the pace of modern civilization but *because* of that fact. After all, Gordon knows he managed to devise an equation for a small but fast-growing corner of a single new industry, . . . and that equation proved so *precise* that it captured the entire zeitgeist. Not even Newton or Maxwell accomplished that. And even after the integrated circuit itself is obsolete, it is possible that Moore's Law will still dominate human existence as what it has always been: not really a law but a commitment to perpetual progress, as Silicon Valley's greatest legacy.

Because he is a plain man blessed with an extraordinary mind, in his own way Gordon Moore is a far more exotic creature in Silicon Valley than any of his fellow tech tycoons, with their flamboyance or eccentricities and their expensive toys. He is also rare in that among the founders of the Valley, he is the only native son, having been born just over the hills in the coast town of Pescadero and raised at the north end of the Valley in the Peninsula town of Redwood City.

Law threads through the Moore family history. If Gordon is a law maker, then his father was a law enforcer—constable, then sheriff, for the entire western coast of San Mateo County, the strip of land never more than a few miles wide, bounded on the east by mountains and on the west by the Pacific, running from just south of San Francisco to just north of Santa Cruz.

Today, with the exception of the city of Half Moon Bay, this stretch of land is largely uninhabited, encompassed either by state parks or snapped up by the Land Commission for preservation. Gordon has joked that Pescadero "is the only town I know of in California that's smaller now than it was fifty years ago."[2]

But in the 1920s and 1930s, this stretch of coastline was more like the South Side of Chicago. Highway 1, which runs along the coast here, was dotted with "restaurants" whose primary offering was bootleg whiskey served to the wealthy who came down from Frisco or over the hill from Burlingame to wet their parched throats. The booze itself came up the highway from central California, or even L.A., or it was delivered by speedboats loaded from boats anchored outside the three-mile limit. A batch of bootleg could be easily brought to shore on one of the many enclosed beaches, piled into a truck, and driven the twenty miles north—and the ship would never have to take the risk of entering the Golden Gate.

Meanwhile, most of this coastal plain was covered by farms—pumpkins, artichokes, melons, anything that loved foggy, windy summers and warm, clear autumns—worked by European and Mexican immigrants who lived in numerous transient camps. Most Saturday nights ended in a knife fight or two.

It was Sheriff Moore's duty—alone—to enforce the law in this lawless region. Gordon Moore's earliest memories include watching his father strapping on a gun and going out in the evening to work. And Sheriff Moore was very good at his work. At six feet tall and about two hundred pounds, he was fearless and would often jump right into the middle of bar fights to break them up. He was also famously clever at finding ways to trick and capture bootleggers or uncover their shipments. "In those days there weren't many shoot-outs," Gordon recalled, "but my father always carried a gun."[3]

Not surprisingly, in San Mateo County, Sheriff Moore was something of a legend, more famous than his son well into Gordon's middle age. Moore: "He was big, but not huge, but he had a way of approaching things where he seemed to be able to settle people down pretty well." Like father, like son.

In 1939, when Gordon was ten, his family moved over the hills to the Bay Area, to the town of Redwood City. His father's success had led to a promotion to undersheriff of San Mateo County, the highest non-elected job in the county. It was a job he held for the rest of his career.

For Gordon, a self-described "beach kid," the transition went smoother than expected. Redwood City then wasn't the crowded Peninsular sprawl under the glass spires of Oracle it is today, but on the edge of a different kind of sea: of cherry, plum, and prune plum orchards stretching for thirty miles into the South Valley and containing several million trees. The cities of the Peninsula, which have long since merged into a single run of housing developments and industrial parks, were then just oases of train stations and tiny retail districts almost evenly spaced like beads on the string of El Camino Real from South San Francisco to San Jose.

In Redwood City, on the shore of this blossomed sea, young Gordon was almost as free as he had been in Pescadero. As had been the case with the previous generation of budding technologists with ham radio (Fred Terman) and subsequent generations of future Valley entrepreneurs with computers and the Internet, the founding generation of modern Silicon Valley—like Noyce with his glider—almost all gravitated to hands-on applied science. Moore was perhaps the ultimate case. His aptitude for chemistry soon produced a chem lab that he set up in the family garage: "Those were the days when you could order almost any chemicals you wanted by mail order. It was fantastic. The things you could buy then were really things you could have fun with."

Like most young chemists, Moore was soon making explosives. As he told a television interviewer, "I used to turn out small production quantities of nitroglycerin, which I converted into dynamite. I made some of the neatest firecrackers, [he held up both hands] and I still have ten fingers!" He then stated the obvious: "Those were simpler times. If I had made the explosives today that I made in those days, I probably would get into trouble."

Having managed to retain both his fingers and eyes, Gordon proved to be an excellent, diligent—and of course, precise—student in public school. In later years, he would become a staunch supporter of

California's public education system, pointing to himself as a prime example of a satisfied graduate.

Like most kids of his generation, Moore worked at a progression of small jobs. From only one was he ever fired: selling subscriptions to the *Saturday Evening Post* door-to-door: "I didn't make it at that. I just found cold-calling on houses to sell magazines was not something I was cut out to do." Even as the chairman of a multibillion-dollar corporation, Gordon eschewed anything that smacked of selling to customers. He left that job to Bob or Andy.

Like Noyce, Moore was just a couple years too young for World War II, graduating from Redwood City High just as the war ended. With his top-notch grades and reputation as a brilliant young scientist, Moore, like Noyce, probably would have been accepted at any college in America. The choice he finally made speaks volumes about his personality.

One of the interesting but rarely discussed phenomena about education in California is that even today most local high-school graduates choose to stay in the state, even if five hundred miles away. Part of this is financial—tuitions at California colleges and universities are comparatively cheap; and part is cultural—Cali kids are loath to experience East Coast or Midwestern winters. So like many before and after him, Gordon Moore, diploma in hand, resolved to stay in the Golden State.

But what school? Even in the 1940s, though the explosion of state-run community colleges and four-year schools was still more than a decade in the future, Gordon still had a wide selection of potential choices, all of which were available to him. Moreover, then as now, there was also a perceived hierarchy among these schools, with Stanford and Cal at the top in Northern California, UCLA and USC in the South. Below them were some of the small private universities, such as (in Moore's area) Santa Clara, USF, and Saint Mary's. And below them were the dozen or so state colleges, many of them evolved upward from teaching colleges.

Moore's choice, San Jose State College, probably raised eyebrows even then. San Jose State was considered a middling college even among the state colleges. Located in downtown San Jose, it looked more like an urban school, and its primary role was to fill the growing number of banks, canneries, and manufacturing companies in the South Bay. Even in its own neighborhood, it was considered a distant second in quality to Santa Clara U just two miles away.

Because he has never written an autobiography or submitted to a full biography, Moore has never made the outside world privy to his decision process in choosing San Jose State. But it appears that the primary reason was financial. Undersheriffs didn't make a lot of money, and even in the 1940s, tuitions at the other local private universities largely kept their enrollment to the children of the upper middle class. Moore has simply said, "I chose to go there because I could commute from Redwood City and didn't have to leave home."

He didn't linger long. Gordon stayed at San Jose State for two years, then transferred to UC Berkeley, likely because his grades earned him some financial support at Cal. But if Gordon Moore's stay at SJ State was brief, it was a period in his life he never forgot—because it was there that he met Betty Irene Whitaker. She was a true Santa Clara Valleyite: she was born in Los Gatos, and her grandparents owned the last great orchard in downtown San Jose. She and Gordon were married in 1950.

He was now a young man in a hurry, with a new bride and not a lot of money. The young couple lived in the married dorm at Cal. Betty worked to support the two of them, while Gordon attended class. He earned his degree in chemistry in 1950, with enough honors to have his pick of graduate schools. Sheriff Moore, who had always let his son live his own life, for the first time tried to influence Gordon's decision; he suggested that his son go to medical school. But Gordon didn't have any interest in becoming an MD. Instead, he chose the California Institute of Technology, and the young couple embarked on what was the biggest move in their lives to date. Gordon recalled, with his usual precision, "I used to claim I hadn't been east of Reno, Nevada, until I got out of graduate school . . . then I discovered Pasadena was east of Reno. But as far as out of the state, Reno was as far as I'd been."

At Caltech, Gordon thrived. The beach kid had long known that he was clever; but now, at the most distinguished school for science in the western United States, it became apparent just how smart he really was. Four years after arriving in Pasadena, Gordon earned his doctorate in chemistry, with a minor in physics. It was the perfect combination for the emerging new world of semiconductors.

Perhaps as a nod to his father's influence, Gordon chose to put "Dr." in front of his name, where it has remained ever since—despite the fact that most other Valley PhDs, including Noyce and Grove, have chosen not to use that honorific regularly.

Describing his time at Caltech, Moore has said with typical modesty, "I lucked out. I had a good thesis topic and a professor who didn't insist on keeping the students around for a long time." Meanwhile, Betty, after a short stint at Consolidated Engineering Corp. in public relations, had found satisfying work at the Ford Foundation.

But now the newly anointed doctor and his wife had a decision to make. Moore: "It's hard to believe that at that time there were really no good technical jobs in California. I had to go east to find a job I thought was commensurate with the training I'd had." Gordon and Betty gritted their teeth and headed to Maryland, where a job was waiting in the Navy flight physics laboratory at Johns Hopkins University.

It was good and challenging work, most of it in the development of new missile technology. Best of all, much of Gordon's research revolved around electronic telemetry systems, "so that let me continue to do things that were reasonably closely related to what I'd done for my thesis."

But within two years, the Moores were ready to go home. As usual, Gordon would later offer an explanation that combined self-deprecation and precision: "I found myself calculating the cost per word in the articles we were publishing and wondering if the taxpayer was getting its money's worth at something like $5 per word. And I wasn't sure how many people were reading the articles anyhow. So I thought I ought to get closer to something more practical—and frankly, I wanted to get back to California. I enjoyed living in the east for a few years, but I really thought I liked it [in the Bay Area] better."

But the initial job search didn't prove promising. In Southern California, Hughes Aircraft was hiring, but in a field for which Gordon had little interest. In the Bay Area, he looked into a couple of oil companies that were hiring for their laboratories, "but those didn't appeal to me either."

Far more interesting was General Electric Company, which at the time rivaled Bell Labs for the quality of its research into solid-state physics. But GE wasn't interested in hiring Gordon for that work. Rather, it wanted him as a researcher into nuclear power. "I wasn't especially interested in that."

Losing hope, Gordon then applied at Lawrence Livermore Laboratory, which had a strong affiliation with his alma mater, Caltech. This was a measure of his desperation to get back to California, because having turned down GE Nuclear, he was now looking at work in developing nuclear weapons. Lawrence Livermore did offer Gordon a job—but in the end, Gordon turned it down. There were limits to what he would do to go home.

Luckily for him and Betty, at that moment there was another homesick Californian. And he didn't need to *find* a job; he planned to start one of his own. And because he was the greatest applied scientist on the planet, William Shockley convinced GE to let him look through its files of job applicants to whom it had made offers but who had turned the company down.

The one name that jumped out was Gordon Moore. It's not obvious that Shockley knew of Moore, as the young man had been mostly buried in rocket research. But for all of his failings, Shockley did have a genius for reading résumés and CVs, and he quickly saw in Gordon a major talent. Moreover, by lucky coincidence, Shockley was also looking for a chemist, as those scientists had proved particularly useful during his time at Bell Labs.

"So he gave me a call one evening. That was the beginning of my exposure to Shockley and silicon."

Gordon and Betty packed up and headed west at last. They arrived on a Monday, and Gordon was officially signed in as employee number 18 at Shockley Transistor. He soon learned that another new hire, Robert Noyce, who had arrived the previous Friday, was employee number 17. Moore: "I've always wondered what would have happened if I'd driven cross-country faster and gotten in on Thursday."

As Gordon remembered it, Noyce stood out from the rest of the new hires from the start. Not because of his intelligence—everyone Shockley hired was intimidatingly brilliant—but because of his experience: "Bob was the only one of the group that had significant semiconductor experience before Shockley. I hardly knew what a semiconductor was. . . .

"Bob came in with much more specific knowledge than us, and he was a fantastic guy in any case."

Despite their desire to make a good impression on their famous new boss, it didn't take long for Shockley to provoke their alienation. On this subject, Gordon's usual polite reserve abandons him: "I don't think 'tyrant' begins to encapsulate Shockley. He was a complex person. He was very competitive and even competed with the people that worked for him. My amateur diagnosis is [that] he was also paranoid, and he considered anything happening to be specifically aimed at ruining Shockley somehow or other. The combination was kind of devastating."

The turning point for Moore, as it was for the rest of the Traitorous Eight, was Shockley's decision, after a few minor incidents in the lab, to implement an ongoing lie-detector program. Modern readers are usually shocked by this and instantly dismiss Shockley as having done the unforgivable. In fact, lie detectors, being de rigueur as the forensic tool of the era, weren't that uncommon in American industry. What *was* unacceptable to the young superstars at Shockley Transistor was that the use of the device seemed arbitrary, punitive, petty, and most of all, permanent. Moore: "*The Caine Mutiny* was popular at about that time, and we saw the analogy between Queeg [the paranoid ship captain] and Shockley."

Though Moore is always lumped in with the mutineers of the Traitorous Eight, he was hardly a ringleader. Left to his own devices, he probably would have stayed another year or two at Shockley Transistor until he found another research job in California, preferably in the Bay Area. Indeed, it is interesting to speculate how the Valley story would have played out if Gordon had instead gone to IBM San Jose or Hewlett-Packard. But the Eight wisely chose to stick together, their strength being in numbers, and Gordon went along as they prepared their exit en masse.

Unfortunately, despite their later status as business legends, none of the Eight at this point were particularly accomplished corporate politicians, and thus their initial strategy was doomed from the start. Moore: "We actually went around Shockley to Arnold Beckman from Beckman Instruments, who was financing the operation, to try to do something where Shockley would be moved aside as the manager of the operation but would remain as some kind of consultant."

They were foolish and naive. "I guess we kind of overestimated our power. . . . That's when we discovered that eight young scientists would have a difficult time pushing a new Nobel laureate aside in the company he founded. [Beckman] decided that would wreck Shockley's career and essentially told us, 'Look, Shockley's the boss, that's the way the world is.' We felt we had burned our bridges badly."

There was now no going back, no repairing that burned bridge. Any boss would be furious at this perceived betrayal, but with Shockley the Eight could only expect the worst kind of reprisals. If Gordon had any doubts about signing on with the Eight to approach Beckman, they now evaporated with the realization that they would all have to walk out of the company together.

In later years, when discussing the Traitorous Eight and his departure from Fairchild to found Intel, Gordon would often laugh and say that he wasn't like the other great entrepreneurs in Silicon Valley history. Those men and women, including Bob Noyce, started companies because they were driven to do so, that this was the canvas upon which they painted their greatest creations. Gordon, by comparison, would then describe himself as a "negative entrepreneur," always joining new start-up companies to get away from his current unhappy work environments. "I'm not a positive entrepreneur."

Perhaps it was this difference between the two men that helped bind them together so closely. Noyce—fearless, clever, charismatic—was a man cut from the same cloth as Gordon's father. Moore—thoughtful, modest, and intellectually ferocious—was much like Bob's preacher father. Almost from the moment they met, these men of opposites seemed to trust each other completely. Each knew the other would never betray him. They deeply admired each other, and both sensed that to succeed, they needed each other's complementary personality traits. Moore's precision kept Noyce intellectually honest, kept him from taking too much advantage of his charms. Noyce, with his reckless confidence, led Moore on one of the greatest and most successful ventures in business history, a trajectory he never would have taken himself.

Now safely cosseted within Fairchild Labs, the day-to-day operation of the company in the hands of smart businessmen like Noyce, Sporck, and Bay, Gordon was free to let his mind speculate, to look into the great sweeping trends, the patterns that he saw better than anyone, to which his mind naturally gravitated. In particular, he had begun to formulate a theory about integrated circuits that he now wanted to thrash out to fulfill a deadline for a bylined article in a leading trade magazine. The paper, whimsically titled "Cramming More Components onto Integrated Circuits," was published in the April 19, 1965, edition of *Electronics* magazine.[4] Its opening paragraphs were among the greatest acts of prediction in tech history:

> The future of integrated electronics is the future of electronics itself. The advantages of integration will bring about a proliferation of electronics, pushing this science into many new areas. Integrated circuits will lead to such wonders as home computers—or at least terminals connected to a central computer—automatic controls for automobiles, and personal portable communications equipment. The electronic wristwatch needs only a display to be feasible today.
>
> But the biggest potential lies in the production of large systems. In telephone communications, integrated circuits in digital filters will separate channels on multiplex equipment. Integrated circuits will also switch telephone circuits and perform data processing.
>
> Computers will be more powerful, and will be organized in completely different ways. For example, memories built of integrated electronics may be distributed throughout the machine instead of being concentrated in a central unit. In addition, the improved reliability made possible by integrated circuits will allow the construction of larger processing units. Machines similar to those in existence today will be built at lower costs and with faster turn-around.[5]

The point was underscored by a *Mad* magazine–type cartoon showing shoppers at a department store ignoring two counters labeled NOTIONS and COSMETICS to crowd around third counter with a grinning salesman hawking HANDY HOME COMPUTERS. But the amazingly accurate nature of that prediction was all but forgotten once readers saw the little graph on the lower left corner of the third page. It had no label or caption, but was merely a grid, with the x-axis reading in years from 1959 to 1975, and the y-axis reading: LOG$_2$ OF THE NUMBER OF COMPONENTS PER INTEGRATED FUNCTION. Those readers didn't have to know that they were looking at the future of the world—even Moore didn't know that yet—but if they were astute enough to study the line on the graph and recognize its implications, it would have made their hair stand on end . . . and shown them the path to become billionaires.

Though in 1965 having only four data points determining a straight line at about 45 degrees, the graph said that integrated circuits, till then just memory chips, were exhibiting developmental characteristics that had no precedent in manufacturing history. In fact, they seemed to be doubling in performance every eighteen months. Moore had intuited that something like this was going on, but only when he sat down to prepare the paper did he realize that he had discovered something stunning that had been right before his eyes for years.

As he would recall later, he pulled out a sheet of standard graph paper and plotted out the performance-to-price ratio of the last three generations of Fairchild integrated circuits. Even though he knew the specs on these devices, Moore was surprised to discover that the performance leaps between generations—especially from the third to the fourth, now in development at Fairchild—were so great, and the hyperbolic curve they created was so vertical, that he had already run out of paper. So he switched to logarithmic graph paper, and when he did, the data points neatly arrayed themselves into a straight line. As he wrote in the article:

"The complexity for minimum component costs has increased at a rate of roughly a factor of two per year (see graph on next page). Certainly over the short term this rate can be expected to continue, if not to increase. Over the longer term, the rate of increase is a bit more uncertain, although there is no reason to believe it will not remain nearly constant for at least 10 years. That means by 1975, the number of components per integrated circuit for minimum cost will be 65,000.

"I believe that such a large circuit can be built on a single wafer."[6]

It has been said that if in 1965 you had looked into the future using any traditional predictive tool—per capita income, life expectancy, demographics, geopolitical forces, et cetera—*none* would have been as effective a prognosticator, none a more accurate lens into the future than Moore's Law. The trend Gordon Moore had identified—and discussed in a series of regular Semiconductor Industry Association speeches over the next two decades that updated his famous graph—was that the world of electronics, from computers to military and consumer products, was increasingly going digital. And thanks to the semiconductor planar fabrication process, it was possible to make the tiny digital engines at the heart of these products ever smaller, cheaper, and denser with transistors at an *exponential* rate.

No human invention had ever exhibited that rate of improvement. In later years, writers would search for analogies. One popular comparison in the 1970s drew from the automobile industry, suggesting that if Detroit had kept up with Moore's Law, cars would go 500 mph at 200 mpg . . . and cost $1.50. But when Moore's Law kept clicking on, long after Gordon's hoped-for achievement in 1975 of sixty-five thousand transistors on a chip, the direct analogies became absurd. Now writers called on the old Chinese legend of the mathematician who has performed a service for the emperor and when asked his fee, replies, "I only ask that you take out a chessboard, put one grain of rice on the first square, two on the second, four on the third, and so forth." The delighted emperor agrees . . . but soon discovers that just halfway down the board, he has already committed to all of the rice in China, and soon thereafter all of the rice in the world.

By 1971, when Moore was chairman and CEO of Intel Corp. and leading the new revolution in microprocessors, he used his SIA speech on his law (already being named after him) to predict that twenty years thence, in 1991, (DRAM) memory chips would jump from one thousand bits (transistor counts had now been all but left behind) to 1 million bits. He was once again proven correct.

By then, Moore's Law had become the defining force of the electronics world. Every company that could, from computers to instruments to even software, jumped onto the rocket that was Moore's Law. It not only offered rapid innovation and explosive growth but, best of all, was *predictable*: you could build your product to take advantage of the power of the next generation of chips and processors—and if you timed yourself right, those future chips would be waiting there for you to use them. Better yet, if you timed your products for the beginning of that next chip generation, you could not only steal a march on your competition with improved performance, but you could also initially charge more for that performance.

And that has been the history of the digital age ever since, as the predictability and power of Moore's Law have led to one technological revolution after another: consumer electronics, minicomputers, personal computers, embedded systems, servers, the Internet, wireless telecommunications, personal-health and medical monitoring, Big Data, and many more. Most of the predictions made for the future in the early 1960s, from personal atomic helicopters to colonies on Mars, failed because the incremental, physical steps (not to mention top-down political will) needed to get there were too great. Instead, we have replaced those dreams with other, real miracles, all of them made possible through the small, ground-up commercial digital revolution set at hyperspeed by Moore's Law.

These days, schoolchildren are taught a simplified Moore's Law, and all of us have assimilated into our consciousness the notion of living (or at least enduring) a life immersed in rapid and relentless change. The world's economy now rests on Moore's Law; it is the metronome of modern life—and were it to suddenly end, that event might throw all of mankind into a kind of existential crisis that could last for generations and lead to a complete reordering of society.

Happily, despite Gordon Moore's early expectation that this pace of change might slow in the late 1970s, Moore's Law continues in effect to this day—into the era of memory chips capable of storing more than one *trillion* bytes. More amazing yet—and rarely recognized because the world has long focused on Moore's straight-line graph on the logarithmic paper, rather than on his original hyperbolic curve—is that everything that has happened in the digital world to the present, from calculators to the Internet to smartphones and tablets, has occurred in the shallow "foothills" of the law's curve. The real curve began to bend almost straight upward only in about 2005.

That suggests that for all of the massive, transformative changes wrought by Moore's Law in the last fifty years, the *real* changes lie ahead—perhaps all of the way to the "singularity," the moment, as defined by scientist Ray Kurzweil, when human beings and computers become one. But even if that never occurs, the latest industry forecasts are that Moore's Law could continue working well into the middle of the twenty-first century, when a single chip (if it still takes that form) might contain the processing power found in all of the world's memory chips and processors today. It is a scenario that is almost impossible to imagine, and yet even today our children play with smart toys that are far more powerful than all of the semiconductor chips in the world when Gordon Moore first formulated his law.

History will be likely to record this miracle, first predicted and then driven by Moore's Law, as one of the greatest achievements in the story of mankind, one that sent humanity on a new trajectory as different as that created by the agricultural and industrial revolutions—and with an even more profound and sweeping impact.

One reason this transformation has been so titanic and is likely to be so enduring is that, as Dr. Moore has reminded the world many times, his is not really a "law" at all. Rather, it is a social compact, an agreement between the semiconductor industry and the rest of the world that the former will continue to strive to maintain the trajectory of the law as long as possible, and the latter will pay for the fruits of this breakneck pace. Moore's Law has worked *not* because it is intrinsic to semiconductor technology. On the contrary, if tomorrow morning the world's great chip companies were to agree to stop advancing the technology, Moore's Law would be repealed by tomorrow evening, leaving the next few decades with the task of mopping up all of its implications.

Instead, Moore's Law has endured because each day, hundreds of thousands of people around the world—scientists, designers, mask makers, fabricators, programmers, and not least, Intel employees—devote their imagination and energy to progressing Moore's Law just another inch forward into the future.

One cannot fully appreciate that what drives Intel and what makes it the world's most important (and sometimes most valuable) company is not just that it is the home of the discoverer of Moore's Law, but that the company has dedicated itself from its founding to be the Keeper of the Law, to bet everything, in good times and bad, on maintaining the health of Moore's Law, even if it has to die to do so.

What began as a simple graph whipped up for a trade-magazine article in the early years of the digital age now haunts the entire world—but nowhere more than at Intel Corporation, and most of all in the office of its chief executive. The endless, unforgiving pursuit of Moore's Law has made or broken the reputation of every man who has held the CEO title.

One of the ironies of Moore's Law is not only that it is not a law, but that Moore didn't entirely discover it. And in that lies a bit of justice. There had long been some resentment inside Fairchild Semiconductor that Bob Noyce, along with TI's Jack Kilby, had gotten all of the credit for the integrated circuit. Certainly it was true that Noyce had sketched out the first practical design for an IC. But actually realizing the Noyce design had also required an equal, if not greater, piece of innovation by Jean Hoerni: the planar process. Hoerni rightly grumbled that history had not given him sufficient credit for his contribution to this revolutionary product. But there were others as well, including most of the Traitorous Eight, who had done their part—and none more than Gordon Moore, who had led the other team in pursuit of the IC design and whose work had made major contributions along the way. Gordon, the quiet gentleman, never made a fuss about it, but it grated on him.

Now Moore had made a discovery even greater and likely to be even more enduring than those of Noyce, Kilby, and Hoerni, more even than the integrated circuit itself—and unlike those of the others, his name was actually on it. And yet it is probably not a coincidence that at about the same time as he was formulating his law, just down the hall, Bob Noyce was announcing the first real application of that law with his now famous price slashing on all Fairchild transistors. After all, what was Noyce's decision to give devices today a price they would have years in the future but his calculation of how the as-yet-unnamed Moore's Law would influence industry pricing? Noyce's application was simultaneous with—or even preceded—Moore's theory. But this time, Gordon Moore got all the credit. It speaks volumes about Bob Noyce and his partnership with Gordon that he never once suggested that he should share credit with Moore. Gordon Moore owns his law not just because he discovered it, but because he made it real at Intel.

Looking back as he approached retirement, Gordon said of his old partner, "Bob Noyce was a fantastic guy. You know, he was extremely bright, very broad ideas, and a born leader. Not a manager, a leader. Bob operated on the principle that if you suggested to people what the right thing to do would be, they'd be smart enough to pick it up and do it. You didn't have to worry about following up or anything like that.

"He was one of those rare individuals that everybody liked the moment they met him. And you know, I worked very well with him. I complemented him a bit. I was a little more organized as a manager than he was, though not an awful lot more."[7]

Gordon was joking with the last remark. Dr. Precision was far more organized than his old partner. And the two men complemented each other to an astounding degree—Fairchild and, even more, Intel are proof of that. As the press of events slowly drew Bob Noyce away from science into business management, it was Gordon Moore who filled the vacuum he left behind—and in assuming the role of chief scientist/ technologist at Intel, he ultimately positioned himself to hold the same de facto title for the semiconductor industry, and then, with his law, for the entire electronics industry.

That said, the two men weren't perfect complements. Together, they made a remarkable team, capable of creating two of the most storied companies in American history. But they both were lacking—largely without regret—in certain traits that were and are key to great business success. Neither man lived or died with the fate of the companies they created. Neither had the killer instinct needed to crush a competitor when it was down. And most of all, neither man was comfortable with the face-to-face confrontation and ruthless calculus of firing employees for the greater long-term health of the enterprise.

At Fairchild, they had found that tough number two in Charlie Sporck, now building a lean, tough National Semiconductor just a few blocks away. Now, at Intel, they found it in the tough, compact form of Andy Grove. As even they would privately admit—and Andy would publicly pronounce that—without Grove, Intel would probably have grown into a middling company, famed for its innovative products but not much more, on a long, slow slide to a respectable but barely noticed oblivion. In a world not driven by a giant, take-no-prisoners Intel obsessed with maintaining Moore's Law, Intel would have been crushed by a more efficient Nat Semi, a cleverer Advanced Micro Devices, a more innovative Motorola, and a more powerful Texas Instruments.

With Grove, Intel beat or destroyed them all . . . but might well have destroyed itself in the process had not Gordon Moore been there to temper Andy's ferocity and round his sharp corners. Gordon was Bob's foundation and Andy's conscience; a remarkable feat of partnership and adaptation, given that he was dealing with two of the biggest personalities of the age. He was the insulator between these two charged characters—and perhaps most remarkable of all, he filled that role (and dealt with his growing fame and wealth) while remaining true to himself.

CHAPTER 11

A Singular Start-Up

With the exception of one event (itself earthshaking only in retrospect), the first five years of Intel's story are largely forgotten, overshadowed on one end by the Fairchild exodus and the astonishing launch, and on the other by the introduction of the Intel model 4004, the world's first commercial microprocessor. Even Intel's official product timeline doesn't begin until 1971—by which time the company had already introduced a half-dozen proprietary new products.

Yet the years in between have their own importance, because it was during this period that Intel had to organize itself as a real company, prove the technological mastery it had been founded upon, get its first products to market before the rest of the semiconductor world caught up, and incidentally but no less important, establish the corporate culture that would define it for a generation to come (assuming it survived).

New start-ups are notoriously bad at keeping track of their own histories. That's a good thing: any new enterprise that has the time to catalog its own story obviously isn't working hard enough to ensure its own success. But for the really successful companies, this obviously plays hob with future historians, who find themselves buried in material from the company's most successful years while scratching for the few documents that survive from the early days. Even the memories of those who were there are often distorted by later events, by the human need to puff up one's own contribution, and eventually by the haze of time. Young companies often behave in ways very different from their later, larger incarnations—sometimes to the embarrassment of later employees and executives. They want to see the past as one straight upward path to triumph, not petty squabbles, questionable deals, and the incompetent decisions of neophytes.

At the Intel Museum, in its current configuration, those first five years take up about ten running feet of exhibits, about 2 percent of the entire museum floor.[1] And even that is dominated by a giant blown-up, backlit photograph taken in 1970 of the entire 200-person Intel staff standing in front of the company's original Mountain View headquarters. The appeal of the photograph—and it is even more popular with current Intel employees than with the museum's visitors—lies in the contrast with what will be.

It's not an embarrassing photograph—not like the infamous one of the eleven "Micro-Soft" übernerds that would be taken eight years hence. Rather, this is, along with the NASA control-room photos of the same era, a classic image of sixties engineers: white shirt, skinny tie, crew cuts, and horn-rimmed glasses. Derided as hopelessly outdated just five years later, the figures in the photograph now look retro-cool after forty years—young hipster engineers at Intel today try to look as casually square as the figures in this photo.

Employees love to stop and scrutinize the photo for a number of reasons. One is that it seems impossible that their $50 billion, 105,000-employee company, with factories and offices around the world, could ever have been so small and circumscribed that every one of its hundred or so employees could gather for a photograph in front of its little glass and wood headquarters.

The employees also linger to scrutinize the figures in the photograph in hopes of recognizing a few. From the beginning, Intel always prided itself on its egalitarianism—the lack of offices, the seemingly indifferent musical chairs of title by Noyce and Moore (and then Grove as well), the informality and easy communications up and down the organization chart. The company was rightly celebrated for this progressive part of its corporate culture. In the world of 1960s American business, Intel's hard-charging yet democratic style stood out, even in Silicon Valley.

Nevertheless, the photo is also a reminder that even a "flattened" organization like Intel in 1970 was still quite hierarchical by today's standards. Thus, in the photo, senior management stands in the foreground, founders Noyce and Moore nearest to the camera, so close they are almost looking upward. Behind them can be seen Les Vadász, almost unchanged in appearance today, and—the figure most likely to elicit a gasp from both employees and knowledgeable visitors, Andy Grove, almost unrecognizable in heavy horn-rims, dark shirt and tie, and unlike almost every other man in the shot, longer hair than today. At his right shoulder, much taller and with almost the same glasses and sideburns then as now, is senior scientist Ted Hoff.

Behind these leaders of the company, stretching back almost to the front door of the building are figures known and forgotten, most long since gone but a few who still remain with the company. Pondering the image, one slowly notices that almost all of the men seem roughly the same age—in their midthirties, the founding engineers of the semiconductor industry, most of them having joined Fairchild at about the same time.

By comparison, the women, whose numbers grow ever larger the farther back in the shot, are almost all in their twenties. Many are secretaries; the rest, identified by their white smocks, work on the manufacturing line. In 1980, there was only one woman scientist at Intel, Carlene Ellis, newly arrived from Fairchild—and her presence is considered a breakthrough in the fraternity that is the semiconductor industry. Seventeen years hence, she will become Intel's first woman vice president; six years after that, the company's chief information officer. But for now, even at open-minded Intel, engineering and management are men's work.

In modern Silicon Valley, at least in engineering, that is still largely the case. The gains that have taken place for women in the intervening years have largely occurred in the nontechnical, staff positions—PR, advertising, marketing, HR, and even sales. If a similar photo were taken of the modern Intel—it would require a soccer stadium—the ratio of women to men would be almost exactly the same as in 1980; the difference would be that many more women would be toward the front of the crowd.

What would be far more noticeable in this modern gathering would be the radically different ethnic composition of the crowd. Other than one or two black faces, 1980 Intel is lily-white. That is not a feature lost on the millions of visitors, most of them schoolchildren, and a majority of them Japanese, Chinese, Korean, or Indian, who stare up at the old photograph. The face of Intel, like the face of Silicon Valley, and indeed, of the global electronics industry, is very different now. Cupertino, the home of many Intel employees, is now almost completely a Chinese American enclave, largely populated by trained professionals from Hong Kong and the mainland. El Camino Real, the path of the pioneering Spanish missionaries and the commercial heart of the Valley, has become an Indian retail community in the stretch that passes closest to Intel's headquarters. And just a few miles south, in San Jose, one large and thriving neighborhood has voted to call itself Little Saigon.

One of the least appreciated but most important achievements by Intel over its forty-plus years is its corporate personality and culture. These were largely in place as early as this photograph and have changed little even as the company has grown by nearly a thousand times in the number of employees and tens of thousands of times in revenues, even as it has spread its operations around the world and incorporated diverse cultures and ethnicities.

What is that personality? In Silicon Valley, there is certainly an Intel "type": technology-focused, competent to the point of arrogance, combative, free of artifice and proud of it, proudly unhip, ferociously competitive, and almost serenely confident of ultimate victory—and all of it wrapped in a seamless stainless-steel box. Were you to create an amalgam of Noyce, Moore, and Grove, you would not end up with the stereotypical Intel employee. Yet were you to parse those Intel employees' traits, you would find all of them in the three founders—Andy's combativeness and plainness, Gordon's competence and confidence, and Bob's competitiveness and vision.

It's a powerful combination, but not a complete one. In Intel's stripped-down and polished soul, there is little of Grove's wit, Moore's humility, or Noyce's maverick entrepreneurship—and that is a pity. It is as if Intel took those pieces from the Trinity it needed to win, not those it might have used to be its best. And in that, one can see the dominant impact of Andy Grove on the company over the two men who hired him.

That personality is made manifest in the culture of Intel, and it has been there almost from the beginning. From the day Grove and Vadász were hired, Intel has always recruited the best technical talent it can find, then dropped them into a work environment that has little pity for human weakness, much less failure. The classic eighty-hour work week that has come to define Silicon Valley professional life was codified (though unwritten) at Intel. So too was the notion inculcated into every employee that they were the best and brightest precisely because they were employed by Intel (an attitude that really wouldn't fully be seen again until Google).

This wasn't the old Fairchild: drunks and eccentrics weren't welcome. Nor were employees who couldn't keep up with the almost superhuman demands made by the company—at least one early Intel engineer had a heart attack from the stress of trying to get to *Sunday morning* office meetings.

This will to win, combined with the notion that the best people can do the impossible, created the most notorious part of Intel's culture: the "creative confrontation." Inside Intel, this process was carefully nuanced and parsed. These confrontations were designed only to focus the conversation on what was the best solution to the problem at hand and what was best for the company, leaving aside personal considerations, emotions, or ad hominem attacks. But to outsiders, they just looked like a lot of yelling and table pounding. Amused observers (grateful they didn't work at Intel) noted that only Intel would try to make getting your ass kicked by the boss into an uplifting and educational experience.

"The operative word here is *performance.* Screw up at National Semiconductor and you're on the street. Do the same at Hewlett-Packard (at least the old HP) and you're taken aside to a private conference room and told you are having a little problem fitting in. At Intel you're screamed at for not exhibiting the proper level of performance and the next day [you] are back on the job working twice as hard as ever to regain the respect of your peers; and the matter is never mentioned again."[2]

This culture, associated with the mighty Intel of the 1980s and 1990s, wasn't fully in place during the first dozen years of the company—that is, during the years when the company was led mostly by Noyce and Moore. Compared to what came later, that company was more idiosyncratic, guildlike, renegade, . . . and fun. Though the popular notion is that this was due to the two founders, that's only part of the story. Equally important were:

Size. Small start-ups, because they are heavily dominated by new hires arriving from other companies, often have trouble creating and maintaining a corporate monoculture. As late as 1975, Intel still had fewer than 5,000 employees—fewer than in many of its current office complexes. Most employees knew each other and the founders, enabling the company to operate largely on mutual trust rather than on the kind of rules and guidelines associated with a giant company filled with mostly strangers.

This small size also amplified influences that would have had little effect in a larger company. For example, in the early 1970s, Intel grew so fast and hired so quickly that it fell behind in the assimilation of these new hires. Many of these were former Texas Instruments employees who wanted to join in Intel's growing success. Without a new culture to adopt, the former TI-ers fell back on their old one, creating a sizable and destructive enclave of Texas Instruments' hierarchal and highly politicized culture inside Intel that took years to root out. This underscores how fragile and vulnerable was Intel's culture during those early years.

Flexibility. Organizational lines remain indistinct because young companies have flat structures and everybody assumes multiple duties as needed to make the enterprise successful. At Intel, where Noyce and Moore enforced their philosophy of equality and access, the organization chart was so flat in practice—no executive dining room, no executive bathroom, no executive parking places, no first-class travel, and anybody could walk down the aisle to the founders' cubicles—that it was almost revolutionary.

By the same token, this flexibility was a key component of the adaptability that was central to the success of Intel during this era. After all, over the course of that first decade, the company completely changed its core technology, its business, and most of its customers. That was possible only because Intel's employees had been equally able to change their jobs, duties, and responsibilities, often multiple times, over the course of those years. Other chip companies attempted the same shifts through a series of mass layoffs and hirings. While that sometimes worked in the short term, it did little for the intellectual capital of those companies—and nothing for employee loyalty. When those companies got into trouble, as all chip companies did in 1974 and almost every four years thereafter, there was little residual goodwill by either employees or customers to keep them afloat.

The era. The modern semiconductor industry was still little more than a decade old. Billion-dollar fabrication plants and fifty-thousand-employee payrolls were as yet years away. The processes and procedures that enabled chip companies to run at peak production, productivity, and profitability were still being developed. Indeed, the ability to maintain adequate yield rates—to produce fully working chips in the volume necessary to both meet customer demand and turn a profit—still remained as much an art as a science.

Just five years before, the process of fabricating chips was so crude that yields were affected by everything from pollen counts to seasonal groundwater levels to herbicides being sprayed on nearby farms (for a few more years, the Valley would have more family farms than housing developments and business parks) to male fab workers not washing their hands after using the restroom. By the birth of Intel, the fabrication process had largely moved into the environmentally controlled facilities that would be the progenitor of the modern surgery-level semiconductor clean room.

But even as these quality problems were being addressed, new ones were constantly emerging under the demands of Moore's Law. The features on the surfaces of chips were getting ever smaller, as were those on the photographic masks being used to create them. The number of transistors on each chip was racing upward, the failure of any one of them threatening the entire chip. Silicon crystals had to be grown in ever larger diameters (initially they were one inch wide; today, fourteen inches) at purities never before seen in the universe; even the water used to wash the wafers had to improve from merely distilled to water purer than that found anywhere on Earth. The list of new challenges went on and on—slicing the wafers, cutting out the chips, metal plating, gas diffusion, wire soldering, gluing to the packaging, testing—and all had to be upgraded, even revolutionized, every two years.

Thus even as Intel and the other chip companies were slowly systematizing the overall processes of semiconductor fabrication, the details of those processes were, for now, still as much craftsmanship and black magic as ever. It could all go sideways at any moment—as it had at Fairchild under Sporck. And that opened the door for the kind of eccentricity and seat-of-the-pants decision making that kept Intel from becoming as systematic and empirical as it dreamed of becoming.

There was another anarchic force defining the semiconductor industry during these years: *lawlessness*. In the early 1970s, the semiconductor industry was the hottest new business sector in America, and it was beginning to produce mountains of wealth. Yet because it was geographically isolated from both Washington, DC, and Manhattan—and because it involved an arcane technology that few yet understood—it was also among the least watched and least regulated of industries. Add to that the fact that the chip business was also basically a large family (or perhaps more accurately, a vast research laboratory) in which everyone had worked for, with, or against everyone else.

The almost inevitable result was that in those early years, the semiconductor industry was characterized by a combination of deep loyalties among competitors and often brutal competition between friends. That competition often crossed the line into what would usually be considered unethical, even illegal, behavior.

CHAPTER 12

The Wild West

N o one understood the potential for skulduggery better than Don Hoefler, who quickly recognized that he could make a living as the Walter Winchell of the semiconductor industry. After leaving *Electronic News*, Hoefler created *MicroElectronics News*, a weekly scandal sheet whose mission was to tear the lid off the clean, crew-cut-and-skinny-tie image of the chip industry in Silicon Valley and expose all of the bad management, personal scandals, and dirty dealings beneath.[1] Hoefler was not a very good reporter; his willingness to go to press with a story from a single source who most certainly had an ax to grind always made his stories suspect—and ultimately led to his destruction via a libel suit. And Hoefler's growing alcoholism, no doubt in part due to his being the most despised man in Silicon Valley, also began to affect the quality of his work.[2]

Nevertheless, a study of Hoefler's newsletter through the 1970s is a glimpse of the dark side of the semiconductor industry's early years that almost never makes the official histories. Here is a world not unlike the oil wildcatters in Texas and Oklahoma fifty years before: smart, fearless men taking huge risks for huge rewards, and not above stealing talent, customers, and even technologies from one another—or sabotaging the competition—to get ahead.

In those pages, Valley companies steal each other's top scientists and inventors, sometimes to get at the competitor's proprietary products, sometimes just slow down a competitor. There are stories—years before the rise of the personal computer hacker—of one company tapping into the design computer system of another to steal designs even before they are completed. There are fights in hallways between founders, outright theft of prototypes at trade shows, layoffs done over the loudspeaker in alphabetical order as employees cower at their desks, and the shipment of dead devices to customers in order to buy time to bring up yield rates.

How much of this was true and accurate? Years later, chip industry veterans would admit that most of it was. Certainly at the time, many of the companies perceived as the worst offenders acted guilty: at National Semiconductor (Charlie Sporck was a bête noire for Hoefler), the mere possession of a copy of *MicroElectronics News* was a firing offense—though that didn't keep company executives from secretly making copies of the newsletter and quietly passing them around.

Intel, if not considered a paragon of probity, then at least of honesty—it's always easier to play it straight when you are the industry leader—rarely appears on the pages of Hoefler's rag. But that doesn't mean that it was insulated from and immune to the tenor of Valley life in the era. The company played rough almost from the beginning, especially with customers—who would quickly find themselves thrust into the outer darkness (i.e., moved from the top of the delivery list to the bottom) if they dared to divide their loyalties. It would one day pay steeply for this ruthlessness. The company also didn't hesitate to raid its competitors for talent, hypocritically doing exactly what the founders had decried at Fairchild. This hypocrisy would peak two decades later with Gordon Moore's public complaints against "vulture capitalists" stealing budding entrepreneurs from Intel Corp.

This wildcatter style of business that defined the Valley, especially the chip business, in the late 1960s and throughout the 1970s, washed up against Intel Corp., but unlike many of its competitors, never defined it. This was certainly due less to any inherently superior business ethics of Intel's founders—once again, industry leaders have the luxury of being above the fray—than to the fact that Bob Noyce and Gordon Moore set out from the founding of the company to avoid the mistakes of Fairchild Semiconductor. It also helped that, given the company's technical leadership, most companies were trying to steal from Intel rather than the other way around. Twenty years into the future, when Intel found itself under threat from other innovators, the company showed that it wasn't above bending the rules.

But if Intel didn't play dirty in the early years like most of its competitors, the environment in which it was forced to compete certainly colored its corporate culture. The Valley's renegade style of that era was reflected in a high-rolling, push-all-of-the-chips-to-the-center-of-the-table style within Intel that the company has never known since.

But this singular Intel style during these early years wasn't entirely due to growing pains, structural forces, or the zeitgeist of the chip industry during those years. As anyone who joined the company at that time quickly learned, Intel also had a unique style of its own—that strange amalgam of the personalities of Noyce, Moore, and Grove—of creative confrontation taking place in the midst of a company–as–giant research department, all while trying to make everything look as effortless as possible. Thus it was oddly appropriate that in the early months of the company, even as Intel was positioning itself as the most cerebral of chip companies, a sizable percentage of the company's senior management was hobbling around on crutches—in Noyce's case from a broken leg from a skiing accident in Aspen; and another manager with a broken ankle from his ninety-ninth parachute jump.

No one piece of reporting has better captured the nature of this early management style at Intel than the Tom Wolfe *Esquire* profile of Bob Noyce noted earlier. It still stands as the most famous description of the early days of Intel and its singular corporate culture:

"At Intel, Noyce decided to eliminate the notion of levels of management altogether. He and Moore ran the show, that much was clear. But below them there were only the strategic business segments, as they called them. They were comparable to the major departments in an orthodox corporation, but they had far more autonomy. Each was run like a separate corporation. Middle managers at Intel had more responsibility than most vice presidents back east. They were also much younger and got lower-back pain and migraines earlier.

"At Intel, if the marketing division had to make a major decision that would affect the engineering division, the problem was not routed up the hierarchy to a layer of executives who oversaw both departments. Instead, "councils," made up of people already working on the line in the divisions that were affected would meet and work it out themselves. The councils moved horizontally, from problem to problem. They had no vested power. They were not governing bodies but coordinating councils.

"Noyce was a great believer in meetings. The people in each department or work unit were encouraged to convene meetings whenever the spirit moved them. There were rooms set aside for meetings at Intel, and they were available on a first come, first served basis, just like the parking places. Often meetings were held at lunchtime. That was not a policy; it was merely an example set by Noyce. . . .

"If Noyce called a meeting, then he set the agenda. But after that, everybody was an equal. If you were a young engineer and you had an idea you wanted to get across, you were supposed to speak up and challenge Noyce or anybody else who didn't get it right away. . . ."[3]

For Tom Wolfe, coming from the autocratic, top-down, command-and-control, executive-dining-room, boss-versus-worker world of Manhattan, Intel's corporate culture was a revelation—and a revolution. And indeed it was: modern descendants of Intel's management style can now be found everywhere, from the overgrown kindergartens of San Francisco social network start-ups to those same corporate giants—now radically transformed by the need to compete in the Silicon Valley–created modern business world—in New York and every other major city in the world.

Yet for anyone who had seen the precursor of this casual management style at Fairchild, Intel's leadership "innovations" just seemed more of the same. Noyce and Moore may have pledged to themselves that they weren't going to repeat at Intel the mistakes they had made at Fairchild, but at least at first glance, the two founders were merely playing to their weaknesses, creating the same chaotic environment that had proven so disastrous at their previous company.

So why did it work this time? What was different?

Andy Grove.

Andy understood more than anyone that his name was attached to Intel—and if he had his way, would be synonymous with it—and there was no way he would ever let *anything* under his watch fail. Noyce and Moore were serious men, and they were willing to risk almost anything for their new company; but ultimately they would survive the death of

Intel. Bob had the integrated circuit; Gordon had his law. But what did Andy have? The authorship of an influential textbook on semiconductors? That wasn't enough for a man of his vast ambitions. No, he knew that his mark on history, good or bad, would be made at Intel. And by god, he wasn't going to lose the company because everyone sat around jawboning at meetings and getting nothing done.

It is obvious now that this was exactly what the Noyce management style needed: *consequences*. Sure, you had equal right to a conference room, and you could stand up and challenge Noyce or Grove or Moore head-on, but when that meeting adjourned you'd better be walking out with a plan or a call to action or an established set of duties for you and every one in attendance . . . or there would be hell to pay. Get into a pattern of holding discursive and inconclusive meetings and you would find yourself standing in front of Andy Grove's desk, and that was the one place in the universe you didn't want to be.

By adding consequences to the process, Grove added the one ingredient missing from Bob's management style at Fairchild. The Fairchilders caught on quickly that there was no punishment for wasting time at endless meetings. On the contrary, the key was to be the last person to make your case to Noyce, so there was every reason to indulge in endless meetings. But with Andy riding herd, everyone was motivated to keep meetings short, concise, and decisive.

A lot of organizations have just such figures in the second position—the major who follows the affable colonel on inspections and screams at the troops, the notorious corporate hatchet man who does the CEO's dirty work and lets the senior executive keep his hands clean, the ultracompetent chief of staff who is unleashed to make decisions in the president's name. Modern corporations have codified that position into the genteel-sounding term chief operating officer, but the role and its duties haven't changed.

In Intel's first decade, this was Andy's unwritten role. He kept the company aligned on a daily basis with the long-term goals set by Noyce and Moore. He imposed consequences on every employee and every action in the company. And he ruthlessly enforced cost accountability on every office at Intel—Grove did not accept excuses for a failure to hit one's numbers. Noyce saw this, and whatever his thoughts about Grove (a bit put off by how Andy got successes but happy with the results), he knew that he had found the right man for the job.

Still, as the years passed, there were many occasions when Andy's behavior toward Intel employees was in direct violation of Bob's management style. Why didn't Noyce intervene? After all, while he may have been loath to confront company employees, he seemed to have no hesitation (sometimes he even seemed to take pleasure in) cracking down on Grove.

One possible reason is that Intel's success meant far more to Bob Noyce than he ever let on, and if it took Grove's tough-guy tactics to get there, Bob was willing to live with them. Noyce always believed in a personal destiny, that he was among the select few blessed with talent, brains, and good luck. He expected to win at everything and whenever possible make it look effortless. With Andy even more than with Charlie Sporck, Noyce would float above the fray, playing the long game, thinking the big strategy, and leave the hard work to Grove, secure in the knowledge that, by hook or by crook, he would get it done.

In a long-forgotten comment, Les Hogan said of Grove, "Andy would fire his own mother if she got in the way."[4] Marshall Cox once remarked, "You have to understand. Bob really has to be a nice guy. It's important for him to be liked. So somebody has to kick ass and take names. And Andy happens to be very good at that."[5]

Noyce needed a success at Intel at least as much as Grove did, only he wasn't so nakedly obvious about it. History has treated Bob's tenure at Fairchild far better than his contemporaries in the electronics business did at the time. Sure, Fairchild corporate's ham-fisted treatment of its golden goose subsidiary was the official explanation agreed upon by almost everyone. But there was no escaping the fact that by the time Fairchild Semiconductor imploded, it was in serious trouble: falling sales, market share, and profits, an inability to get new designs built and shipped, a frustrated senior staff. And much of the blame for this—a belief held even by Grove, the top operations guy at Intel—fell upon Bob Noyce. Nobody wanted to say this about "Saint Bob," but Noyce, the shrewdest man in Silicon Valley, knew it. He needed a win—a big one, as big on the commercial side as the IC had been on the technical side—if he was going to live up to the world's (and his own) image of him.

Intel was going to be that win. It had it all: the technology, the image, the money, the team, and, Noyce slowly discovered, in Andy Grove it had the plow horse to pull the company to victory.

It certainly hadn't started out that way. When Grove was hired at Intel, the expectation was that he would ultimately assume leadership of Intel's technical operations as Moore moved up to help Noyce with the larger business operations and strategic planning of the company. Moore: "Andy, we thought originally, would end up as something like our chief technical guy. And he started out worrying about the operations, the setting up of the manufacturing and the engineering activities.

"But it turned out that he just loved the idea of assembling an organization and seeing how it worked. He 'got over' his PhD essentially and started worrying about management kinds of problems. His horizons expanded when he got to Intel. Looking at the whole operation, seeing what it took to set up manufacturing. Andy is extremely well organized. He was at the opposite end of the spectrum from Bob as far as managerial capabilities are concerned.

"The three of us worked very well together."[6]

That last is classic Gordon Moore. In fact, the strengths and weaknesses of the three men dovetailed together beautifully in the running of Intel. There was very little disagreement among them about what was best for the company and the direction Intel should go. Indeed, their very differences seemed to work toward greater business accord than is typically found among partners who are much more alike in sensibilities. But as is also true in life, an ant and a grasshopper can make for a powerful team, but the grasshopper often takes the ant for granted as someone who likes to be a drudge, while the ant comes to despise the grasshopper for flying above all of the hard work of daily existence while still taking most of the credit.

Those forces were at work in the executive offices of Intel from the very beginning, but in those hectic early days, the founders were too busy to concern themselves with them. And soon, an opportunity would present itself that would be so immense, so conflicting, and so transformative that it would overwhelm every other internal conflict for years. But when Intel emerged from this era, those old frictions would return, with a vengeance.

CHAPTER 13

Bittersweet Memories

I n the spring of 1969, Intel introduced its first product, a memory chip built with bipolar technology. Even though it apparently wasn't meant as one, this device proved to be a brilliant act of misdirection to the rest of the semiconductor industry, lulling it for the second part of Intel's one-two punch.

This first product, the model 3101 Static Random Access Memory (SRAM), was a 64-bit memory chip that used Schottky TTL (transistor-to-transistor logic) architecture to achieve speeds twice as fast as the Fairchild 64-bit SRAMs already on the market. SRAMs were becoming very popular devices in the computing world because they could be put, alone or in clusters, into existing computer models to provide what IBM was calling a "scratchpad" for the temporary storage of data for easier user access. Traditional computer magnetic core memory didn't lend itself to this application, and SRAMs had an advantage over other memory chips in that they didn't leak power and did not need to be periodically refreshed: as long as they had electric current they could store information almost forever. That said, because they required a half dozen or more transistors per cell, SRAMs were expensive to build (yield was a perpetual problem), were slower, and had lower densities than their counterparts.

In practical terms, all of these factors typically meant that any company going into the SRAM business had to be prepared to donate several years to get into full production. The shock created by the model 3101 was that it appeared so quickly, almost impossibly quickly.

In talking to the media at the founding of Intel, Noyce had given the impression that it was a company that, rich with talent and money, had all the time in the world to study the latest technologies and markets and then settle upon a strategy. As Noyce said, Intel wanted "to take a snapshot of the technology of the day, trying to figure it out, what was going to work, and what would be the most productive." It would not be rushed; rather, it would take advantage of this unique moment as a corporate tabula rasa. "We didn't have some commitment that we had to keep some manufacturing line going for, so we weren't locked into any of the older technologies."

Thus when Intel suddenly announced the model 3101 chip less than eighteen months after the company was founded, the industry was flabbergasted. How had Intel even gotten a fabrication facility up to speed in that brief time, much less designed, built, and announced a whole new product?

To Gordon Moore, who knew exactly what had happened, it was still something of a miracle. "Start-ups are an exciting time," he has said. And in those first days of Intel, the key was to build upon what was already in place, rather than the usual start-up practice of starting from scratch:

We found a facility where a small semiconductor company had once operated. Silicon Valley was full of small semiconductor companies, many of which could trace their origin to Fairchild. The building had many of the utilities that we needed and allowed us to avoid some of the start-up costs and delays that would otherwise have been required had we moved into a completely empty building. We hired a staff of young, highly capable people. At Fairchild we had learned that one of the advantages of a start-up is the opportunity to train managers. We hoped that we could recruit a staff that would grow in capability as fast as the new company grew.

Getting the company's manufacturing operations under way also had the hallmarks of a veteran management team. Moore:

We divided the labor of establishing the technology and developing the first products among the senior people. The technology group pledged to develop the rudimentary processing before the end of the year so that we could proceed with product development. This pledge became a bet. To win the bet, the process people would have to get the equipment to operate reasonably well, make a stable transistor, and demonstrate a couple of other technical milestones by the end of 1968, only five months after start-up.

Moving into a building with no equipment on August 1 left relatively little time to purchase the process equipment, install it, get it running, and demonstrate at least a rudimentary process. The last of the three milestones was achieved on December 31, but only with the very dedicated effort of the entire staff.[1]

The young company thus began the new year with its factory in place and the basic design of its first product near completion. No one outside the company had any idea that Intel was moving so fast. Industry observers had some notion that Intel's founders would likely go into the memory business; after all, that had been Fairchild's great strength. But they had no inkling in which memory-chip business Intel would plant its flag. The model 3101 seemed to answer that question: Intel would be a bipolar SRAM builder. Silicon Valley geared up for a fight over the SRAM business.

Why static RAM? From the beginning, Intel had predicted that the biggest emerging new opportunity for semiconductor devices would arise from the replacement of magnetic-core memory with semiconductor chips designed for either read-only memory (ROM), to hold software; or random-access memory (RAM), to hold raw and processed data. SRAMs were difficult to build at high yields because they required at least six transistors per cell and were comparatively slow in operation. But Intel knew it could make up some of this lost architectural speed with the high-speed underlying physics of bipolar technology.

When the 3101 was introduced in spring 1969, all eighteen Intel employees (three of them on crutches) gathered in the little corporate cafeteria and toasted each other with champagne. The company was on its way—and had given notice to the rest of the industry that it would be a serious competitor, a technological innovator, and a manufacturer capable of achieving high enough yields to get to market in the shortest amount of time. Meanwhile the fear induced in the industry by Intel's incredible speed to market was mitigated somewhat by the realization that at least the company was targeting a predictable market in SRAMs.

But it was only a commercial head-fake. In a matter of months, Intel shocked the world by next introducing first a classic Schottky bipolar ROM (model 3301), but at an industry-leading 1,024 bits, and then, to the amazement of all, the model 1101 SRAM—in metal-oxide semiconductor (MOS) technology.

Regarding the 3301, as with many products in the company's early history, the impetus for this device came from an outside vendor wishing to place a contract with the company for a cutting-edge—and thus highly competitive—design. This was the advantage of being considered the brain trust of the semiconductor industry: all of the really interesting ideas went to Intel first.

As for the model 1101, it stunned the semiconductor industry not because of its performance, though it was impressive compared to the competition, but because of its unexpected process technology. Just when everyone had settled into the assumption that Intel would be just another bipolar company, suddenly it announced a mature product in MOS.

Why MOS? Here's Moore's technical description: "MOS transistor technology, called silicon-gate MOS, is the metal *electrode* terminal through which electric current passes between metallic and nonmetallic parts of an electric circuit. In most familiar circuits current is carried by metallic conductors, but in some circuits the current passes for some distance through a [non-metallic conductor such as a silicon film]. This new MOS process offered the advantages of self-registration (meaning that one layer of the structure is automatically aligned with those previously applied to the wafer), so the devices could be smaller and perform at higher frequencies. It [also] gave us greater flexibility in designing the interconnections among the various layers."

Put simply, MOS wasn't as fast (nor as the military knew, as radiation-proof) as bipolar, but it more than made up for that limitation by being much easier to design and fabricate, and it could be made much smaller and denser. Just as important, as Moore noted, MOS chips used a new design feature in their transistors called a silicon gate. This feature, part of the manufacturing process, simplified the placing of the layers on the surface of a chip by using the electrode region of each transistor to act as a mask for the doping of the surrounding source and drain regions.

That's a complicated description of what would prove to a powerfully simplifying process in chip making. Just as important, it made the transistors on the surface of MOS-based chips self-aligning, and that in turn radically improved yield rates over comparable bipolar chips. Ultimately, that improved quality and reduced size would transform the computer industry, making possible comparatively low-cost minicomputers for small business and setting the stage for the personal-computer revolution. The inventor of silicon-gate technology, a young Italian physicist named Federico Faggin, at that moment working at Fairchild, would one day be among the greatest inventors of the century—and would soon change the course of Intel's history.

As already noted, making the shift from bipolar to MOS was very difficult, and even by the mid-1970s, some chip companies still hadn't fully made the transition. But Intel not only made the jump to MOS, incorporating silicon-gate technology in the process, but it seemed to do so almost effortlessly, even as it was also leading the industry in bipolar memory chips. If Intel's inaugural chips impressed the world of customers, vendors, and trade-press reporters, it was this easy move into MOS that earned the most respect.

In little more than a year, Intel had already put itself in the front ranks of the chip industry. And yet it still had one more card up its sleeve. Once again, the source was an outside vendor looking to gain an edge.

In this case, the ambitious potential customer was Honeywell, then still one of the world's largest mainframe computer companies. In October 1959, the biggest firm in that industry, IBM Corp., had introduced the first small business computer, the model 1401, and the rest of the Seven Dwarfs (as they were called in light of their comparative size to Big Blue's Snow White) spent the sixties scrambling to catch up. Honeywell's response, announced in 1963, was the model 200, and well into the 1970s, the company continued to upgrade its design.

It was during one of these performance upgrades, in 1969, that Honeywell approached Intel to see if it could create a special type of memory chip, called dynamic random access memory (DRAM).

DRAMs were rapidly becoming the hot rods of the semiconductor industry. Initially fabricated in bipolar, they required very few transistors per computational cell. This made them both extremely fast and of a transistor density greater than any other integrated-circuit chip. Thus it was DRAMs then and now (though they have largely been replaced by flash memory chips) that set the performance pace for every other device in the business and have been the centerpiece of almost every graph of Moore's Law. But like hot rods, DRAMs were built for speed, not durability. Unlike SRAMs, which retain their content as long as they are supplied with power, DRAMs slowly leak their power and are thus "dynamic" in that they needed to be periodically refreshed.

What Honeywell wanted from Intel was a DRAM that used just three transistors in each cell, a pretty radical improvement in design for the era (modern DRAMs use just one transistor and a capacitor). Honeywell went to the right place, but it made a fundamental mistake: not trusting the new company to do much more than fabrication, it imposed on Intel its own design.

Intel did as it was asked, and in early 1970 it delivered to the computer maker the Intel model 1102, a 1-kilobit MOS DRAM. Honeywell was happy with the result, but Intel was not; the design suffered from performance problems. So Gordon Moore ordered the creation of an internal skunkworks—remarkable in itself, given that the company still had fewer than one hundred employees—to design a wholly new 1-kilobit MOS DRAM with no overlap with the Honeywell design so as not to create a conflict of interest.

Intel wasn't just going for a win with this new chip, but to sweep the table. As he had done at Fairchild, Noyce set a pricing target of less than 1 cent per bit—meaning that if Intel could pull it off, this new memory chip would be not only the technology but the price leader. That would be an unstoppable combination.

It was not a simple challenge. The masks for this new chip were designed by Joel Karp and laid out by Barbara Maness (making it perhaps the first important semiconductor chip ever cocreated by a woman).[2] In the end, it took five revisions of those masks before Intel finally reached an acceptable production quality. Yield rates have always been important in the chip business, but for Intel it was particularly important to get this device right, because it would be the first-ever commercial DRAM chip.

The device, designated the model 1103, was introduced in October 1970. Its timing was perfect: bipolar chips, with their speed and toughness, were ideal for military and aerospace "mil-spec" applications and thus found a home with the Defense Department in fighting the Vietnam War and facing down the Soviets in the Cold War and with NASA for the space program. Now at the beginning of the seventies, with two of the three winding down, MOS technology, with its greater levels of miniaturization and lower cost, was ideal for the emerging world of consumer electronics. The semiconductor industry, now led by Intel, made the turn and has never looked back.

By the end of the decade, DRAMs became the "currency" of the semiconductor business—so important to the computer, calculator, and video-game industries that a global economic war resulted as the Japanese electronics industry used its own DRAMs as the spearhead of its assault on the US electronics industry.

With the 1103, Intel effectively finished bracketing the world's memory-chip business—just as those devices were about to become vitally important throughout all of electronics. In less than two years, the company had taken design leadership in bipolar SRAMs and MOS SRAMs and had pioneered DRAMs. Furthermore, each of those devices had set a new standard in capacity, showing that Intel could not only play in these different vertical markets, but was also the leading innovator in each of them. In the time it took most semiconductor companies just to build their factories and bring their production up to speed, Intel had all but swept the table in an entire industry sector. Any chip company that wanted to compete in the memory business now knew it would have to go head-to-head with Intel, and many chose the safer path of pursuing logic chips and other parts of the semiconductor world.

Despite flying out of the gate long after many of its competitors, in just two years, Intel had won the first memory war in the semiconductor industry. But in one of the company's back labs, a small team was at work on a new type of chip that would so profoundly change the digital world that it would put even the transistor and integrated-circuit revolutions in the shadows. It would change the nature of the modern world and become the most ubiquitous product in history. And it would force Intel to make its most important and difficult decision—and in the process, make this remarkable early period in Intel's history so unimportant that it is usually dropped from the company's own official story.

PART III

The Spirit of the Age (1972–1987)

CHAPTER 14

Miracle in Miniature

There was one more customer, this one even more unexpected than the others, heading Intel's way.

The first great semiconductor-based consumer electronics market was desktop calculators. Tabulating machines were as old as the abacus.

It was inevitable that electronics would come to the calculating industry. And that process began in the late 1950s with the introduction of the first "electronic" calculators—although actually most still used vacuum tubes or electric relays. The pioneer in the field was Casio, in 1957, serving notice to the world (largely ignored) that the Japanese electronics industry, until this point a minor player in technology, even an object of derision, was determined to own the emerging world of consumer electronics. Soon other Japanese companies, including Canon, Sony, and Toshiba, had jumped into the game, along with veteran players like Olivetti, Monroe, and Smith-Corona.

With the 1960s and the arrival of the first low-cost logic and memory chips from Fairchild, TI, and Motorola, the race began to build small, light desktop calculators featuring integrated-circuit chips. Friden introduced the first transistor calculator in 1963. Soon after, Toshiba introduced the first calculator with semiconductor memory; in 1965, Olivetti announced the first programmable desktop calculator—the precursor to the personal computer.

All of this was very good news for the semiconductor industry, as it opened a potentially vast new marketplace away from the tightly controlled and cyclical mil-aero industry. And its growing demand for cheap RAM memory chips certainly played a role in Intel's early product decisions.

By the late 1960s, setting the precedent for the many consumer-tech booms that followed, the calculator business enjoyed almost vertical growth, as prices held steady, performance improved, and retailers, then customers, flocked to this hot new home/office appliance. And again setting the trend for future consumer-tech bubbles, the calculator industry began to, first, attract scores of new competitors hoping to cash in on the big profits to be made and, second, segment into different submarkets.

There may have been more than one hundred calculator companies by 1970, and while the newcomers were rushing in to compete with cookie-cutter four-function machines, the established and ambitious firms were already racing off to add more functions, make them small enough to be handheld, or drive manufacturing costs down so far as to commoditize their basic models, shrink profit margins, and drive competitors from the field with massive investments in marketing and distribution. It was mostly American companies, especially Hewlett-Packard, that took the first tack; American and the most veteran Japanese calculator companies, like TI and Casio, the second; and giant Japanese companies like Canon the third. That left everyone else high and dry when the inevitable industry shakeout came.

One of those doomed also-ran companies was Japan's Busicom. At least that was its name in 1969. Capturing the fly-by-night nature of the calculator boom during that era, Busicom had also been known in its brief existence as ETI and Nippon Calculating Machines. Busicom was hardly the biggest competitor in the calculator industry, but neither was it the smallest. And if it had neither the business acumen nor the investment capital to compete and win, what it did have was a kind of risk-taking entrepreneurial courage found in few of its competitors, especially its Japanese counterparts. And by 1969, when most of its competitors had begun to realize that they were being left behind and would soon face financial oblivion, Busicom refused to fold but instead decided to bet everything on one crazy throw of the technical dice to leap ahead of the industry leaders.[1]

By then, it was generally recognized in the tech world, particularly in semiconductors, that it was theoretically possible to use the new MOS technology to create a single chip that contained multiple functions.

That is, it might be possible to design and fabricate one chip on whose surface different regions could be dedicated to logic/calculation, read-only memory (ROM, to hold permanent firmware programs), a cache of random-access memory (RAM, to briefly hold input data and output results), input/output management, and power control. These regions could then, in theory, be linked together on the chip, resulting in a single chip that could perform all of the tasks on the multichip motherboard of say, a calculator. Theorists dubbed this miracle chip a *microprocessor*—a "CPU on a chip"—because it assumed all of the tasks, at the single-chip level, of the whole board of chips currently used as the central processor of a calculator.

There was no one moment that history can point to and say that *this* was the moment, the epiphany, that led to the invention of the microprocessor. Gordon Moore would say, "There was no real invention [in a technical sense]. The breakthrough was a recognition that it was finally possible to do what everyone had been saying we would someday be able to do."[2]

At the beginning of 1969, the microprocessor crossed that line from theory to possibility, and a number of companies, from giant RCA to computer networker Four-Phase to mighty IBM, began inching their way toward it. But realizing this possibility, everyone realized, was going to be daunting. There seemed to be an endless list of obstacles to be overcome—design, mask making, fabrication, heat dissipation, programming, packaging, testing, yield rates, and more. While the concept seemed doable, no company was rushing forward to spend its treasure on a unproven concept without any obvious customers.

It was the sheer impossibility of inventing and installing such a magical device as the microprocessor that appealed to Busicom. And what did it have to lose? The market was already beginning to implode, so why not bet everything? Whatever chip company took the contract would be assuming all of the risk. If they pulled it off, then Busicom would take the technological leadership of the entire calculator industry—with a product that could be produced at a lower cost and at a premium price over the competition. And if the chip maker failed, the vendor would take the financial loss, and Busicom could still stagger along in hopes of another miracle.

Even Busicom didn't think that this pioneering microprocessor could be built as a single chip—that was still a bridge too far. No semi-conductor company in the world would assume the risk of that kind of technological leap. But perhaps a small constellation of ten custom chips: that would still be a major breakthrough in both size and cost, and it would still give Busicom a jump on the competition. Perhaps it could convince an American chip company to accept that slightly lower level of risk. Busicom decided to start with that new chip company in California that was getting all of the publicity about its innovative new products: Intel Corp.

By now, a couple semiconductor companies had already begun to see some real results in their casual (and secret) quest for the micro-processor. The defense contractor Rockwell had already prototyped a primitive processor. And at Fairchild, that young Italian physicist, Fed-erico Faggin, who had invented silicon-gate technology, was reportedly conducting his own research.

Meanwhile, among industry theorists, a debate had begun over where calculator-chip technology should go next. The dominant camp held that the future lay in developing custom ICs for each customer. That, they claimed, was the best way to increase the performance of those calculators while still retaining proprietary control over the un-derlying chip designs. The minority camp argued that designing new chips for each and every calculator company would ultimately prove—as Moore's Law made them more and more complex—prohibitively ex-pensive. Instead, they promoted a very different solution, one that de-rived from the computer industry: general-purpose chips that could then be programmed for different applications and clients.

In retrospect, the latter strategy is obviously better. This is not just because, as the minority camp suggested, it would soon become in-sanely expensive to produce all of those custom chips, but also because it would allow these chips, when reprogrammed, to be sold outside the confines of calculators into other consumer electronics industries. But keep in mind that, at that time, there were no other consumer electron-ics industries, and chips then were still so simple that customization was comparatively easy.

Busicom may have been willing to take a risk with its proposed new ten-chip CPU contract, but when it came to the architecture of those chips, it was much more conservative. It went with the custom-chip camp—and that's what it put in its contract.

Why did Busicom choose Intel to build the chip set for its new calculator?

The easy answer is that Intel was basically the last semiconductor company left that hadn't yet signed on to build chips for a calculator manufacturer.

But there was another, more subtle, reason that more than any other event in the story of Intel explains why Bob Noyce is so important to Silicon Valley history and why his presence made Intel so special.

It begins in Japan, where Noyce was all but worshipped for his co-invention of the integrated circuit—which to Japanese engineers, struggling to make their place in the world of digital electronics, represented the signal intellectual achievement of the age—and for his leadership of Fairchild, which represented to the Japanese everything they wanted to emulate in American entrepreneurship. It became a badge of honor, first at Fairchild and then at Intel, for Japanese executives to visit Silicon Valley and have an audience, no matter how brief, with the great Noyce . . . and then return home and tell everyone how they had been transformed by being in his presence.

Most of this adoration was lost on Noyce himself. Bob saw this hero worship as just an extreme case of Japanese formality and politeness . . . that is, until he embarked on his first trip to Japan. There, with Intel's product manager Bob Graham, Noyce saw the real depth of the awe and veneration in which he was held by the Japanese electronics industry. (Intel's corporate attorney later said the Japanese thought Noyce was "a god.") And that attitude had even turned into profit: Intel marketing VP Ed Gelbach remembered that senior Japanese engineers would approach Bob and say, "We designed this [Intel chip into our product] just because of you, Dr. Noyce."[3]

But none of the hero worshippers Noyce met did as much as Tadashi Sasaki, an executive at Sharp, one of Japan's biggest electronics companies and a major global player in the calculator business. Sasaki had followed Noyce's career ever since the Shockley days, and his early knowledge of both the integrated circuit and the planar process had proved crucial to Sasaki's rise at Sharp. So he believed he owed his career success to Dr. Noyce and was desperate to meet him. And like a nervous young swain on his first date, Sasaki even had a conversational topic that he thought would intrigue the great man: one of Sasaki's engineers had recently approached him with an idea to put all of the functions of a computer on a single chip.

But the meeting didn't work out as planned. There is no record of Noyce's response to Sasaki's story about the computer-on-a-chip—the notion really was in the air all over the world—but we do know that Sasaki was deeply disappointed that he couldn't give his hero the one thing that Bob Noyce really wanted: chip orders from Sharp. There just wasn't any business available.

Desperate now not to let the great man down, Sasaki says that he proposed a dinner with Noyce, Graham, himself, and a colleague and old college friend, Yoshio Kojima, president of Busicom. We don't know what was discussed at that dinner—there is, in fact, no record that it took place—but we do know the result: not long after, Busicom approached Intel with the chip deal. Sasaki would even later claim that he funneled 40 million yen from Sharp to Busicom just to make the deal happen.[4]

Bob Noyce may not have been the greatest businessman, but the simple fact is that nobody else in the semiconductor industry could have provoked that chain of events simply through his reputation and charisma. Noyce got Sasaki, who got Kojima; Busicom got Intel, . . . and Intel got the world.

Tall and gangly, with horn-rimmed glasses and a goofy grin, Marcian Edward "Ted" Hoff was already showing a special gift for scientific invention even as a boy in Rochester, New York. As a senior, he earned national attention as one of the finalists in the Westinghouse Science Talent Search, a celebrated science fair. Hoff then attended Rennselaer Polytechnic Institute, two hundred miles away in Troy, and went home in the summers to work at General Railway Signal Corporation. It was there, while still an undergraduate, that Hoff applied for his first two patents.

He then went off to California and Stanford University, where he earned a master's degree in 1959 and a doctorate in electrical engineering in 1962. With his dissertation he co-invented (with his professor) the "least means square filter," a device still in use that adapts to its setting in order to reduce interference.[5]

Ted Hoff was a man built to invent new things, and with his new PhD in EE in hand, he eventually gravitated to the local company most likely to need his gifts, joining Intel as employee number 12 in 1968.

Intel wasn't thinking much about a CPU-on-a-chip in those early days; it had other concerns, like actually building its first product. But Ted Hoff was. And the more he looked at the problem, the more he came down with the generalists. He even knew where to look for the architecture needed to build such a multipurpose chip set: the brand-new industry of minicomputers. These new, smaller computers had carved out a valuable market niche, but not just by being smaller and cheaper than their predecessors. Unlike mainframe computers, which processed input in large batches, minicomputers operated continuously in "real time" as the raw data were inputted.

Ironically, Hoff had first learned about the power of minicomputers by using them to help design integrated circuits both before and during his brief time to date at Intel. And the machine that had impressed him most was the Digital Equipment Corp. PDP-10. Hoff was especially taken by how this machine was essentially a raw computing engine controlled by a comparatively simple instruction set that could tackle incredibly complex tasks through the use of switchable sophisticated "application" software programs. Here was the architecture Hoff was looking for to build his general-purpose processor.[6]

After the deal with Busicom was cut, Hoff volunteered to manage the new client—in particular the design team it was planning to send to work in Intel's headquarters. But if he planned to work with Busicom to build the general-purpose, programmable computer-on-a-chip, or just a few chips, he was soon disappointed.

By the time the three-man Busicom design team arrived, that company's comparatively straightforward (albeit customized) ten-chip strategy had morphed into a twelve-chip monster that added all sorts of new features to the planned calculator. To accomplish that, each of these new chips would have to feature three thousand to five thousand transistors—*five times* the number found on typical calculator chips.

Moreover, the Busicom team planned to do this design on-site in Santa Clara, with a minimum amount of help from Intel. According to the terms of the contract, Intel would then be paid $100,000 to manufacture this chip set, then be paid $50 per set on a minimum order of sixty thousand sets.[7]

From Noyce's point of view, this was a sweet deal. The Busicom team, featuring a talented young design engineer named Masatoshi Shima, would do all of the heavy lifting, with Intel providing some unused lab space and, when requested, some advice and assistance. Other than a tiny amount of time from project director Hoff, Intel didn't expect to expend much labor or capital on the project. In other words, Intel would do to Busicom—put the risk on the other party—what Busicom had planned to do to Intel.

There was only one problem: almost from day one, Ted Hoff was "kind of shocked at how complex this was," and he was convinced that the Busicom design would not only fail, but catastrophically, and that might leave Intel high and dry on the deal. Some of this may have been the bias of a man who had already bought into his own vision, but it was also the appraisal of a top-notch scientist who could read a diagram.

Ted Hoff now found himself in a terrible bind. Intel was struggling to get its new memory devices out the door, but he was convinced that the Busicom team was blowing one of the great product opportunities of the era. Worse, he knew how to fix it and perhaps save Intel in the process: "[Intel] was a start-up company, and a lot of us had hopes for its financial success, so I didn't want to let a major effort go into something disastrous."[8]

So despite the expectations that he would spend only a few minutes each day in "oversight" over the three Busicom researchers, Hoff was soon spending large portions of each day with the newcomers, trying to turn their trajectory. Hoff: "I had no design responsibilities for the project, but soon I was sticking my nose where it didn't belong."

But the harder he pressed, the more the Busicom team persisted. Hoff: "So I made some proposals to the Japanese engineers to do something along these lines [the general-purpose architecture]—and they were not the least interested. They said that they recognized that the design was too complicated but they were working on the simplification and they were out to design calculators and nothing else. They simply were not interested."[9]

Among themselves, according to Shima, the Busicom team thought Hoff's vision of a simple general-purpose chip set with a lot of programming was interesting but insufficiently fleshed out. To Shima, Hoff's design had a "lack of a system concept, lack of decimal operation, lack of interface to the keyboard, lack of real-time control and so on."[10] The Busicom team was there to build a real product that would fit in the company's calculator, its last shot at survival, not to try out some unproven, incomplete concept.

On the other hand, Hoff was already burning up too much of his precious time on the Busicom project. There was no time left to work out all of the nuances on the small chip-set concept that Shima and his crew demanded. So Hoff made a crucial decision: he would let the Japanese team go on with their project as planned and go back to his official role of hands-off adviser. But at the same, Hoff resolved to take the situation—and his solution—directly to Bob Noyce.

"I think we can do something to simplify this," Hoff told Noyce. "I know this can be done. It can be made to emulate a computer."

Noyce was sympathetic—he always was with subordinates—but as even he was the first to admit, he knew almost nothing about computers. He had come up in the analog world, and even though he was now building digital chips used in computers, the world of programming was still beyond him. That presented a serious obstacle to Hoff's argument. The good news was that Noyce was an incredibly quick study. He quickly began to grill Hoff with a series of questions—How could a calculator be like a computer? How does a computer operating system work? et cetera—that Ted Hoff recognized as the classic signal of Noyce's support.

Once again as with integrated circuits, Noyce found himself at a crossroads, facing a crucial decision about the future of his company, his industry—and as it turned out, the world. And once again he made the same decision. At Fairchild, developing the integrated circuit, Noyce had essentially created a blue team and a red team, one under Hoerni and the other under Moore, and let them pursue the design along parallel tracks. Hoerni won, inventing the planar process along the way. This time, with the computer-on-a-chip, the microprocessor, Noyce decided on the same strategy, but this time he was even more explicit. The Busicom team would keep going on their project, and Hoff would advise them as needed. Meanwhile, to Hoff he said, "Why don't you go ahead and pursue those ideas? It's always nice to have a backup position."[11]

Unleashed, a thrilled Hoff went off to come up with his own processor design.

The only problem was that, strictly speaking, Ted Hoff reported not to Bob Noyce but to Andy Grove. In going directly to Noyce, Hoff had just committed a corporate organizational no-no—and in committing the company to support Hoff's new skunkworks project, Noyce had just undermined the authority of Grove, his operations manager. This was not going to end well, especially with an individual as volatile and wary as Andy Grove.

Just as bad, Noyce had unleashed Hoff just when he was needed to help get the model 1101 memory chip, whose yield problems seemed insurmountable, out the door. Intel was having product problems, the company had yet to show a profit, and it was at risk of going broke— and at that moment, its CEO had signed off on a side project by its top scientist to produce an unproven product design with a tiny chance of success for a market that might never exist.

It is interesting to speculate why Noyce did this. Perhaps it was the same old nonconfrontational Bob, the boss who couldn't say no, bringing the same weakness he exhibited at Fairchild to his new company. Or as Leslie Berlin suggests, Noyce saw himself in Hoff and refused to crush the younger man's ambitions the way Shockley had tried to crush Noyce's creativity.

Perhaps it was both. But Bob Noyce also had an unequaled record for hiring the very best talent, including most of the men who created the semiconductor revolution and ran its greatest companies. He did so by being an astute judge of people and having the courage to trust that talent to make the right choices and to create great things. At this world-historic moment, Noyce had that faith in Ted Hoff. And if Hoff, knowing all that he did about Intel's current financial health, still wanted to pursue his own processor design, Noyce was going to back him all of the way, no matter who objected.

Because of his vision of a general-purpose computer-on-a-chip and his successful advocacy of that idea inside Intel, Ted Hoff is usually credited—sometimes alone—with the invention of the microprocessor. But it wouldn't have happened if Bob Noyce hadn't gone renegade on his own company. How many CEOs have ever done that? The microprocessor decision alone puts Noyce among the great corporate leaders.

For the next three months, Hoff worked mostly alone on his microprocessor concept in the brief intervals he had between putting out the many technical fires that threatened Intel's very existence. Because the project was not only secret but a potentially huge internal scandal, the only person he ever spoke to about it was Noyce, who despite his own manifold distractions served as sounding board, adviser, critic, cheerleader, and devil's advocate. Hoff: "He was always very encouraging, always very helpful, always had an idea." It was a bravura performance by both men.

Along the way, Gordon Moore also heard about the project, perhaps because Noyce decided to confide in his partner. Always curious about any new tech breakthrough, he investigated—and liked what he saw. But, as he recalled later, no doubt because he was dealing with the 1103 yield crisis, "my enthusiasm may not [have been] so obvious at times."[12]

By August, recognizing that his skill was in hardware architecture, not in the software needed to run a general-purpose processor, Hoff brought in twenty-eight-year-old Stan Mazor, a soft-spoken and self-effacing young software engineer who had recently jumped from Fairchild. Mazor's job would be to characterize the software architecture for what would now be a four-chip design and to write the instruction set for the master arithmetic processor chip, now designated the Intel 4004.

Within two weeks, the pair had blocked out the architecture of the 4004.

Later that month, Noyce sent a letter to Busicom president Kojima to tell him both of his concerns about the Busicom team's design and about the alternative Intel design. His trip to Japan had taught Bob the value of diplomacy in Japanese business culture, so he began his critique by first saying, "I don't criticize the design" of Busicom's calculator . . . and then went on to do just that. He noted that the sheer complexity of the planned circuitry for the calculator meant that there was "no possibility that we could manufacture these units for [the planned] $50/kit, even the simplest kit." Frankly, Noyce continued, they'd be lucky to build these big, complicated chip sets for even $300 per kit.

Noyce concluded by asking, "Is it reasonable to proceed with this development on the basis of this design, or should the project be abandoned?"[13]

There must have been a flurry of subsequent correspondence flying across the Pacific, because by September 16, Bob Graham, who was now privy to the secret project, sent a note to Kojima—once again decrying the complexity of the Busicom design . . . but now, for the first time, proposing that Kojima and staff consider an alternative Intel design. Within the month, Busicom sent a contingent of its technical experts to Mountain View.

It was good timing, because also that month, October 1969, the Busicom team finally hit the predicted wall.

Hoff: "We held a meeting in which the managers of the Japanese company came over here for a presentation by their engineering group and by a group representing our firm [i.e., Hoff and Mazor]. At that time we presented our idea: that our new approach went well beyond calculators and that it had many other possible applications.

"They liked that and they went for our design."[14]

The Intel team members were stunned by how quickly and easily Busicom had rolled over. Did the Busicom managers like Intel's four-chip design? Certainly. Did they like the idea of having to run those chips with software? Probably not. Did they care that this general-purpose design would open opportunities in other markets? No. In the end, it was that price point. Because of Noyce's letter (and who would doubt the words of the Great Noyce?), Kojima knew that even if his team succeeded, the resulting calculator would be priced out of the marketplace. He had no choice.

Intel, in what Hoff would modestly characterize as "a bit of a coup," had convinced a desperate Busicom to turn on its own top designers—who had proposed more than a dozen familiar chips stuffed with 2,000 transistors each—to go with the Intel design of four radically new chips containing no more than 1,900 transistors each, plus a lot of software yet to be written. Only the prestige and persuasion of Bob Noyce, combined with the increasingly troubled reports from the Busicom design team, could have pulled off that deal.

And then, having nailed down what would prove to be the most important piece of intellectual property in the history of the digital age, Intel seemed to walk away from the project. Four months later, the contract finally made its way through the legal departments of both companies. But still Intel didn't move. It was only when, a month later, in March 1970, Busicom sent a polite note to Bob Noyce inquiring about the status of the project and requesting an update (and proof of progress) that Intel finally got off the dime and fired up the computer-on-a-chip project.

Why the delay? Because Intel was in trouble, as captured in one internal company timeline of the era that read "Management Near Panic." As difficult as the model 1101 had been, the 1102 was proving to be even worse. All hands, including Hoff and Mazor, had been mobilized to get the Honeywell-backed memory chip to market, or at least to function with any consistency. The situation had gotten so bad that even as Busicom was pestering Noyce about the status of *its* chip, Intel, losing money fast, was going through a wrenching layoff of twenty employees—nearly 10 percent of its entire staff.

There was even worse news: throughout the 1960s, the chip industry had enjoyed an almost continuous period of growth. One after another, huge new markets appeared: transistor radios, the military, the space race, UHF television, minicomputers, calculators. But for now, at least, all of these markets were played out or (in the case of calculators) peaking with no equivalent new booms in sight. The semiconductor industry, increasingly the leading indicator for the rest of the electronics world, was not immune, and soon every one of Intel's competitors was seeing its own flattened financials.

But that was little consolation for Intel, which wasn't just burning through its initial investment, but headed toward a net loss for the year. From above, chairman Art Rock was using every board meeting to remind the founders to be more disciplined in their spending and ruthless in their cost cutting. From below, Andy Grove was demanding support from Noyce and Moore in his belief that the company had to focus all of its resources on getting 1101s built and out the door, on fixing the 1102's yield problems to get the deliveries done to Honeywell, and on taking the 1102's core to build the one new memory chip, the 1103, that might actually find customers in a slowing economy. Noyce and Moore listened, but to Rock's dismay and Grove's fury, went on their merry way.

It was at this moment that Bob Noyce decided that he'd better not wait any longer but must finally drop the bomb on Grove. He went into Grove's office, and according to Berlin, "sat down on the corner of his desk, a move that immediately raised Grove's suspicions. Anytime Noyce affected a faux-casual air, Grove knew he was not going to like the message. Noyce looked at Grove more from the corner of his eye than face-on. 'We're starting another project,' Noyce said with a little laugh." Grove later remembered thinking, "Go away, go away, we don't have time for this. What kind of company started new projects when its very survival was at stake?"[15]

The answer is: a company that can't resist a great technological breakthrough.

As the quote suggests, this was a nightmare moment for Andy Grove. His greatest fear about joining Intel had been that it would be a company without business leadership, that Moore would bury himself in the lab with the latest technological problem, . . . and that Noyce would be Noyce: more concerned about being liked than successful, influenced by the last person who spoke to him, averse to confrontation, and chasing glory at the expense of profits. And now, less than two years after the founding of Intel, Noyce seemed to be already up to his old tricks. The company was dying and Saint Bob was chasing phantasms.

But Andy knew he had no choice. He had thrown in his lot with Intel, and the chances of his finding another job at that level anywhere else in the chip industry, especially in this economy, were slim. Besides, at this point in his career, it was Moore who believed in him most, and after Gordon (ironically), Bob Noyce. To the rest of the chip world, he was still a former Fairchild scientist who, to date, hadn't been successful in running operations at Intel.

So as best as he could, Grove took decisive action to get this new nightmare off his desk. The Busicom project had stalled, he concluded, because its two Intel team members, Hoff and Mazor had been drawn elsewhere: Mazor to work on the memory chips, Hoff to be almost continuously on the road calming customers. Meanwhile, the Busicom team had gone home to Japan. The computer-on-a-chip project was now an orphan, so Andy decided it needed a new father. And he had the perfect candidate.

CHAPTER 15

The Inventor

S mall, impeccable in appearance, soft-spoken, and affable, Federico Faggin was almost the antithesis of the growing stereotype of nerdy engineer. Yet this twenty-nine-year-old son of a philosophy professor from Vicenza, Italy, who had just moved to the States two years before, was already well on his way to becoming one of the great inventors. Behind him was the invention of the MOS silicon gate, the heart of billions of chips to come, and ahead of him was the invention of the touch pad and screen, which would make possible the mobile computing and smartphone revolutions. For now, though, he was working in Fairchild's lab, having arrived just after the company's diaspora. Like everybody else at Fairchild, he was watching with curiosity and envy the hot new competitor just a few blocks away.

It wasn't that Faggin was being treated badly at the Les Hogan Fairchild. On the contrary, the company was more structured and professional than ever before. The economy might be declining, but Fairchild was once again showing profitability, and whatever bloodletting and purging might be going on along executive row, life in the Fairchild labs was good.

But for someone like Federico Faggin, being comfortable wasn't enough. He wanted to be at the very forefront of technology innovation in the semiconductor industry. He had earned it. And he craved the excitement and industry leadership that appeared to have moved to Intel.

So when the call came, he jumped. Then, amazingly, when he arrived, Faggin was told that he was to be in charge of Intel's new project to build the mythical computer-on-a-chip. It was a dream come true—until he saw the state of the project. As Faggin would recall:

"Presumably, Hoff and Mazor had already completed the architecture and logic design of the chip set, and only some circuit design and chip lay-outs were left to do. However, that's not what I found when I started at Intel, nor is it what Shima found when he arrived from Japan.

"Shima expected to review the logic design, confirming that Busicom could indeed produce its calculator, and then return to Japan. He was furious when he found out that no work had been done since his visit approximately six months earlier. He kept on saying in his broken English, 'I came here to check. There is nothing to check. This is just *idea*.' The schedule that was agreed on for his calculator had already been irreparably compromised."[1]

Remembering this, Faggin shook his head in disbelief. "So there I was: behind before I had even begun."

So he set out to make up for lost time. He handed off some of the design work on the main processor chip to Shima, whose expertise was logic design. Meanwhile, says Faggin, "I worked furiously 12 to 16 hours per day" on a gigantic technical problem that hadn't been anticipated by Hoff in the 4004 design: how to make the circuits very small and fast—i.e., using low power so as not to cook the surface, and yet strong enough to drive a readable signal off the chip.

Intel had made the right choice, because at that moment there was only one semiconductor physicist in the world who knew how to do it: the inventor of the MOS silicon gate. Faggin:

"First, I resolved the remaining architectural issues, and then I laid down the foundation of the design system that I would use for the chip set. Finally, I started the logic and circuit design and the layout of the four chips. I had to develop a new methodology for random-logic design with silicon gate technology; it had never been done before."[2]

This little side invention would prove to be almost as important in the history of semiconductors as the silicon gate: Faggin "bootstrapped" each gate atop its partner transistor, something heretofore considered impossible. The technique would soon be adopted in almost every type of chip for the next forty years. But at that moment, he didn't even have the time to step back and admire his achievement. Rather, he plunged on.

By the end of summer 1970, Faggin and Shima had completed the design for the four-chip set and had the underlying technology in hand.

It was to be officially called the 4000 family—underscoring the fact that Intel still didn't think of it as a single solution, a precursor of the computer-on-a-chip, but as a group of four distinct chips. Thus the nomenclature continued down to each of those chips:

1. The model 4001 was a 2,048-bit ROM memory chip designed to hold the programming.
2. The model 4002 was a 320-bit RAM memory chip to act as the cache of data to be operated upon.
3. The model 4003 was a 10-bit input-output register to feed the data into the main processor and remove the results.
4. Most important of all, the model 4004 was a 4-bit central-processor logic chip.

In a few years, when Intel actually did built single-chip microprocessors, the industry would begin to call these pioneering chip sets by the name of the processor chip. Thus today when we speak of the four chips of the Intel 4000 family, we typically refer to them simply as the Model 4004 microprocessor.

(These bit/byte listings may be somewhat confusing. In fact, they are apples and oranges. Bytes [composed of four to sixty-four or more "bits"] when we are talking about memory, are a measure of the capacity of that memory. A modern 1T disk memory holds approximately one trillion bytes in the form of numbers, symbols, or letters.

Back in 1970, Faggin and Shima were developing a read-only memory chip capable of holding about 256 bytes of instructions, and a 40-byte random-access memory chip, designed to briefly [less than one second] hold about 320 bits of raw data just before they went into the processor, and then hold the now-processed data just after it emerged and before it was sent off to external memory.

On the other hand, when we speak of bits in the context of logic chips or processors, we are talking about the size of each unit [byte] of data being processed. The 4004 processor chip was capable of handling data of up to four digits—making it a four-bit microprocessor. By comparison, a modern Intel microprocessor operates on 64- or 128-digit data. Volume in logic chips and processors is measured in the clock speed or frequency at which that chip can operate on the data. The 4004 ran at 740 kilohertz [740 kHz], or about 740,000 operations per second. A modern Intel microprocessor runs at nearly 4,000 MHz, or about four billion operations per second. This astonishing leap in performance in just a few decades is the very essence of Moore's Law.)

Faggin now had his chip-set design, but he still had to build it. In October 1970, prototype fabrication began. First up was the 4001. Faggin and Shima took the finished prototype, hooked it up to a specially created tester, and . . . it worked. Perfectly. The two men cheered each other and notified Hoff and Intel management.

Next came the 4002 and 4003. They exhibited the same flawless performance as the ROM chip—one of the most successful prototype runs of a new design in electronics history. Faggin was thrilled, but not entirely surprised. He had proven the effectiveness of his bootstrapped gate/transistor design, and most everything else was a proven (albeit relatively new) design.

Then just a few days before New Year's, the very first prototype 4004 processor chips arrived from fab. This was the one that had Federico scared. Not only was it the first chip of its kind, with an unproven architecture, but because its complexity was much higher than any of the other three chips, Faggin was afraid he might miss some bugs. With much trepidation, Federico placed the 4004 in the testing unit and fired it up.

Nothing.

The chip was completely dead. It didn't just emit a weak or confused signal; it didn't emit anything at all. Faggin's heart broke. And it only got worse when he put each of the handful of new chips into the tester. Every one was stone dead. Such a result for a new device, especially a new chip, was hardly uncommon in the electronics industry in those days. But after the spectacular results with the first three chips, both Faggin and Shima had set their hopes high. Now Faggin, after an almost superhuman effort, could only look forward to spending the rest of his lost holidays and probably most of January on a long postmortem analysis of the dead chips to figure out what went wrong.

And then just hours later, to his astonishment, Faggin looked through his microscope and instantly identified the source of the 4004's catastrophic failure: the prototype fabrication folks had failed to put down one entire layer of circuitry on the surface of the chips. Faggin sighed with relief: it was a manufacturing mistake, not a design error.

He went home to celebrate an unexpectedly happy New Year. Faggin: "Three weeks after that disappointment, a new run [of 4004 chips] came. My hands were trembling as I loaded the 2-inch wafer into the probe station. It was late at night, and I was alone in the lab. I was praying for it to work well enough that I could find all of the bugs so the next run could yield shippable devices. My excitement grew as I found various areas of the circuit working. By 3:00 a.m., I went home in a strange state of exhaustion and excitement. . . . All of that work paid off in a moment of intense satisfaction."[3]

For the next few days, Faggin continued to perform intense verification testing on the 4004, and other than a few other minor bugs, he was immensely satisfied with the great results. Faggin reported the news to Bob Noyce. By February—it was now 1971, almost two years after Intel had signed the contract with Busicom—the 4004 was ready to go into full production. Stan Mazor had also delivered the ROM codes, the software that would drive the 4004, to be entered into the 4001. Everything was ready to go.

A few weeks later, in mid-March, Intel shipped the first 4000-family chip sets to Japan, to Busicom.

The microprocessor age had begun. And though the company didn't yet know it, Intel had also begun its ascendancy to become the most important company in the world.

CHAPTER 16

A Sliver of History

In the decades that followed, as it became increasingly understood that the microprocessor wasn't just a more sophisticated chip, but a revolutionary invention that was at least the equal of the transistor and the integrated circuit, a kind of low-key war began for credit for the invention of this device.

While the microprocessor was still just a high-priced novelty used in a few premium electronic devices, ownership had been easy to share. But when microprocessors began to be produced by the billions and became the engines of personal computers, the Internet, smartphones, and a thousand other devices in automotive, industrial, consumer, and medical applications, ownership became infinitely more valuable. And soon politics—personal, corporate, and even national—began to influence the honors and recognitions.

Within the next fifteen years, all of the players in the 4004 project except Ted Hoff would be gone from Intel, not all of them on good terms. With Intel now synonymous with microprocessors and quickly becoming one of the most valuable companies in the world, some sort of official history was necessary, and with it some sort of official designation of the 4004's "real" inventor (emphasis on the singular).

Intel chose Ted Hoff. Soon company documents and documentaries, even the new Intel museum, gave Hoff almost full credit for the 4004.

But geopolitical forces were at work as well. The battle between US and Japanese electronics—and in particular, semiconductor—industries was at a white heat, with both sides guilty of making undiplomatic comments about the other. The Japanese, having long labored under the image of being makers of cheap, disposable electronic products, were luxuriating in their new reputation for quality, innovation, and competitiveness. But they needed heroes—men to match Packard, Noyce, Moore, Gates, and Jobs. They found them in Sony's Morita, NEC's Osafune, and among scientists, Masatoshi Shima, by now with Faggin at one of Intel's biggest competitors.

Shima welcomed the attention. He had long bristled at his treatment by Intel—and even more at the official history written by Intel, in which he had been largely ignored. Shima believed that he deserved most of the credit for the 4004, as he had not only done much of the design work, but as prototypers at Intel later admitted, Shima's logic design, done manually at the transistor level, made for a very understandable "floor plan" for fabricating the chip.

In Shima's view, the only real revolution in the 4000 family was the 4004 processor chip, and he had been its chief designer. Ergo, he more than anyone deserved to be called the inventor of the microprocessor. The Japanese electronics industry, weary of being accused of stealing the chip business from its American originators, was more than willing, with the support of the Japanese media, to go along with this version of the story. Soon Shima, bolstered by this support back home, was threatening to sue any reporter or writer who departed from this version of the story.

As for Faggin, who among the group was arguably the one person—if only one person had to be chosen—to be given the credit for the 4004 chip set, he would soon engender the animosity of Andy Grove and be wiped for decades from the official history.

The most charitable distribution of credit for the 4004 and the invention of the microprocessor typically spreads it across all four individuals: Hoff the visionary, Faggin the creator, Shima the builder, and kindly Stan Mazor, too polite to get into the fight, the programmer. That is increasingly how the world treats the invention of the microprocessor, and that is why President Obama awarded the President's Medal of Invention to the three Americans in the group. It is of a piece with how history has treated the integrated circuit (Noyce and Kilby) and the transistor (Bardeen, Brattain, and Shockley)—and if the Nobel is ever given for the microprocessor (it is long overdue), those will most likely be the four names attached to it.

And yet there is a very strong argument to be made that there is a fifth figure who also deserves credit for the microprocessor: Robert Noyce. Throughout the entire development of the 4004, Noyce had been there. At a time when the company was in serious financial straits, Noyce had given Hoff permission to follow his own dream, to create a secret skunkworks, and then had cajoled, challenged, and pushed Hoff to get things right. Noyce had then faced down his own management team, which wanted to kill the project and make better use of precious company resources. And he had led the effort to retain the rights from Busicom to sell the resulting chips to non-calculator customers. And now for the next few years, Noyce would dedicate himself to two simultaneous campaigns: first to get the microprocessor accepted inside Intel as a new product line, and second to get it adopted by the world for use as "embedded" intelligence.

There were other scientists around the semiconductor industry who were thinking about putting all of the functions of a computer on a single chip, so even if Ted Hoff had not put forth his idea, Intel would likely have gone into the microprocessor business in the years to come. Shima was brilliant, but Intel had his equal on the payroll, as it did for Mazor. Faggin was much less replaceable, but the industry would have matched in years what his genius did in weeks. But without Bob Noyce getting the order from Busicom for the 4000 Series project, backing Hoff in his secret project and diverting resources to make it succeed, approving Faggin's hiring, . . . and then, as we will see, walking the halls of Intel and flying around the world fighting for the acceptance of this new technology, a real working microprocessor might have taken as much as a decade more to be invented, and certainly not at Intel. Though he is left off the rolls of inventors of the microprocessor, Noyce is the indispensable man.

CHAPTER 17

Dealing Down

E ven as the Intel skunkworks team was struggling to get the 4000 Series chips designed and prototyped, the majority of the company was dealing with an existential crisis. It involved the company's big new memory chip, the 1,024-bit model 1103, the first commercial DRAM and the one built on the Honeywell core. It was designed to be a big commercial seller and keep Intel alive through the current industry downturn.

But that plan would work only if Intel could get the 1103 out the door, and by autumn 1970, the company was failing to do just that. As Andy Grove described it, "Under certain adverse conditions the thing just couldn't seem to remember."[1] That wasn't good for a memory chip, especially one designed for high-end, mission-critical applications. Some fogged up under their glass covers; others shut down when shaken. Others just quixotically turned themselves off at random moments. Ted Hoff, perhaps to compensate for being distracted by his other project, even wrote a twenty-eight-page memo explaining both the 1103's operations and its bugs.

The history of the semiconductor industry is replete with stories of cutting-edge new devices producing shockingly low yields, burning themselves up in use, or just being dead on arrival at customers. Fairchild had been responsible for more than its share of these disasters. But Intel, from the beginning, had presented itself as being not only the technological innovator, but also the quality manufacturer of the semiconductor industry. Now it had an untrustworthy new product on its hands—and a desperate need for revenues. Seeing no other option, the company broke its own rules.

In October 1970, just as the 4000 Series prototyping began, Intel formally announced the 1103 with an ad that read:

The End:
Cores Lose Price War to New Chip

To industry insiders, these words were a ringing alarm. Intel was announcing that for the first time in computing history, the traditional internal storage medium, the tiny magnetic rings and interwoven wire of core memory, had now been matched in performance and price by integrated circuit memory, hundreds of times smaller and more power efficient, fast, . . . and at a comparable price. And for those who had been following the industry closely, also implicit in this announcement was the fact that the curve of Moore's Law for memory chips had already crossed the horizontal line of core memory development and would soon be rocketing upward.

What those knowledgeable readers didn't know was that Intel still hadn't figured out the 1103's bugs but had decided to ship anyway. No one was more upset by this decision than Andy Grove, even though he had been part of it. He was rightly terrified that the chips would arrive DOA and his phone would start ringing with furious customers canceling all of their contracts with Intel.

Instead, the strangest thing happened, a result so unlikely that it can only be credited to sheer chance. Intel has long been known for its innovative excellence, for its ability to learn from its many mistakes, and for its competitive ferocity—but this is one case where Intel, on the brink of disaster, just plain got lucky.

According to Gordon Moore, the reason the company dodged a bullet with its equally dodgy first 1103 shipments was the eccentric nature of its market. The engineers who worked in the computer memory business faced operational breakdowns and blackouts with core memory that were remarkably similar to those experienced with the 1103. Both were affected by the presence of nearby electronic devices, and the periodic refreshes of the 1103 that sometimes gave it amnesia were stunningly like what was called a *destructive read* in core memory.

For mainframe computer memory engineers, this familiarity of failure bred not contempt but trust. The fact that the 1103 was cranky and didn't work quite right perversely made these engineers more comfortable about adopting this wholly new technology. Recalled Moore, "All these things made the 1103 more challenging and less threatening to engineers [at computer companies]. We did not plan it to happen that way, but I think if [the 1103] had been perfect out of the box, we would have had a lot more resistance from our competitors."[2]

Unfortunately for Intel, as is often the case with luck, the company learned the wrong lesson from getting away with selling a buggy product—and it would be punished for it twenty years later.

But in the meantime, the company's luck continued to hold. The very importance of the 1103—it would establish the path to minicomputers and then personal computers—made it a vital new technology for every computer company in the world. Yet that potential demand, combined with the device's unpredictability, also made it crucial to these same computer companies that there be a constant supply of the chips with, hopefully, ever better performance. That is, they wanted a "second-source" supplier.

The first company to recognize this opportunity was Microsystems International Limited, the manufacturing wing of Bell Canada. The company approached Intel—in particular, Bob Noyce. Noyce may not have been a very good businessman on a day-to-day basis, but when it came to masterstrokes, he rose to the occasion. He became a fearless gambler. His reply to MIL was characteristic: he told the Canadians that if they wanted to be the second source for the 1103, they would have to "pay us our net worth, and we'll tell you what we know."

It was a crazy play, one based less on any empirical analysis than on a gut sense of what MIL would accept and an eye on how much money Intel needed to survive. It was just the kind of executive improvisation that drove real businesspeople like Andy Grove (needless to say, Noyce had not conferred with him) crazy with both frustration and, when it inevitably succeeded, a kind of awe.

MIL agreed to everything: $1.5 million for the second-source rights, plus royalties on sales through 1972, in exchange for Intel turning over the "cookbook" for the 1103 and sending a technical team to Ottawa to set up the MIL fabrication laboratory. Then, if that fab met MIL's production goals, the company would send another $500,000 to Intel. As Noyce wrote in a note to Intel's investors: "[Microsystems'] payments under the agreement will substantially improve our already good cash position, [and] the implied endorsement of our process excellence will also be a valuable marketing aid for us."

When Andy Grove heard the news, he went ballistic. He would later say, "We were seeing one or two [working "die"—i.e., chips] per wafer of 1103s," a shockingly low yield rate. "It's hard to describe what shit we were in with that product." It was taking every scientist and engineer Intel had just to accomplish that. And now, the MIL contract meant that "I would have to improve yields without a manufacturing manager or chief technologist," as they were both heading off to Ottawa.[3]

There was another reason for Andy's anger: he was the senior operations executive at Intel, and he considered himself, if not the peer of the two founders, at least near enough to their level to be privy to all major decisions made by the company. Now in what might be Intel's last important decision, he had been left standing on the sidelines, kept in the dark about where the company was going. Even the man he admired most, Gordon Moore, had treated him like hired help. It is remarkable that Andy didn't quit on the spot.

Instead, Grove approached Noyce several times to talk him out of the idea, but Bob waved him off. Finally, in the summer of 1971, as the engineering team was preparing to leave for Canada, Grove gave it one last shot. He marched into Noyce's office (by coincidence, Moore was already there in a meeting) and made his case "as aggressively as I was capable—which was very." He told the two men that there was no way that the company could simultaneously increase the 1103's yield at home *and* send top talent to Ottawa to get production running up there.

"This will be the death of Intel!" Grove shouted in warning. "We'll get the $1.5 million, and then we'll sink."

Noyce may have been legendary for his aversion to confrontation, but for some reason, with Andy Grove, he seemed to operate under a different set of rules. He seemed to have Andy's number, and that was to never be conciliatory, to never compromise. So instead of agreeing with Grove, as Noyce likely would have with anyone else, Noyce just patiently listened to Grove's rant, then said slowly and carefully, "We have decided to do this. You need to put your energies into figuring out how to do this."

Grove left, as he described it, "with my tail between my legs." Once again, why he didn't resign remains a mystery.[4]

In the end, the MIL deal not only worked but worked brilliantly, and the Canadian company paid Intel the full $500,000 bonus. The 1103 team returned from Ottawa to a heroes' welcome—and a free trip with their wives to Hawaii for a three-day party at a resort on Kauai. After a fair amount of liquor, the men all stripped down to their boxers for a quick dive in the ocean. Noyce, the bemedaled diver, led them in—only to discover that the tide was out. He emerged laughing from the water covered with sand and scrapes . . . to the cheers of his staff.

Intel had every reason to celebrate. Back home in the Valley, as word leaked out about Intel's fiendishly clever MIL deal and its eventual payoff, competitors were put on notice that despite their Fairchild reputations, Intel's founders were now sharp business dealmakers. The success of this memory chip also marked the moment that Intel became a real business, with revenues and profits. "That Microsystems deal really saved their ass," said Marshall Cox, who watched the deal from Intersil.[5] It was the kind of sweetheart deal one expected more from Charlie Sporck, down the street at National Semiconductor.

The MIL deal also did a lot to rehabilitate Bob Noyce's image as a businessman. Other than his dollar-per-chip announcement at Fairchild, it may have been the best single business maneuver of his career. Whereas he had been able to pilot Fairchild brilliantly as long as it was successful, once that company went into a tailspin, he had been unable to pull it out of its dive—a rescue that Les Hogan had been able to accomplish pretty quickly upon his arrival.

But now at Intel, faced with an industry downturn, falling revenues, and a serious yield crisis in its flagship new product, Noyce had broken almost every rule of management. He had divided his team to take on a long-shot project that was secret even from his chief of operations, then divided his staff again to help set up a second-source supplier that would quickly become a dangerous competitor if he failed. And yet Bob Noyce had pulled it off—and in the process he had not only doubled Intel's capital, but as the company would eventually realize, had mentored the invention of one of the greatest products in the history of electronics.

Throughout Silicon Valley and the rest of the electronics industry, even as belts tightened, jaws dropped. Anyone who thought Intel's one fatal flaw would be Noyce's fecklessness was now disabused of that hope. And they knew to their dismay that even as they were hunkering down to weather the impending hard times, Intel was now sitting on enough cash to maintain a high level of R&D investment and come roaring out when good times returned.

Bob Noyce was already a living legend for the invention of the integrated circuit; now his reputation as a world-class business leader was established forever as well.

CHAPTER 18

The Philosopher's Chip

B ut what to do with this new invention, the microprocessor?
More than a year had passed between the signing of the deal between Intel and Busicom and the delivery of the first 4000 Series chip sets. In that time, as expected, a lot had changed in the desktop calculator industry—as Bob Noyce quickly discovered.

In February 1971, he embarked on a tour of Japan designed to use his reputation to convince Japanese computer, calculator, and device makers to design the new 1103 chip into their upcoming products. It proved to be very successful: within a year, 15 percent of Intel's revenues would come from Japanese customers.

A second purpose of the trip was to visit Busicom and brief president Kojima on the impending delivery of the first 4000 chip sets. But if Noyce was expecting a warm welcome, he got a big surprise. In the intervening months, the desktop calculator bubble had quickly deflated. Prices were falling fast, and competitors were retreating to defensive redoubts, typically reducing their offerings to low-cost four-function machines with which they could price-bomb their competitors. Busicom, always an also-ran in this industry, was hardly an exception. And its planned high-function, onboard-printer calculator with its four-chip controller now looked like an expensive anachronism.

Bob Noyce had arrived expecting hearty congratulations, perhaps even acclaim, for what Intel had accomplished. Instead, he was met by a stone-faced Yoshio Kojima, who informed him in no uncertain terms that the 4000 Series was overpriced and that the contract now needed to be renegotiated.

A frustrated Noyce returned to California and met in turn with each of the key figures on the project and explained what had happened. Hoff told him, "If you can't get any other concession, just get the right to sell to other people." Faggin told him the same thing.

Then, sensing that their historic effort might now be at risk if Noyce faltered, each of the inventors then took their case to the rest of the company. Hoff: "That's when all of us—Mazor, Faggin and I—went to our marketing people and said, 'Get the rights to sell to other people!' And there was a reluctance on the part of the marketing to do that. Intel was a company that made memories; these were custom chips. Marketing was very much afraid of the computer business, and they came up with arguments as to why we shouldn't go into it."[1]

Still, they found one ally: Intel's new marketing consultant and publicist, Regis McKenna. Just a year before, McKenna had hung out a shingle for his own marketing/PR agency—soon to become the premier such firm in high tech—and signed on Intel as one of his first clients. After learning about the 4000 Series and the situation with Busicom, McKenna decided to insert himself into the debate. Shrewdly, he presented the one argument for the 4000 that would work with Intel: that the chip set might ultimately sell more memory chips.

In the end, though, it came down to Bob Noyce. And once again, Noyce proved to be the microprocessor's greatest defender. He agreed to talk with Kojima.

By May 1971, Busicom had signed away exclusive rights to the 4000 chip set for everything but calculators. It was one of the biggest business mistakes of the century. Had Busicom held on to ownership, it would have owned the world's only working microprocessor design. It would have made billions on the licensing alone in the years to come. Instead, increasingly desperate, by the end of 1971, Busicom even negotiated away to Intel its exclusivity on the microprocessor in calculators. Busicom could have ruled; instead it died, swept away like scores of other calculator companies in the great industry shakeout.

Busicom's mistake would prove to have implications far beyond that hapless calculator company. Over the next few decades, Japan would again and again try to compete in the all-important microprocessor business, and it would consistently fail. And that failure would ultimately determine whether it won or lost in the semiconductor war onto which it would soon embark against the United States.

Back at Intel, the technology debate had moved on from whether it was possible to reduce a computer's central processor to a handful of chips to where that now-proven technology should go from here.

One notion was to keep shrinking the chip set, down to two, perhaps even one. That was certainly where the technological imperative and Moore's Law suggested it should go. But there was a big problem in taking that direction: until now, even with the four chips of the 4000 set, the individual chips were still dedicated to one task set: logic, RAM, ROM, and input/output. Any further reduction would require putting two or more of those functions on a single chip, a difficult and largely unproven task in which different regions of the silicon surface would be dedicated and designed for their distinct roles, with complex interconnections fabricated between them. As appealing as this true microprocessor might be—and its arrival in the next few years seemed all but inevitable—it was still too risky a step at this point, even for Intel.

But there was another alternative. That was to temporarily stick with the four-chip set and focus upon improving its performance and applicability. This was a safer strategy, but it was not without risk. And in fact, even this safer strategy was almost stillborn.

Less than two months after Intel signed the revised contract with Busicom, the company decided to turn the 4000 Series into an official product line to market to commercial customers. After all that Intel had been through to get to this point, including fighting for ownership of the 4000 Series, one might think that this decision to go commercial would have been automatic. In fact, it was anything but; even Ted Hoff lobbied against the idea.

Hoff's reasoning was this: the 4000 Series was a revolutionary product, but the same technical limitations that made it appealing for use in a low-cost calculator—a four-bit word length, a low-capacity ROM, a comparatively slow clock speed—severely compromised its usefulness in other applications, notably big computers. Ironically, one of Hoff's arguments against the 4000 Series now was precisely the one he'd used for the design in the first place: the general-purpose architecture required customers to do a lot of programming, which a lot of prospective commercial customers weren't yet prepared to do. Meanwhile, as Hoff and others inside Intel asked, who was going to train and support Intel's sales force as it went out into the field and tried to sell this revolutionary new product? What customer would take such a risk on a radically new, unproven product with no second source?

But if one of the 4004's (it was at about this moment in history that the entire chip set began to be named for its all-important processor chip) parents was prepared to make it an orphan, its other parent was not. Federico Faggin believed in the commercial viability of the 4004. He pointed to the device he had created to systematically test newly minted 4004s coming out of fab and, noting that it was controlled by a 4004 itself, suggested that this was proof of the chip's market potential.

He won the argument. But Faggin's victory proved Pyrrhic: three weeks after Intel made its decision, company employees opened the latest issue of *Electronics*, the industry's flagship trade magazine, and were shocked to see a new advertisement by Texas Instruments. It showed a huge chip ("huge" being a quarter-inch square) over a headline that read:

CPU ON A CHIP

Intel had just spent more than a year, burning up its precious talent and treasure, to build a revolutionary new device, and had just completed a blistering internal debate about whether to even take this product to market . . . and now, before it had even started the race, it had already lost it.

The ad's copy was even more dispiriting. It seemed that one of the world's biggest semiconductor companies and one of Silicon Valley's greatest competitors not only had the first true microprocessor, but apparently even had the supersmall line width large-scale-integration (LSI) fabrication technology in MOS to produce this microprocessor in volume.

Then the final stab to the heart. The copy then went on to note that TI's new CPU on a chip was being developed for Computer Terminal Corporation (today's Datapoint). This bit of news was particularly painful, because CTC had awarded the contract to Intel first, but then Intel had bungled the project.

It happened like this. CTC had first approached Intel with its own computer-on-a-chip idea at about the same time as Busicom did. In CTC's case, it wanted Intel to take its current model 2200 computer terminal motherboard, with its nearly one hundred bipolar logic chips, and reduce that number to a couple dozen MOS chips. Like Busicom's, CTC's goal was to reduce both size and cost without compromising performance.

Ted Hoff had looked at CTC's request and realized that not only could Intel do everything the contract asked, but it could even reduce those scores of bipolar chips to a *single* MOS chip if Intel were to use that new silicon-gate technology devised by Fairchild's Federico Faggin (yet another reason for the latter's recruitment by Intel). Hoff approached CTC with this stunning solution, and unlike Busicom, CTC bought into it immediately.

There were two other crucial differences between the Busicom and CTC deals. The first was that, whereas Hoff inserted himself heavily into the Busicom project and its resident team at Intel, now he was already busy with what would be the 4004, so he handed off the leadership for the new CTC chip, to be called the 1201, to one of his subordinates, Hal Feeney. Feeney was more than capable of designing and building the 1201, especially with Faggin's help, but he had neither the reputation nor the political influence (i.e., access to Noyce) at Intel that Hoff had. And so a few months later, when the 1103 memory-chip crisis hit, Grove didn't hesitate to pull Feeney off the 1201 project and throw him into the fight to save the company.

And there, in project limbo, the 1201 had languished, until CTC, finding Intel unresponsive, reawarded the contract to TI.

Now, eight months later, the shocking Texas Instrument advertisement sent Intel scrambling. This time it was Andy Grove, his competitive fire burning, who did an about-face from his earlier opposition to the 4004 in order to beat TI to market. He went to CTC and successfully fought to keep the original contract alive. Even if he didn't necessarily see a future for microprocessors at Intel, he still intended to win this race.

Meanwhile, Federico Faggin, still testing his prototype 4004, had already been assigned to salvage the 1201. Feeney, the only person at Intel who actually knew what had happened to the original 1201 project, had also been assigned to be Faggin's partner. The project had been moving along at a solid pace; now the TI ad made it a priority.

The two men had spent until spring 1971 figuring out what Faggin had learned in the design of the 4004 that could be used in the 8-bit architecture of the 1201. (The fact that the 1201 was 8-bit rather than 4-bit was the biggest argument in its favor: it could find a home in commercial applications for which the 4004 was too limited.) When the bombshell TI ad hit in June, the team shifted into high gear.

Then another shock. In July, CTC came back to Intel and announced (shades of Busicom!) that, with a new recession under way and the price of logic chips plummeting, it wanted out of the contract. This news might well have ended the 1201 project—and with it, Intel's foray into the microprocessor business—but thanks to Noyce's connections in Japan, Intel soon learned that watchmaker Seiko, about to pioneer the next consumer electronics business, digital watches, was looking for a powerful logic chip like the 1201. This potential deal with Seiko was so good that Intel not only let CTC out of the contract, but didn't even charge for the work done to date. Intel in turn got the complete commercial rights to the 1201, which it soon named the Model 8008.

Once again, it was time for Faggin and his team to rush a new product to market. This time, however, they would do it without Ted Hoff, who had been ordered to hit the road to industry trade shows and technical conferences to talk up Intel's new chip sets in hopes of educating the marketplace . . . and with luck, drum up a few customers to justify Intel's now enormous investment of time and money.

Unfortunately, Hoff wasn't trying to sell just a new product, but a whole new paradigm. And as is usually the case when you are trying to convince an audience to accept a radical innovation, almost by definition the idea is so far from the status quo that many people simply cannot get their minds around it.

Recalled Hoff: "People were so used to thinking of computers as these big expensive pieces of equipment that had to be protected and guarded and babied and used efficiently to be worthwhile and cost-effective. So there was this built-in bias in people that *any* computer had to be treated that way. I remember one meeting in which there was all of this concern about repairing microprocessors, and I remember saying, 'A light bulb burns out, you unscrew it and throw it away and you put another one in—and that's what you'll do with microprocessors.' But they just couldn't accept doing that with a computer."[2]

To be fair, those were $400 chip sets Ted Hoff was talking about in 1971, so the notion of tossing out and replacing faulty ones would have been a bit less appealing then than now. Moreover, it wasn't as if Intel had unanimously ratified the idea of the microprocessor either. Hoff himself, after all, had lobbied against marketing the 4004; and even at the top of the company, Andy Grove, though he respected the technological achievement, still questioned if there really was a market for these new chips—and with good reason: nobody at Intel had yet articulated just exactly who would buy the chip sets and for what purposes.

But the biggest obstacle to the adoption of these new processors was simply a failure of imagination by potential customers, one that wasn't helped by Intel's own indecisiveness. Computer scientists and programmers had spent their entire careers working on giant mainframes that required information to be loaded (via cards, tapes, and eventually terminals) and processed in large batches; the output typically was delivered via large, noisy printers that operated like giant typewriters. These mainframe computers were the size of studio apartments, required their own climate-controlled rooms, cost millions of dollars, and required a team of operators. Just fifteen years before, the biggest breakthrough in data storage, the one-ton IBM RAMAC magnetic disk drive, had to be delivered to customers via a specially equipped Boeing 707 jet.

Obviously these big computers had evolved over the previous two decades from slow, giant monsters to somewhat smaller and fleeter early generation minicomputers—the size of a couple refrigerators and costing "only" a few hundred thousand dollars. Computer scientists had anticipated this kind of evolution in power, size, and price and had updated their skills to keep up.

But the devices that Ted Hoff described to them were so radically different in every way that if he had prefaced his presentations by saying they had come from outer space, no one would have been surprised. After all, he could hold the four chips in the palm of his hand, and at four hundred bucks per set, they still cost less than just one terminal for a mainframe computer. And there he was announcing that these little silicon caterpillars would soon replace the Big Iron to which they had long ago become accustomed. On top of that, Hoff pointed out, these chip sets also processed data in real time, so you could pour data in and draw results out twenty-four hours per day, and not just in an expensive-to-run, air-cooled processing center, but right out into the natural world. And you could even use some of the established computer programming languages to make them work.

That was the good news, if they believed it. The bad news to these old mainframe jockeys was the sheer legacy problem of all of those expensive machines, IT centers, millions of lines of code, and thousands of operators. Even if these little chips could replace all of that (and few believed they could), who would want to? Hoff famously interviewed a promising candidate for the 4004 team—and was turned down because the candidate didn't think Intel had enough IBM mainframe computers for a company with any kind of future.

Worst of all, even Intel's marketing and sales people, who would have to sell these chip sets in large volume if the company was going to succeed, didn't believe in them. As marketing director Bob Graham told Hoff: "Look, [computer companies] sell 20,000 minicomputers a year. And we're latecomers to this industry. If we're lucky we'll get 10 percent. At 2,000 chips a year, it's just not worth all of the trouble."[3]

As that comment suggests, inside Intel as much as outside, there was a general inability to see beyond that enticing but ultimately limiting term, computer-on-a-chip, to see the microprocessor as anything more than a tiny, bolt-in replacement for the central processor unit of a big computer. Seen that way, it was little more than a novelty: incredibly small and cheap but also comparatively slow and unscalable. Even if you could build a computer with, say, an array of microprocessors somehow linked together, you'd still need all of that Big Iron and millions of lines of code to make it work.

So why throw out a quarter century and billions of dollars in developing the modern computer simply to go solid-state? Frankly, if that was your goal, why wouldn't you (as computer companies eventually did) just buy a bunch of dedicated logic and ROM chips?

A major decision, a reckoning, was coming for Intel and the rest of the electronics industry. But for now, at least for the most aggressive chip companies, like Intel, Texas Instruments, and Motorola, the only choice was to plunge ahead and hope that the market would sort itself out in time.

At Intel, this meant Faggin and his team going all out in the development of the model 8008, even as senior management hit the road and tried to convince the world to take the 4004 seriously. For the tour, Hoff and (sometimes) Noyce were joined by Intel's new marketing director, Ed Gelbach. Faggin aside, Gelbach's hiring was perhaps the most fortuitous of this era at Intel—and yet another example of Intel's uncanny ability to turn potential disasters into victories.

Intel's first marketing director, Fairchild veteran Bob Graham, had been Bob Noyce's close associate and right-hand man almost from the very first day of Intel's existence. He had played a valuable role during the negotiations and subsequent contract management with Busicom. Moreover, he was a good friend and fishing buddy of Gordon Moore. But his adamant resistance to the microprocessor had damaged Noyce's and Moore's support for him. Most of all, and this was the fatal touch, Graham had crossed swords with Grove too many times. Andy had been no fan of Graham at Fairchild, but now that he had to work with him on a day-to-day basis, that disrespect had turned to contempt. As was often the case with Grove, the personal was the corporate, and he had convinced himself that Graham was impeding not only his progress, but that of Intel itself.

This perpetual feud with his boss, especially one as unrelenting as Andy Grove, was taking its toll on Graham as well and beginning to impede his ability to get anything done.

In spring 1971, this internecine war between Grove and Graham came to a head. In May, Grove marched into Moore's office and informed him that he thought he might quit Intel because "it was far too painful for me to continue to do what I needed to do at work *and* fight Bob Graham." The experience was so stressful for Moore that he was reduced to his last-ditch stress reducer, twisting and untwisting a paper clip.[4]

Knowing that Moore wasn't much better than Noyce at confrontation, Grove walked out of the office unsure whether Gordon would actually do anything about his ultimatum or whether, after an interval of nonaction, he would just have to leave the company. In fact, Moore almost immediately walked down to Noyce's office and gave him the news.

Now it was Noyce's decision. He didn't like being blackmailed, nor did he like the prospect, one he had always resisted, of having to fire one of his employees—especially one he respected as much as Bob Graham. But this wasn't Fairchild; at Intel, Bob was in to win—and he knew that would never happen without a strong chief of operations to back him up. He had learned that at Fairchild and he needed one now more than ever at Intel, where there wasn't yet the same level of infrastructure. Noyce understood, at an elemental level, that Intel couldn't succeed without Grove. So Graham had to go.

Noyce spent June 1971 in a quiet search for Graham's replacement. He finally found him at Texas Instruments in the form of marketing expert Ed Gelbach.

Gelbach was a very different character than Graham. Graham was like Noyce—engaging, customer-oriented, and visionary, while Gelbach was much more like Grove—tough, pragmatic, plainspoken, and empirical. Noyce certainly must have seen the resemblance and known that bringing Gelbach aboard would please Andy and make him less likely to walk. Gelbach's current job at TI meant that he was familiar with memories and microprocessors, making the transition almost seamless. After some heavy negotiations, in which Gelbach showed that he would be an asset when playing hardball with customers, an agreement was reached.

That left the matter of firing Bob Graham. Because Moore was off on a prescheduled fishing trip in early July (which "made it a lot easier for me," said Moore later), the task fell to the ever reluctant Bob Noyce.

Noyce biographer Leslie Berlin has speculated that "Graham was probably the first person that Noyce personally fired, a remarkable situation given that Noyce had been a senior manager for more than a decade." That's entirely possible; certainly it appears to have been Noyce's first-ever face-to-face firing. In the end, Noyce simply girded himself, walked into Graham's office, and announced, "Shit. We have an incompatible situation here, and you're going to have to go."[5]

"What that must have taken for Noyce to do," said Grove years later. "It must have been like cutting out his liver."

Grove, of course, was thrilled. For him, it was a termination hat trick. He got rid of an enemy; Intel, he thought, would now have a tougher and more disciplined marketing operation; and perhaps best of all, Andy was left alone in the pivotal slot between the founders and the rest of Intel.

As for Graham, the termination left him more relieved than angry. The battle with Grove, combined with Intel's already killer work demands, had been taking a terrible toll on his own mental health and that of his family. When he called home with the news, his wife said only, "Thank God."

Though he would soon be all but forgotten in the glittering story of Intel Corp., Bob Graham's contribution to the company was substantial. Beyond his work in dealing with Busicom and nailing down Intel's ownership of the microprocessor, Graham also devised the company's first marketing campaign, "Intel Delivers," which not only served the company for a decade before the arrival of "Red X" and "Intel Inside," but also served as the platform for those celebrated campaigns.

Even more important, Graham tied Intel's product development to its internal engineering development pushed by Moore's Law. The result was Intel's seemingly relentless rollouts of new product upgrades approximately every two years. It was this pace—286, 386, 486, Pentium, et cetera—that ultimately crushed the competition. Graham, of course, got little credit for this in the official history.

Nevertheless, getting fired from the company may have been the best career move of all for Bob Graham. The relationship with Grove would have never healed, and as he grew in power, Andy would have fired Graham anyway. Instead, Graham took his gifts as a marketer to Applied Materials, the great builder of the equipment used to build semiconductor chips, and turned the company around by skillfully replacing the notion of simple "hardware" with that of complete hardware/software/training/service "solutions."

There is no little irony here. The guy Andy Grove didn't think was up to the marketing job at Intel was soon, through his brilliant management, providing Grove with the equipment he desperately needed to keep Intel competitive. And as we will see in the story of Operation Crush, at the moment in its history when Intel was facing the single greatest threat to its survival, the company would (perhaps not coincidentally) turn to a "solutions"-based marketing model to produce an even more celebrated turnaround.

Graham would achieve a similar turnaround at another semiconductor equipment company, Novellus, where he again reversed himself, moving from "solutions" back toward ever more productive hardware—once again vital to Intel and the US semiconductor industry when the battle with Japanese chip companies began.[6]

Graham died at age sixty-nine in 1998, and his services to marketing in the semiconductor equipment industry were considered so great that the industry trade association, Semiconductor Equipment and Materials International (SEMI), created in his memory the Bob Graham Award for the highest achievements in marketing and sales force management.

Andy Grove never credited Graham for his contribution to Intel during (or after) his employment there.

Was Bob Graham's firing good for Intel? History says that indeed it was. The company was heading into a difficult era, one that threatened to tear it in two. And as much as it would need a talented marketer to sell its revolutionary new products, it needed even more a numbers-oriented marketer who could convince the equally numbers-oriented engineers and scientists inside Intel to follow his lead. And in Ed Gelbach, Intel found exactly the right person.

CHAPTER 19

Product of the Century

·

Down in Intel's labs, the game of musical chairs taking place on executive row was of little importance. What counted was the challenge at hand: turning the compromised 1201 chip set into a world-beating 8-bit 8008. Happily for Faggin and his team, the 8008's development went smoothly, no doubt in part because they didn't have to come up with a whole new design but merely fix the mistakes of its predecessor. This luck held right through the prototype phase.

Then, just as Faggin and his team were about to pop the champagne, something went wrong. It was a fugitive problem, invisible during the early testing, but now with constant use, it began to show up as intermittent failures in the 8008's operation. "It took me a week to solve the problem," he recalled. "It was a nasty one, at the crossroads of device physics, circuit design, and layout."[1] In particular, Faggin discovered that the design of the 8008 allowed electrical current to leak out of some of its memory circuits. Skipping sleep, Faggin redesigned that chip's entire layout—a fix on the fly that would have been all but impossible a decade later with the immensely more complex layouts of future processors.

The Intel 8008 was introduced in April 1972. That was just five months after Intel formally introduced the 4004 in the November 1971 issue of *Electronic News*—an astonishing speed to market, one that would never be matched in the microprocessor business again. And it wasn't as if the 8008 was just a turbocharged version of its predecessor. It was, in fact, a major breakthrough of its own, with 50 percent more transistors (3,500 versus 2,300) and nearly *eight times* the clock speed (800 kHz versus 108 kHz).[2]

The trade press noticed. The 8008 earned some of the best reviews of Intel's brief history. Unfortunately, those good notices weren't sufficient to fire up a still skeptical market. It seemed as if Intel was trapped in a vicious circle, one in which it would be increasingly lauded for products that nobody wanted to buy.

And yet something had changed. It wasn't obvious at first, as Noyce, Hoff, and eventually Faggin hit the road to try to sell the advantages of the 8008 as they had, unsuccessfully, just a few months before with the 4004. Then, suddenly and without warning, orders began pouring in . . . not for the two chip sets but rather for the various design tools and testing boards (such as in-circuit emulators, which enable engineers to imitate the performance of microprocessors and figure out how they might be used in their systems) that Intel had just begun to create to support these new products. The first of these, the Intellec MIC-8, had just been released. So even though nobody was yet buying the two Intel chip sets, it was obvious that a lot of designers out there were looking at them.

That realization was underscored once Intel's executives got out in the world before prospective customers. It was the company's first experience with a phenomenon that would become increasingly common in the chip world as the customer base shifted from industrial to commercial customers: the *learning curve*.

Military, aerospace, and industrial customers typically had the technical teams and the budgets to adopt and incorporate new technologies quickly. But that wasn't necessarily the case with commercial companies, and it was even worse with consumer products. In that market, as the Busicom experience had shown, profit margins were often so slim that even if the company wanted to adopt a revolutionary new technology, it still might not have the money or talent to do so.

Thus at the very least, especially with transformative products like the Intel 4004 and 8008, it was going to take time for customers just to get their heads around the idea of a handful of chips replacing a wall of magnetic ring cores or a motherboard or two filled with logic and memory chips. Then they had to see if any of their competitors were looking at the same solution and thus validating the technology. Then they had to look at the cost of redesigning their products to incorporate this new technology. And *then* they had to look at the cost of building these new, microprocessor-powered designs, rewriting their operating code, and figuring out the new marketing and pricing on the result.

That took time—in fact, about a year from the first reports about the 4004 in the trade press in mid-1971 to the summer promotional tour for the 8008 by Intel's executives. It was a stunning experience for the participants: "The entire electronics industry seemed to undergo an awakening. Intel's efforts had finally had their effect. Suddenly, as if overnight, engineers they visited understood the meaning of the microprocessors. They had read the articles, heard the speeches, talked to their peers, and as if as one, jumped aboard the silicon bandwagon."[3]

In fact, their embrace of this new technology almost went too far, as if they had spent so many months with no expectations that they were now trying to make up for lost time: "It was a little disorienting. Where a few months before, the typical audience to a speech by Hoff or Noyce had sat in disbelief, now suddenly they seemed to expect too much. Doubt in one direction had almost instantaneously shifted to overexpectations in the other. Now potential customers demanded to know why a $400 chip couldn't do everything a $50,000 minicomputer did."[4]

Intel, which had worked so hard to ignite market interest in its new microprocessors, now found itself at risk of being consumed by the flames.

Nobody took this response more to heart than Federico Faggin. He had worked so hard to make the two chip sets work, even inventing important new technologies in the process. Now, as he lectured at a series of technical seminars throughout Europe, he found himself meeting not indifference but seemingly endless, passionate complaints about the inadequacies of his chips. Even more depressing, Faggin had to admit to himself that many of those complaints were actually valid: "When I returned home, I had an idea of how to make a better 8-bit microprocessor than the 8008, incorporating many of the features that people wanted; most importantly speed and ease of interfacing."[5]

Faggin was anxious to get going and filled with notions about how to turn the four-chip 8008 into a single-chip device—the world's first true computer-on-a-chip—and also to make it even more powerful than its predecessors.

The requisite miniaturization, which to the outside world seemed almost miraculous, was in fact in many ways the easiest part. Moore's Law had clicked over once again, so now the features that could be etched on this new chip could be as small as 6 microns wide—almost twice as small as the 10 micron features on the 4004 and 8008. That made it possible to not only move the 3,500 transistors from the four chips of the 8008 onto this single chip, *but also to add another 1,000 transistors.*

And if all of that wasn't enough, Faggin believed he could crank up the clock (processing) speed of the new chip, from the 800 kHz of the 8008 to 2 MHz—meaning that, with the right application, this new chip might perform more than one million calculations per second. That was minicomputer speed—the ultimate response to all of those newly unsatisfied computer engineers.[6]

And there was more. Being Federico Faggin, he wasn't content just to upgrade the microprocessor. Once again, he wanted to *revolutionize* it. For this new chip, he also wanted to increase the number of legs (pins) on the packaging to boost input-output speeds, add more instructions to the now-onboard ROM memory, and adopt an important new transistor architecture, called n-channel, to replace the standard p-channel.

In early 1972, at about the time Intel formally introduced the 8008 to the public, Faggin approached senior management with his proposal for this new chip, which would be called the Model 8080. So confident was Federico that the company would adopt his proposal that he even recruited Shima to join the new design team.

But to Faggin's dismay, the company turned him down. He was left to cool his heels while Intel tried to find a market for the three microprocessors it already had. Despite the apparent shift in attitude by the marketplace toward the new Intel microprocessors, the company was still unsure whether there would ever really be a sizable demand for these devices and whether they were worth the distraction (and money bleed) away from the company's proven core business of memory chips.

This doubt had been there from the start. The Busicom contract had been seen as precisely that: a custom project worth enough money to temporarily divert resources away from Intel's main operations. And while Noyce and Moore had, with prompting from Hoff and Faggin, gotten the commercial rights to the 4004, they still hesitated about actually getting into the business. As for Grove, he was putting out so many fires in manufacturing Intel's DRAMs and SRAMs that: "Microprocessors meant nothing to me. I was living and dying on two points of yield in memory."[7]

In the end, Ted Hoff almost had to put his career on the line to fight for his invention. When Noyce expressed indecision about *ever* bringing the 4004 to market, Hoff almost shouted him down, saying, "Every time you delay [announcing the microprocessor] you *are* making a decision! You're making a decision *not* to announce. Someone is going to beat us to it. We're going to lose this opportunity."[8]

What Noyce didn't tell Hoff at the time was that some of the resistance to embracing the microprocessor was coming from Intel's board of directors. It, too, was looking at the company's struggles getting its current crop of memory chips out the door and properly concluded that the last thing Intel needed was to divide its focus or its efforts on a wholly new and unproven line of products. As with many of the greatest products of the digital age—from the transistor to the iPhone and beyond—there were many more thoughtful and pragmatic reasons *not* to get into the new business than to make the risky leap.

Once again, Bob Noyce came through. Marshaling Moore, chairman Arthur Rock, and the newly hired Ed Gelbach, Noyce overwhelmed the handful of directors resisting the idea. From now on, Intel Corporation would be a microprocessor company.

But taking that endeavor from a boutique operation to a real business required more than just brilliant technologists like Hoff and Faggin—or even the enthusiastic (if distracted) support of company senior management. It also needed *missionaries*—true believers with the skill to devise a compelling message and then do the hard work of selling that message to employees, the press, investors, and most of all, prospective customers.

Once again, Intel's luck held. In 1971, the company found two such advocates for the microprocessor. Ultimately, their contribution to the advent of the microprocessor and Intel's leadership of that industry was as great as either the company's top management or its design team.

One of these advocates was Ed Gelbach. He had arrived from Texas

Instruments with almost unequaled expertise in marketing chips and a vision of the future of microprocessors learned at Intel's biggest competitor in that business. As veteran journalist T. R. Reid would write in his history of the integrated circuit: "As it happened, Gelbach had started in the semiconductor business at Texas Instruments; like everyone else at TI, he was steeped in [chairman] Patrick Haggerty's view of the world. Gelbach realized immediately that Intel had reversed the course of the industry by producing a general purpose chip. 'General purpose,' Gelbach saw, was just another way of saying 'pervasive.' The real markets for the new device, he [saw], would be completely new markets; with this one-chip central processor—known today as a microprocessor—the integrated circuit could 'insert intelligence into so many products for the first time.'"[9]

As he traveled around the world promoting the 4004 and as he watched the development of the 8008, Gelbach became more and more convinced that the microprocessor was the most exciting invention of the era and that his new employer needed to own that technology at all costs.

The second important advocate was not an employee but a consultant: Regis McKenna. McKenna had arrived in Silicon Valley from Pittsburgh in 1962 in search of a career in marketing. He and his wife, Dianne (a future Santa Clara County supervisor) purchased a home in Sunnyvale they would still be living in a half century later—receiving visits even from the president of the United States.

Skilled in marketing but as yet knowing little about high tech, in 1965 Regis joined the marketing department at General Micro-Electronics, the semiconductor company founded just a year before by the three former Fairchilders to pioneer MOS chip technology. That experience would prove valuable when Regis went to work with Intel.

McKenna spent two years at General Micro, then jumped to National Semiconductor, arriving at almost the same moment as Charlie Sporck and his crew from Fairchild. Once again, this was a lucky break, because under Sporck National Semi quickly turned into a fast-moving and hugely aggressive contender in the chip world. With the company introducing new products at breakneck speed and relentlessly attacking one market after another, Regis, as the company's marketing services manager, earned an education under fire as he quickly developed one product marketing strategy after another.[10]

With Moore's Law already creating a generational change every couple of years, by the end of the sixties, Regis knew as much about marketing high tech—particularly chips—as anyone alive. Shrewdly, instead of staying in the lucrative but dreary world of Charlie's National, in 1970 Regis decided to set up his own agency. Though officially a marketing consulting firm, Regis McKenna Inc. quickly found (to Regis's endless dismay) that many of its clients were so inexperienced that it had to get down in the trenches and help with PR, speechwriting, and advertising. Regis liked the big picture, but at least for the first few years, the blocking and tackling of public relations paid the bills.

Intel, of course, knew Regis—the company had watched his work at one of its biggest competitors—and soon (1971) it was knocking on McKenna's door.

Though he is now more associated with Steve Jobs and Apple—an association resulting from his close relationship with Bob Noyce—Regis McKenna has an importance to Intel over the next three decades that is almost immeasurable. Within a decade, Intel's management trusted him so completely that reporters, when calling Intel to talk with an executive, would find themselves transferred by the operator directly to Regis's office in Palo Alto. The reporter, expecting Andy Grove or Les Vadász to pick up the phone, would instead be stunned to hear: "Hi, [the reporter's first name], this is Regis. What can I do for you?"[11] Few companies have ever given as much control over their interaction with the outside world as Intel gave Regis McKenna.

This remarkable relationship was set at the very beginning. Almost from the moment he signed the consulting contract with Intel, Regis found himself engaged with the company's microprocessor conundrum. As he would describe the situation, "You couldn't promote the things [if] people didn't know what they were."

McKenna quickly discovered that the marketplace wasn't just confused by the concept of the microprocessor, but was actually frightened by its implications. "Many feared this new technology. Some ignored it.

Others dismissed it. Many of my engineering friends scoffed at it as a gimmick." His solution? "The market had to be educated."[12]

In that belief, he made common cause with Bob Gelbach, and soon they had enlisted Intel management as instructors. McKenna: "At one point, Intel was conducting more seminars and workshops on how to use the microprocessor than the local junior college's total catalog of courses. Bob Noyce, Gordon Moore and Andy Grove became part of a traveling educational road show. Everyone who could walk and talk became educators. . . . It worked."[13]

At the same time, and in order to drive this message home, Ed Gelbach created his own independent marketing operation just for microprocessors (skunkworks seem to have characterized the early story of microprocessors at Intel) to address the challenge. To run this operation, he hired a smart young Dartmouth/Caltech/Stanford PhD newly out of Hewlett-Packard named Bill Davidow. Davidow's father had made a small fortune in the Great Depression by obtaining the rights to reference books and then reselling them in low-cost editions in retail stores. That gift for marketing had been passed on to the son. In the years to come, Davidow's impact on the marketing of high tech would be as great as McKenna's.

Ostensibly assembled to figure out how to effectively bring the 4004 and 8008 to market, this troika quickly realized that their real task was to save the microprocessor business at Intel.

They knew the world was immensely intrigued by this new invention. What they didn't yet understand was how to convert that interest into actual sales. Said Davidow: "There were just lots of people who wanted to read about microprocessors, independent of whether they were buying any."[14]

The microprocessor marketing team fed this interest by producing endless manuals (the one for the 4004 was ten times longer than the company's usual ten-page manual for its other products), technical articles, data sheets, white papers, et cetera. Tellingly, Intel shipped many more manuals for the 4004 microprocessor to the marketplace than it did the actual chip sets. This at least kept the curious and the tire kickers engaged a little longer and bought time for the team to come up with a more persuasive message.

In the end, that solution came in the form of a whole new set of products from Intel designed to serve its microprocessor family. One was a brand-new type of memory chip, the erasable programmable read-only memory (EPROM), invented at the company by Dov Frohman, a Dutch-born Israeli whose parents had died in the Holocaust (he had been given to the Dutch Resistance for protection). Frohman, eventually an Intel vice president, would be instrumental in constructing Intel's famed Tel Aviv research facility.[15]

Frohman's creation was essentially a ROM chip with two revolutionary features. First, it was *nonvolatile*—that is, it could be turned off and still retain all of its data. Second, that data could be quickly erased by exposing the chip to ultraviolet light (usually in the form of a mercury-vapor lamp). The EPROM also had a distinctive look because, instead of the usual metal cap over the chip in its mount on the lead frame, the EPROM had a tiny quartz glass window to let the light beam shine through when needed.

Replacing the ROM chips with EPROMs in Intel's early microprocessors meant that the chip sets could be delivered to customers with their onboard programming (firmware) already in place. Now customers wouldn't have to create and then constantly upload code into their Intel microprocessors to make them function.[16]

The second solution came from Davidow's team. These were the first of the Intellec in-circuit emulators. In conjunction with the new EPROM, they made it possible to test new applications for a microprocessor and then store and transfer that code to thousands of the same microprocessors embedded in the customer's products. That went far beyond anything available from the competition. As already noted (and pointed out by Davidow himself in the years to come), these Intellec emulators and testers, had they been outfitted with keyboards and displays, would have beaten IMSAI, Altair, Apple, and the other makers of the earliest personal computers by nearly five years. Intel, however, had more pressing concerns.

In 1972, Intel was just four years old. Its annual sales were just $18 million—one tenth the size of Fairchild at the time—and it was still not profitable. Its entire staff could have filled a single large conference room. And probably most important, the company already had a revolutionary product line—MOS memory—that had already gained traction and was now growing with impressive speed. The idea of pursuing two divergent product lines, with two almost entirely different markets, was crazy enough, but to have one booming and the other still not finding a home and then to choose to pursue *both* seemed like certain business suicide.

And yet . . . Ted Hoff was coming home from the speeches he was giving around the world and reporting a sea change in attitudes—not least because he had slowly modified his pitch based on audience feedback. Davidow's design tools were picking up in sales, suggesting that potential customers out there might be designing the 4004 and 8008 into products. Gelbach was using every tool of persuasion he knew to nudge his bosses to a decision—even as he was warning them that TI and other companies were coming.

Meanwhile, Regis McKenna, with Gelbach's blessing, had prepared a series of notebooks for presentation to Noyce and Moore that offered a parade of potential new applications (and thus markets) for Intel's microprocessors beyond just computers. McKenna's list would one day be the subject of considerable amusement for its goofy ambition—even Regis laughed when recounting that "it ranged from automatic toilet flushers to cow milking machines, airport marijuana sniffers, electronic games and blood analyzers"—but it served its purpose of getting Intel's leadership to look beyond the computer motherboard. One member of Intel's board listened to the presentation and asked, "Don't you have any customers you could be proud of?" But the board of directors too voted its support.

Finally and ultimately most important, Federico Faggin and his team were in the lab waiting to begin work on what promised to be the world's first single-chip microprocessor, a device so revolutionary that it could literally change the entire world of electronics. No one understood this better than Gordon Moore—and in the end, no matter what their opinions on the matter, it would be his decision that trumped either Noyce or Grove. It only helped that, to date, Federico Faggin had not only never failed them, but had delivered more than he promised.

With the acquiescence of the board, Intel management gave the microprocessor group a green light both on putting the 8008 into full production and on building the new single-chip microprocessor. This new device would be called the Model 8080 microprocessor . . . and it would be the most important single product of the twentieth century. The 8080, its descendants, and its competitors would so profoundly affect the world that human society would look markedly different before and after their introduction.

Faggin, Shima, and the rest of the design team immediately went to work. From the beginning, they felt themselves under the gun and anxious to make up for lost time. Three months had now passed since the introduction of the 8008 and there were growing rumors of similar products in the works at the big semiconductor manufacturers. The one consolation was that Faggin, as usual, had come up with even more improvements during those lost months. And even Stan Mazor was freed up in time to make some additions and adjustments to the firmware.

Unlike the two earlier products, the 8080 was created smoothly and without any major hitches. It took six months, the first tests on the new chip taking place in December 1973. There were only a few errors, which were quickly fixed. In March 1974, just nine months after Faggin and his team embarked on the product's design, Intel Corp. formally introduced the model 8080 microprocessor to the public.[17]

It was an extraordinary feat of design and engineering that has had few equals in the history of the chip industry. But it was matched by an equally impressive and innovative marketing effort by Gelbach and his crew, with support from Regis McKenna and his agency. Intel had shown the world and its competitors that despite financial hardship and the distraction of another, even more successful business, it was a company that could move at lightning speed, could be as innovative as even the biggest technology companies, could form up and work as a team from the top of the company to the bottom, and could *learn*—from customers, competitors, even outside observers—as it went, improving its product, support, and messaging almost by the day.

Intel would show this skill over and over in its history, but never better and never for bigger stakes than in the creation and introduction of the 8080. Within three months, Texas Instruments would introduce its own microprocessor, with Motorola and others soon to follow. But by then, armed with its three-microprocessor family and especially its new flagship 8080, Intel had taken market leadership. And over the next forty years as the microprocessor became the "brains" of thousands of products throughout scores of major industries, and against seemingly endless challenges from the cleverest competitors, Intel would never give up that leadership.

PART IV

The Most Important Company in the World (1988–1999)

CHAPTER 20

Crush

On Wednesday morning, December 4, 1979, a worried group of Intel executives moved into a meeting room at Intel headquarters . . . and they hardly left for the next three days.[1]

The executives were the best marketers Intel had: microprocessor director William Davidow, board products general manager Jim Lally and his lieutenant, Rich Bader, regional sales manager Casey Powell, and Intel's outside marketing/PR guru, Regis McKenna. They had been ordered into that conference room by Intel COO Andy Grove and told not to emerge until they had found a solution—a *marketing* solution, because, incredibly, Intel had run out of engineering break-throughs—to the biggest existential threat the company had faced in its decade of existence.

Grove had been provoked to this desperate move by a memo he had received a few days earlier from Powell. It described in painful detail how Intel salespeople around the world, long confident that they had the best devices on the market, were suddenly and inexplicably losing sales—more precisely, "design wins" into future products—to the company's most dangerous competitor, Motorola.

Eighteen months before, Intel had unveiled the Model 8086 16-bit microprocessor. Not only was it a huge technical advance over its equally distinguished predecessor—the 8-bit 8080, the first true single-chip microprocessor—but its architecture would prove to be the defining standard of generations of processors that dominate the digital world to this day. Intel's ads for the unveiled 8086 showed a sunrise and the phrase "The Birth of a New Era." The company wasn't kidding.

Intel quickly snapped up every design win in sight, including the growing number of hot new personal companies.

The 8086 rocked the semiconductor world on its heels, but not for long. The semiconductor industry had been a brutally competitive free-for-all now for almost two decades—and any company that couldn't claw its way back from such a blow was probably long dead. And though Motorola had reeled, within six months it had counterattacked with its own 16-bit microprocessor, the Model 68000. Any doubts that Motorola was too big and old to be a scrapper anymore instantly evaporated: the 68000 was a masterpiece, one of the greatest microprocessors ever designed.

That's what had shaken Grove, and that's why his best marketing execs were now hunkered down in a conference room. The Moto 68000 wasn't just the equal of the 8086, but as even the die-hard Intel loyalists had to swallow hard and accept, it was *better*. Beaten to market, Motorola had brilliantly played out the classic strategy of the player late to the game. Like a stock car driver, Moto had let Intel cut the path ahead of them both: innovating the next-generation processor design, ironing out the bugs, spending the money needed to educate the marketplace, and getting customers thinking about how best to use these new chips. Then Motorola simply designed a processor that featured the best of the 8086's features and improved on the worst.

This strategy wasn't easy to execute. Despite all of the advantages that accrue to the "second entry" in a new market, the biggest challenge is speed: can you design and build that product, then get it out into the marketplace before the pioneer steals most of the customers and captures the highest profit margins before prices start to fall? That's why, while many second-entry companies desperately attempt this strategy, few succeed at it.

Motorola not only pulled it off—a lesson for generations of tech companies to come—but did it so successfully that the 68000 hit the marketplace early enough to steal some of Intel's best sales prospects at peak profit margins.

This news was ignored at first by Intel headquarters, but the field sales force took a battering. "The message wasn't getting through to management on the West Coast," said a salesman at the East Coast sales headquarters. Recalled Intel marketing executive Bill Davidow, "It was demoralizing to have one customer after the next lecture you about your employer's failures and your competitors' strengths. Many customers actually relished the opportunity to stick it to the famous Intel."[2]

Finally, on November 2, the head of the East Coast office, in Hauppauge, New York, blasted an eight-page telex to Santa Clara explaining in painful detail the unfolding disaster. Even this message might have been dismissed had not a second memo arrived almost simultaneously from another salesperson, this one in Colorado, describing the same emerging disaster there.

Almost overnight, Intel went from complacent to terrified. Except for chief operating officer Andy Grove. True to form, he just got mad. This was, after all, a company that defined itself as the technology leader, not just in semiconductors, but in the entire business world. That was the company's greatest conceit—*we are the best engineers in the universe*—and now Motorola had exploded that sense of superiority. Worse, Motorola wasn't alone: there were rumors in the trade press that another competitor, Zilog, was racing to complete its own 16-bit processor, the Z8000. And with Federico Faggin, the scientist who had led the invention of the microprocessor at Intel, now at the helm of Zilog, no one doubted that the Z8000, too, would be better than the 8086.

This defeat would not only cost Intel customers and market share, but even worse for the long-term health of the company, it would also make Intel's proud employees look in the mirror and ask if they really *were* the best around. In a company that ran and recruited on pride, that might prove crippling.

The natural response of Intel, as it was before and after, would have been to seek out an engineering solution. That was, after all, the company's core philosophy and its greatest strength. But by the composition of that December meeting, it was obvious that Intel had already investigated such a response and concluded that it was too late to design an upgrade to the 8086 in time to respond to the Motorola and Zilog challenges. No, Intel would have to play outside its core competency and figure out some novel way to *convince* customers to buy a product that was an also-ran on the specifications chart. But before it could do that, the company had to be honest with itself about its failure. It was a bitter pill to swallow.

"The executive staff meeting the following Tuesday couldn't have been more unpleasant," Davidow would remember. Traumatic too, because years later he still couldn't remember whether he "either volunteered or [was] asked" by Grove to lead Intel's response.

Bill Davidow, who would later write a classic text centered on this experience, *Marketing High Technology*, knew that his first task was to get his arms around the magnitude of the problem by polling all of Intel's sales force. Then he needed a team with the right talent to tackle this threat based upon what he had learned. It took several weeks, but by the time Davidow's handpicked task force met early that December, he was prepared to present to the team "a precise analysis of what Intel was up against."[3]

"The first thing the group did was agree on the problem. That wasn't hard. There were three of us in the race: Motorola was going to be first, Zilog second, and Intel was headed for obscurity."[4]

Once the awful truth had been looked in the eye and the existential risk to Intel directly faced, the old company arrogance and bravado could finally reemerge: Intel wasn't going to play just to survive; it was going to bet everything on a win. Davidow: "All of us agreed that if we whipped Motorola, we would win. For that reason we made our goal not simply regaining market share but restoring Intel's preeminence in the market."[5]

Preeminence meant design wins, the goal not just because they meant short-term revenues, but because they were the gift that kept on giving. Once a company designed a particular microprocessor into its new product, it was wary of replacing that chip in future versions or upgrades of that product for fear of having to make secondary changes in packaging, power supply, software, and manuals. Thus, especially for a successful product or the first of a family of products, a single design win might result in renewed contracts for years, even decades.

The team quickly decided it needed to set a minimum quota of design wins for each Intel salesperson around the world. Only by presenting a united front rather than depending upon a few sales superstars could the company recapture market dominance. So what should that quota be? One win per salesperson per month seemed like a realistic, doable number. The team agreed to set the quota at that number.

It was only after it had reached this agreement that someone actually did the calculation and determined that this would mean Intel would have to land two thousand design wins by the end of 1980. That was an insane number. Even the most optimistic industry analysts were not predicting that Intel could land even half that number; most predicted a third as many.

In one of the most important decisions in company history, the marketing team voted to stick with the original number. Two thousand design wins in the next year—or Intel would die trying.

But that still left the question, With what would Intel capture these design wins?

To answer it, the team spent a day just on competitor analysis. What was it that really made the Motorola and Zilog microprocessors better? In the end, the team concluded that while the hardware design of the two competitors' chips was a little bit better than the 8086 in terms of speed and processing power, those advantages were not decisive in the minds of most potential customers. Rather, the real advantage held by the 68000 and the Z8000 was with a comparatively small but very prominent subset of users: "software-oriented" companies—that is, enterprises that focused on user solutions rather than hardware.

Most of these software-oriented companies had seen their first great success designing software programs for minicomputers to help those once revolutionary machines to become viable replacements for the previous generation of mainframe computers. Now, looking at the hot new generation of computing companies being led by Apple Computer, these software companies quickly recognized that a new generation—personal computers—was about to become the Next Big Thing. So the rush was on to *port* (convert) those popular minicomputer programs to run on dedicated personal computers and other "smart" instruments and devices.

But this conversion process wasn't going to be easy. The early generations of microprocessors may have been miniature miracles, but to software-oriented companies, they represented a big step down in performance from "real" computers. Software couldn't just be translated and simplified but very likely had to be rethought and rewritten in some fundamental way.

Understandably, that worried these companies. Not only would it take time and money, but in the end they might well end up with a kludgy, buggy, and unpopular solution. Or just as bad, they might arrive in the marketplace too late and lose all of their existing customers, who themselves were undergoing the same evolution. In other words, these very important companies couldn't go back, couldn't stay where they were, and were terrified of making the crossing into the future. And that's why, even though many of these companies might have preferred to go with Intel and its unequaled reputation in the chip industry, in their minds the safest course was to adopt the more "software-friendly" architectures of Motorola's and Zilog's processors.

For the Intel team, the shape of the problem was now much clearer but no less intimidating. Given the market's perception of what made a good product offering in the new era of 16-bit microprocessors, even if Intel threw everything it had into the creation of a new and better 8086, it would still likely be too little and much too late. No, the Intel team would have to figure out a way to work with what the company already had. In Davidow's words, "We decided what we needed was a new product that better fitted the needs of the customer base. We would have to *invent* one."[6]

In retrospect, one of the shrewdest executive decisions Andy Grove ever made was in how he constituted that team. In a company like Intel, which always gave priority to its engineers and scientists, creating a team out of marketing and sales professionals—experts in the kind of squishy, nonempirical stuff that Silicon Valley never really trusted—was an uncharacteristic move; one that probably would have drawn resentment from the rank and file if they had known about it. After all, entrusting your fate to flacks and promoters was not the Intel way of doing business.

In fact, it was a brilliant choice, because in all of Intel, only this group had the ability to look outside the usual box of technical answers and into the hearts of Intel's customers. No one had ever accused Intel of empathy before, and rarely after, but during those three days Davidow and his team managed to put themselves into the shoes of their customers and finally began to appreciate the fear of the unknown those customers were feeling and their desperate need to be protected from making suicidal technical decisions.

It slowly dawned on the team that what these customers were looking—begging—for was not the most powerful or even the cheapest microprocessor. Nor were they giving the highest priority to specs or even, surprisingly, to "software-friendliness," even though they said so. No, what these chip customers really wanted was a *solution*. To them, microprocessors were merely an answer to the question: how do you take the computing power of big computers and put it into small boxes to make those new low-cost products smart? The less thinking and worrying that they had to do to accomplish that task, the better. Yeah, it was great that the 68000 was a little smarter than the 8086, but what really mattered was that it was easier to shove into the new boxes without a lot of extra work on software. End of story; contract signed.

Here in the twenty-first century, we live in a world in which the term *solution* is part of everyday business conversations, where it shows up endlessly in advertising and corporate mission statements. Every MBA is taught Peter Drucker's famous dictum: "Nobody pays for a 'product.' What is paid for is satisfaction." Ted Levitt expressed the same idea in an equally famous dictum: "People don't want quarter-inch drills—they want quarter-inch holes."[7] In that Intel meeting room on the eve of the 1980s, this thought was a revelation. It gained momentum when one of the salespeople noted that there was a rumor on the street that Motorola's early customers for the 68000 were having trouble using the chip.

From the early transistors to the current microprocessors, semiconductors had been designed by engineers, sold by engineers (in mufti as salespeople), bought by engineers, and designed into larger products by engineers. They all spoke the same language, worshipped at the same altar of empiricism, read the same trade magazines, and went to the same trade shows. It was a perpetual tech race, and whoever led at the moment in terms of performance specifications (and to a lesser degree, price) was declared the winner and also scooped up most of the market, while the losers licked their wounds and vowed to win the next race.

Now, with this recognition of an irrational dimension to the microprocessor wars—one characterized by fear of failure, by embarrassment at a lack of technical acumen, by lack of money, even by laziness—the Intel team members suddenly were looking at their industry in a whole new way. They had just identified an enormous competitive opportunity: the way to victory was to redefine *product*.

Davidow: "Everyone on the task force accepted the harsh truth that Motorola and Zilog had better devices. If Intel tried to fight the battle only by claiming our microprocessor was better than theirs, we were going to lose. But we also knew that a microprocessor designer needed more than just the processor and we had our competitors beaten hands down when it came to extras. We had been playing to [our] competitors' strengths, and it was time to start selling our own."[8]

Needless to say, the team next embarked on a hurried inventory of those "extras." They quickly had a list of *company* advantages. It included Intel's reputation and pedigree. The company did, after all, have the most sterling image in the semiconductor industry—perhaps with Hewlett-Packard and IBM, in all of tech—for innovation and quality. The two founders were industry legends—Noyce for the integrated circuit, Moore for his law—and together with Grove were generally recognized as the strongest management team in semiconductors. And the temporary stumble of the 8086 aside, Intel had a long history of putting out the first and best devices in each new product generation. The team also were convinced (though the outside world didn't know) that Intel had the best long-term strategic plan, the 8086 being the founding chip to what could be decades and generations of future microprocessors built on its basic architecture. It wouldn't be a one-off seller like Zilog's; rather, Intel would continue its history of building strong and enduring relationships with its customers.

The team cast its net out further. Besides its image, what products and services did Intel have that made the 8086 a better solution for nervous customers than its competitors? This led to a second list, this one of *product* advantages. It included the fact that Intel was a company of specialists—the entire company was devoted to the world of microprocessors—not generalists making everything from memory to car radios, like Motorola. Better yet, with the 8086, Intel was offering a more complete package of supporting chips, including a math coprocessor—Zilog didn't have them, and Motorola's weren't nearly as good. Intel also offered its customers a collection of tools specifically designed to help them design products around its microprocessors. The best of these, the Intellec in-circuit emulators, would later be recognized as the first proto–personal computers. Tellingly, Intel had been so focused upon using these systems to help customers that it never considered putting a display on the boxes and making them into consumer products.

By Friday, it was becoming clear to the team that it really did have a response to the Motorola challenge, that it could create a wholly new "product" merely out of the other products and services that Intel already had on hand, once they were combined into an overarching solution. They even had a bold, in-your-face, name for this new initiative: Operation Crush.

Years later, when marketing students read about Operation Crush, there was always the lingering question whether the Intel team actually believed that they had created a new product out of the 8086 and its constellation of hardware, software, and support or whether they merely thought customers would believe so. The answer isn't obvious; about the most we can see at this distant remove is that the Crush men convinced themselves that they believed in this new product and that they could make that case both to Intel management and to the business world.

That the Crush team managed to take a losing hand and turn it into a hugely competitive advantage—one that changed high-tech marketing, packaging, and branding forever—is often seen as something of a miracle. But the real miracle may have come a few hours later. That's when the meeting adjourned, the exhausted but happy members headed home, and Davidow delivered the plan to Andy Grove.

Intel was already justly famous for the speed at which it operated, from design to fabrication to the creation of new generations of products. But Grove's response to the plan turned heads even at Intel. By the following Tuesday, December 10, he had studied and approved the plan. A week later, while company departments were beginning to hold holiday parties, Grove called in from the field more than a hundred company sales engineers and presented the plan to them. By the holiday break, Operation Crush had gathered up more than a thousand Intel employees, making it as big as the original 8086 project. Intel was betting the store.

It wasn't just the sales force that mobilized: "Numerous committees were created to work out the details. New sales aids were created to reflect a system, rather than a product, viewpoint. System-level benchmarks were prepared; technical articles written; existing customers were convinced to write their own articles; new data sheets were prepared, as was—remarkably given the short time window—a completely new product catalog. [Regis] McKenna devised a new advertising campaign. Within the next few months, more than fifty customer seminars were presented throughout the world, as was a users' conference. In each case, whatever didn't work was abandoned and something new tried."[9]

Intel, born in the belief that superior technology would overcome any obstacle to success, was now discovering the value of marketing and branding. It would be a crucial lesson.

For now, though, the company was hardly out of the woods. It had managed to convince itself that it had invented a whole new product out of nearly thin air; but it had not yet proven that it could convince its competitors of the same thing. New orders were beginning to pour in, but not yet at a rate that would meet Intel's target of two thousand design wins per year.

Then a small miracle. Motorola, serenely confident that it had captured the 16-bit microprocessor market, was caught completely flat-footed by the "new" Intel 8086 announcement. Worse, when it saw the small but real shift in orders to Intel, it panicked and did the worst thing possible: it tried to mimic Intel's strategy. The company quickly put out an amateurish systems catalog for the 68000.

It was the worst kind of mistake, made for all the right reasons. Motorola thought that it could defuse Intel's claims by showing that it too had a "solution." Instead, Motorola shifted the competition to a battlefield where Intel could win, then went further by showing that, in this new battle, Motorola had the weaker hand. In the long history of competitive duels in high tech, from IBM versus Burroughs to Microsoft versus Netscape to even Facebook versus MySpace, no company has ever held such a strong position and then destroyed itself so completely in the face of a weaker opponent.

Davidow: "Had Motorola chosen to remain aloof from our challenge, I think Intel would have been in deep trouble."[10] And once Motorola stooped to slug it out, Intel quickly put the challenger on the ropes. As for Zilog, it was too small and undercapitalized to respond and quickly retreated to its natural niche market.

It took a while for Intel to ramp up. By midyear it was still behind the pace it needed to reach two thousand design wins, but the trajectory was becoming clear. Intel's sales force, smelling victory and record bonuses, began to get excited . . . and competitive. Operation Crush, with Andy Grove's full support, now began to take some of the early revenues on the increased sales and reinvest them in new additions to the 8086 product constellation.

One of these was a budget version of the 8086 called the model 8088. Despite the name, it was in fact a strange little choked-down version of the more expensive original chip. Like the 8086, the 8088 was a 16-bit processor—that is, the data it operated upon had sixteen bits of complexity. But unlike the 8086, with its 16-bit external bus (the pipe that carries data into and out of the processor), the 8088 was slower, with a cheaper 8-bit bus.

Confident now that they had a winner, Intel's sales engineers began to take risks, calling on unlikely potential customers who wouldn't even have been classified as viable leads a few months before.

One of the takers of the biggest risks was Earl Whetstone. Earl was one of Intel's top salespeople, but that didn't make him any less immune to the pressure of Crush's quota. Searching everywhere to fill his one-per-month quota, he decided to call on the biggest, unlikeliest whale of them all: IBM Corporation.

It was widely assumed that Big Blue had no interest in buying from outside vendors. It didn't need to: with a market valuation of $40 billion, it was twenty times larger than Intel, and in fact, IBM's *internal* semiconductor operations (making chips to sell to itself) were larger than those of Intel or any other independent chip maker. Moreover, IBM had dominated the computer market, the most important sector of the electronics industry. In the 1950s, after crushing competitors like

Burroughs and Univac, IBM captured dominance over the mainframe computer industry, and held it into the 1980s. Meanwhile, in the late 1960s, it had invented the minicomputer and managed to dominate that industry as well, despite brilliant challenges from Digital Equipment Corporation, Data General, and Hewlett-Packard. When Earl Whetstone came calling, the phrase "Nobody ever got fired for buying IBM" still dominated the IT departments of the world's corporations.

But all of this success had not come without cost. Indeed, the seventies had been a miserable decade for Big Blue. In 1969, in response to IBM's extreme dominance of the computer industry after the collapse of its major competition, the US Justice Department filed an antitrust suit against IBM for violation of section 2 of the Sherman Act. The case would drag on for thirteen years, throwing the entire decade under a cloud for IBM. Even worse, in a separate but related case, 1973's *Honeywell v. Sperry-Rand,* a federal court ruled that a patent-sharing agreement made between IBM and Sperry in 1956 was invalid—thus voiding the original electronic computer patent and throwing the rights to the computer into the public domain.

The government case against IBM would endure until 1982, when the feds would finally drop the case for "lack of merit." And though the Justice Department would lose the case, it had its intended effect. Most famously, IBM "unbundled" its computer hardware from its software—establishing a new trajectory for the company for the decades to come. But in the near term, the ever present threat of being broken up by the government made Big Blue extremely careful about entering into new markets that it might dominate. Thus it watched the explosion of the new personal computer market, yet (to Apple Computer's relief) it remained inexplicably on the sidelines.

This ongoing legal threat also made IBM increasingly paranoid and secretive. After all, any memo or the minutes of any meeting might be subpoenaed by the Justice Department. By the late seventies, the company had become almost walled off to the outside, grinding out upgrades to its minicomputers and dragging along the vast installed base of its world-dominant 360/370 mainframe computers.

As if this were not enough, Big Blue was also suffering a crisis of leadership. At the beginning of the decade, CEO Thomas Watson Jr.

suffered a heart attack and retired, ending the family dynasty that had ruled the company for a half century. The resulting power vacuum created two warring camps. The Old Guard wanted IBM to keep its head down, stick to its core businesses, and ride out the antitrust suit. The Young Turks believed that the company should not be diverted by the government threat but should use its technical and marketing prowess to start staking out claims in new markets.

One of the Young Turks was Philip Donald "Don" Estridge. He was just a project engineer at IBM, but as he watched the first generation of personal-computer companies—Commodore, Atari, Apple, and a hundred others—he became increasingly convinced that IBM was not only missing a major opportunity, but perhaps even its grip on the future of computing. In 1978, he approached his superiors with the idea that IBM should get into the personal computer business, if only to have a beachhead in case that new industry ever gained real traction. He was rebuffed. IBM was in the business of real computers, he was told, not consumer novelties.

But Estridge didn't give up. Big Blue was beginning to notice all of the attention paid to Apple and was wondering if its minicomputer competitors were noticing the same thing, so to shut him up, Estridge eventually was allowed to take five other engineers from his department and set up a small skunkworks at IBM's facility in Boca Raton, Florida. To keep it from becoming a disruptive force, the parent company was keeping this little maverick team at arm's length, so IBM broke with precedent and let Estridge go outside the company for components, parts, and eventually software programs.

It was at that critical moment, in what in retrospect seems almost a seam in history or a moment of incredible luck and timing, that Earl Whetstone called on IBM's Boca Raton division, pursuing a rumor that Big Blue was working on a black-box project that might involve microprocessors. Whetstone fully expected to be rebuffed in the lobby and sent away.

Instead, shockingly, he found himself met and welcomed by Don Estridge. Whetstone, expecting the old IBM arrogance and "not invented here" attitude, was thrilled to find Estridge both warm and receptive to new ideas. Better yet, Estridge's interests seemed wonderfully aligned with the message of Operation Crush. He was more interested in service than performance, in Intel's long-term commitment to its processors over the course of multiple generations rather than in any small advantages in current design. And he was especially impressed with Intel's new, Crush-influenced product catalog, with its emphasis on all of the add-ons in the 8086 family. In subsequent conversations with Whetstone, Estridge began to emphasize the lower-price lower-power chip, the 8088, as being of particular interest. He also liked the fact that Intel offered supporting chips, like the math coprocessor, for the 8086/8.

Whetstone couldn't believe his good luck. He even went along with the growing weirdness of the relationship between IBM Boca Raton and Estridge's team. Whetstone:

"Everything was very secretive. When we went in to provide technical support, they'd have our technical people on one side of a black curtain and theirs on the other side, with their prototype product. We'd ask questions; they'd tell us what was happening and we'd have to try to solve the problem literally in the dark. If we were lucky, they'd let us reach a hand through the curtain and grope around a bit to try to figure out what the problem was."[11]

As the world would learn two years later, on August 12, 1981, this stealth project was the IBM PC personal computer, one of the most successful electronic products in history.

So successful was Operation Crush that Intel completed 1980 not just meeting its goal of 2,000 design wins, but shattering it: by year's end, it could celebrate an astonishing 2,500 new wins. It had turned the industry around. Now both Motorola and Zilog were struggling to stay in the game, while the 8086 was rapidly becoming the industry standard.

But it was that one design win with IBM, though Intel didn't know it yet, that counted most. Recalled Gordon Moore: "Any design win at IBM was a big deal, but I certainly didn't recognize that this was more important than the others. And I don't think anyone else did either."[12]

Whetstone: "A great account was one that generated 10,000 units a year. Nobody comprehended the scale the PC business would grow to—tens of millions of units per year."[13]

Whetstone went on to become Intel's group vice president of sales.

IBM at last recognized Estridge's gifts. He was made vice president of manufacturing and an adviser to IBM's board of directors. Steve Jobs offered him the presidency of Apple Computer. And it was said that he was on track to become a future CEO of IBM itself. But then, tragedy: on August 2, 1985, Delta 191 crashed at Dallas–Fort Worth Airport, killing Estridge and his wife and leaving their three daughters orphans. Estridge, forty-eight, did live long enough to see his vision vindicated and the IBM PC become the world's top-selling computer.

Though no one knew it at the time, Intel's sale of the 8088 to IBM to become the central processor unit of the PC signaled the end of the microprocessor wars almost before they began, with Intel the clear and dominant winner. From the moment the IBM deal was inked, every other semiconductor company was either doomed to oblivion or acquisition (like Intersil, American Microsystems, and others), or left rushing to capture niche markets (like Zilog and National Semiconductor).

Only Motorola survived intact, largely because of a curious event beyond its control. The newborn Apple Computer used a clone of its microprocessor in the Apple I and II computers.

To understand why Intel's contract with IBM meant "game over" for everyone else, it's important to appreciate what happened next. With the Intel 8086, Estridge and his team in Boca had the box and the central processor, but they still needed the software to run them. That's when a third figure entered the story. Bill Gates had dropped out of Harvard and joined his friend Paul Allen to create Micro-Soft Inc. (The company name was changed to Microsoft in 1976.) Thanks to his mother's connections, Gates also got an invitation to make a presentation in Boca Raton.

IBM was interested in Microsoft's word processing software (which would become Word), but more important, it was in desperate need of a good operating system. Gates recommended DR-DOS, created by Digital Research in Monterey, California. Big Blue followed his advice, but when negotiations there collapsed, IBM came back to Gates and asked Microsoft to develop a comparable operating system (OS). Gates and Allen quickly snapped up a nearby Seattle company that had the right code, renamed it MS-DOS (and after a few generations of upgrades, Microsoft Windows), and delivered the product to a very happy IBM.

Everyone knows what happened next. In an effort to catch up on the six-year market lead held by Apple Computer, IBM chose to pursue a very un-IBM strategy, once again driven by the remarkable Don Estridge. Whereas Apple sued any company that dared to clone its computers or design applications for its OS, IBM allowed not only cloning, but the sale of MS-DOS to other computer makers. It even opened up the source code to allow software companies to design their own proprietary application programs to sell to businesses and consumers.

It was a brilliant stroke. Soon the market was filled with PC clones, and the aisles of electronics retailers were dominated by endless shelves of Windows-compatible applications and, even more important, computer games. Furthermore, thanks to IBM's half century as the world's dominant supplier of business machines, its PCs quickly found their way into corporate offices—a place where even Apple had failed to make much more than a beachhead.

Against this onslaught, Apple Computer, which had held as much as 90 percent of the personal computer market, staggered and watched its market share plummet. Even the introduction of the stunning Macintosh computer in 1984 failed to do much more than slow the fall. Within a decade, Apple's market share had fallen to single digits.

Even more than Big Blue, the biggest beneficiaries of the IBM PC were Intel and Microsoft. They sold not only to IBM, but to all of the other IBM clones out in the world and in time to makers of other Windows-based products, from game consoles to smartphones. So intertwined were the fates of these two companies that soon—to the distaste of both—they were regularly called by the portmanteau nickname Wintel. The advantages of this relationship (most of all, being the industry standard) and the disadvantages (being hostages to each other's mistakes) would largely define the actions of both companies for the next quarter-century.

Looking back on Operation Crush and the 8088 deal with IBM, it is easy to dismiss both as a combination of cleverness and luck. Cleverness, because Davidow and his team basically took an uncompetitive product, wrapped it with a few spare parts and a lot of hype about services, and made it look like something shiny, new, and superior to the competition. Luck, because Motorola foolishly took the bait and let Intel set the terms of the battle. Then even more luck because Earl Whetstone happened to call on IBM Boca Raton just when, for the first time, the company was allowing the design-in of products from outside vendors. Then the final bit of luck: IBM decided to let the PC be an open platform and thus set the stage for Intel microprocessors (and Microsoft's operating system) to become the industry standard in computing for a generation.

But that judgment fails to recognize that at key moments along the way, Intel made brilliant decisions and showed the power of its culture and executive team. For example, in many companies, then and now, the field sales force would have been wary of sending bad news, much less an angry demand for action, back to corporate headquarters. Moreover, this news probably would not have made it all the way to the chief operating officer nor would it have resulted in such a quick call to action as was made by Andy Grove.

As for the team itself, few companies—especially a hard-core tech company like Intel—would have put their fate in the hands of a bunch of sales, marketing, and PR people. Nor would a team thus composed have felt safe enough to come up with such a radical solution.

But the most important and unlikeliest difference between Intel and most other established companies is what happened after the task force delivered the Crush concept to Andy Grove. He accepted it and executed it throughout the company in a matter of days.

In his memoirs, Regis McKenna said that this last was the single most important step in Intel's success during the Crush era: "Some time later, when I told a former Motorola executive that it took only seven days to develop the Crush program, he told me that Motorola could not have even organized a meeting in seven days."[14]

In military terms, by moving so quickly, Intel had gotten inside Motorola's Boyd cycle or OODA (observe, orient, decide, act) loop, such that every time Moto reacted to one of Intel's moves, Intel had already embarked on the next one. Motorola's feeble response to Intel's catalog betrayed a befuddled company reduced to reacting to its opponent's moves.

By comparison, Intel recognized the jeopardy in which it found itself and responded by betting everything it had on regaining primacy in the microprocessor industry. Intel had made similar bold choices before this, and it would make many more after, but in Operation Crush, the company for the first time revealed its true character: fearless in being willing to push all of its chips to the center of the table, confident to the point of arrogance, filled with a sense of entitlement as guardian of Moore's Law, and determined to dominate the microprocessor industry in the process. But the characteristic least noticed by the world (or admitted by Intel) was the company's unexpected willingness to admit it had made a mistake. Then it moved heaven and earth, even playing from weakness, to rectify the situation.

This reality of Intel has always been at odds with the myth of Intel—and that's for the best. Intel has always presented itself to itself and to the world as all but infallible, an enterprise that has embodied intelligent planning and rational action in one long march from its legendary founding to its current world-bestriding leadership. The reality, to anyone who makes a close inspection of the history of the company, is that Intel has probably made more mistakes than any company in high-tech history. And there are two good reasons.

The first is that, from its founding, Intel has taken one great risk after another for nearly a half century, more than any other tech company, even legends like IBM and Apple. More risks mean more mistakes, no matter how well run an enterprise, and at its peak in the 1990s, Intel may have been the best-run company in the world.

The second is that most tech companies, when they make major mistakes, simply don't survive them. Tens of thousands of electronics companies have been built and died since Intel was founded. Some of those doomed companies, especially during the dot-com bubble, barely survived a few months despite starting out with far more venture capital money and much bigger market opportunities than Intel had at its birth. The greatness of Intel is not that it is smarter than other companies (though it may well be) or that it is too clever and competent to make a false move (we've just seen a stunning example of the very opposite) but that it has consistently done better than any company, perhaps ever, at *recovering* from its mistakes.

That Intel has survived all of these decades and stayed at the top of the most complex and dynamic industry imaginable is testament to this remarkable ability—by a singular combination of ferocious competitiveness, uncompromising management, and seemingly bottomless reservoirs of employee commitment and energy—to not only recover from mistakes but actually capitalize on them. To the amazement of observers and the dismay of competitors, whenever Intel has stumbled, it has not only quickly picked itself up and brushed itself off, but has then charged off again even harder. The 8086 stumble led to Operation Crush, to the triumph of the 8088, and to the Intel Inside campaign, the most successful branding campaign in high-tech history.

That's how you come back a winner. And Intel has done it again and again throughout its story. It is in the company's DNA. You can hate Intel for its sometimes hardball—and borderline illegal—tactics. You can resent the company for its arrogance as the high priesthood of Moore's Law. And you can complain about its consistently high-handed treatment of customers and every other stakeholder. But in the end, even competitors admit that this is a company that placed itself at the center of the global economy, assumed the toughest and most unforgiving role in the business world, clawed its way to the top and remained there for decades through all of the vast changes that took place in high tech and the world around it—and never once took its task lightly, never shirked its immense responsibility, and always took seriously its stewardship of Moore's Law, the essential dynamo of modern life.

Intel is the world's most important company because it assumed that role and made it work in a way that perhaps no other enterprise on the earth could. Intel is a special company because it decided to be special and never settled for anything else. And thanks to Operation Crush, Intel, more than any Valley company before or since, had *learned to learn*. That ability would save the company many times in the years ahead, even as other companies succumbed to their mistakes.

CHAPTER 21

Silicon Valley Aristocracy

The first five years of Intel's history were defined in part by a head-long technical race to take industry leadership in memory chips, then microprocessors, perfecting MOS along the way. They were also defined at least as much by the day-to-day struggle to build a financially viable and structurally sound company capable of dealing with the demands of surviving explosive growth.

Managing growth is the most underrated part of building a successful high-tech company. Observers tend to celebrate (or criticize) the march of new products as central to the story of an electronics firm. And certainly, thanks to Moore's Law, any company that doesn't regularly upgrade its product line to keep up with the biennial clicking of the semiconductor clock will face eventual oblivion. But there are many companies out there in the tech world that have enjoyed decades of profitable existence by *never* being the industry innovator but rather trailing behind, where the margins may be smaller, but so is the risk. They compete on some combination of quality, reduced price, and high volume.

Hewlett-Packard did just that in the twenty-first century under Mark Hurd, largely abandoning sixty years of legendary innovation to become the first electronics company with revenues greater than $100 billion per year. There are scores of semiconductor companies around the world whose names few outside the industry know, companies that have survived happily for many years after licensing obsolete chip designs from the likes of Intel and Motorola and selling them at ever lower prices for ever less sophisticated applications.

For example, it has been said that the Intel model 8051, an embedded processor chip introduced by the company in 1980 and cloned by as many as twenty chip companies today, is the single most popular product in history. It has been produced by the billions for applications that began in personal computers (such as the IBM PC keyboard) and now, priced for a few pennies, in cheap cameras and children's toys.

As Gordon Moore himself has noted, in a business world defined by his law, the first company into a brand-new market is less likely to be the winner than the point man, the one who hits all the mines and trip wires and serves as a warning and example for the more successful firms that follow. Classic examples of this "first-entry failure" syndrome include Altair and IMSAI in personal computers (and very nearly Apple with its Newton before Steve Jobs's return), Silicon Graphics, Netscape, the first online auction sites before eBay, similar search engine companies before Google, and most spectacularly, MySpace. Some couldn't get the paradigm right; others merely set the table for bigger and more aggressive companies to come roaring in and steal the business. But even these scenarios often arise from a single cause: the inability to grow fast enough and with enough discipline to deal with the challenges that confront being a superstar company—a doubling of employees every few months, endless moves to bigger facilities, a shortage of cash to keep up with expansion, overextended suppliers, and at the top, the vagaries of wealth and fame.

Some great companies escape one or two of these traps, but none escapes all of them. Intel certainly didn't—ultimately, it would hit almost every one. But unlike nearly all other companies, Intel not only managed to survive the passage, but learned from each failure and grew stronger as it went. This survival instinct—the all-important skill in the unforgiving world of high-tech business—to pick oneself up whenever one fell, and, wiser for the experience, charge back into the fray, can be credited most of all to Andy Grove. In important ways, Grove is the difference between Intel and all of those failed companies. For more than thirty years, whenever Intel fell (sometimes when it was Andy's own fault), it was he who, through sheer force of will, pulled the company back to its feet and told it in which direction to charge.

In those first years, Intel was beset by one after another of these challenges, at best to its long-term success, at worst to its very existence. Intel's problems with maintaining a viable yield on its first memory chips have already been described. So too the up-and-down relationships with Busicom and Honeywell. But there was much more.

In 1970, while the company was struggling to get sellable yields on the 1101, complete the 1103, and design the 4004, Intel simply ran out of space at its Mountain View facility. The company had to move to a bigger and newer facility. Intel found the site for its new home on twenty-six acres of farmland in Santa Clara, about a half mile from National Semiconductor, at the corner of Central Expressway and Coffin Road (Intel would soon convince the city to change the latter, unappealing street name to Bowers Avenue after a historic local orchard family). The land was purchased in 1970, and Intel's first building, Santa Clara 1, was erected and ready for occupancy a year later.

It was an unlikely location from which to change the world. Now the very heart of ever growing Silicon Valley, in 1970 it was to Valleyites of the era basically nowhere. Even then, the center of the Valley was between Hewlett-Packard in Palo Alto and Fairchild in Mountain View, the latter almost ten miles north of Intel's new home. In fact, Bowers and Central Expressway wasn't even close to the traditional Santa Clara of the mission and university. It was out in the old strawberry and flower fields near Bayshore Highway 101, and the closest structures were long greenhouses managed by Japanese Nissei farmers returned from the internment camps after World War II, their late-night glow about the only sign of human life along the pristine new Central Expressway. In autumn, tumbleweeds would break off from their fragile trunks in vacant lots and roll in the wind across the expressway, piling up as tall as a man against the cyclone fencing. It was a final glimpse of the Old West even as Intel was setting out to create the new one.

The most important feature of the move to Santa Clara was not the obvious one—that Intel now had a shiny new, state-of-the-art building and considerable room to grow—but rather one that was much more subtle. It was that for the first time, the company was away from the old Fairchild neighborhood, with the Wagon Wheel and former compatriots and new competitors. Had Intel stayed in Mountain View, it might have suffered some of the same fate as had Fairchild before it. It's easy to imagine Intel employees, worried about their job futures during the difficult days with the 1103 and the microprocessors being heavily recruited over equally heavy drinking at the Wagon Wheel or Dinah's Shack.

Instead, the Intel team was isolated in the bleak industrial district of Santa Clara, miles from Mountain View and equally distant from downtown San Jose, even from the coffee shops on El Camino Real. For line worker and senior executive, the daily fallback choice for lunch was the company cafeteria. Combined with a growing movement by senior managers out of private offices and into cubicles on the office floor with everyone else, eating with the employees reinforced a philosophy of equality through unadorned engineering excellence that came to characterize Intel in the years to come.

Was this celebrated culture real? Yes, as far as it went. As in most meritocracies, in Intel's culture, some people were more equal than others, the hierarchy built around engineering and business acumen rather than titles and degrees. And as one might expect, this spartan style was often honored in the breach. There is one well-known though perhaps apocryphal story about Intel that spread around the Valley in the early 1980s about the wife of a top company executive who showed up at Intel headquarters in an expensive Mercedes to pick up her husband, who, when he saw the car, began berating his wife with the words, "I *told* you to *never* bring the good car here." Until that moment, he was known for driving a cheap used sedan.[1]

Ultimately, the value of the move to Santa Clara was that it enabled Intel to find itself, to experiment with different management styles, personnel policies, and even public images. It would have been a lot harder to do that in its old milieu—and because it was precisely this corporate character that enabled the company to deal with the huge challenges coming its way, it can be said that the Santa Clara move was critical to Intel's ultimate success.

The move to the new facility was an early glimpse of just how competent Andy Grove could be at running the daily operations of a fast-growing company. That old Valley rule about shorting a company's stock when it builds a corporate headquarters is based upon experience. When a company moves into a new headquarters, it often becomes distracted from business by political fights over who gets a corner office, which department gets the best location, whether there should be an executive dining room, et cetera. And then, of course, there are all of the dislocations of moving laboratory and fab equipment from one location to another . . . and dealing with all of the problems of cooling, power, and space that result.

That Intel had a considerable amount of time—twelve months—to make the move wasn't necessarily an advantage. Without a firm hand in control of the move, that extra time could have been wasted on even more jockeying for offices and company power politics. But this was the kind of challenge in which Andy thrived: specific tasks, a precise timeline, and the fate of the company on the line. The move went without a hitch—and barely a dip in production. There is, in fact, a photo of a grinning Grove, complete with horn-rimmed glasses and the beginning of an afro, standing in the astonishingly primitive assembly area of the Mountain View plant, posing with two male technicians and four female assemblers in lab coats.

Though the back wall is stacked with Bekins moving boxes, three of the ladies are still working at their lab stations testing finished chips. Behind the ladies sits one more Bekins box, this one with its top still open, and the words "1103's for Marketing or Bust" written on the facing side. What the camera has captured is Grove's remarkable feat of keeping production going in Mountain View right up until the very moment operations shifted to Santa Clara.

It was a bravura performance by Andy Grove—and its success wasn't lost on the two founders.

CHAPTER 22

Public Affairs

I n the story of successful companies, there is no event so difficult, exciting, and transformational than "going public," the first public sale
of company stock.

It's difficult, because the company—especially if it is a tech company—is typically young and at the zenith of its period of fastest growth.
At this moment, when it is most likely overextended already and can
ill afford the loss of a single employee, the company's senior management has to divide itself in two. It must assign half to the seemingly
impossible job of running the company even better than before, while
the other half runs an exhausting series of sprints: finding and working with an underwriter, writing the company prospectus, and then
racing around the world meeting with scores of potential investors on
the IPO road show.

It is exciting, because waiting at the end of controlled chaos is the
prospect of finally enjoying the payoff for all of the hard work, lost
months, and stressed marriages and friendships of building the company. And this payoff, to which the world has become accustomed in
the years since, can be worth billions of dollars to the founders of the
company and millions even to secretaries, consultants, technicians, and
others who were shrewd enough or lucky enough to get stock options
early in the company's history. The excitement at this prospect, already
intense, grows to an almost unbearable intensity in the days leading up
to the IPO, especially as the exact day is rarely known in advance.

And it is transformational. From the moment the company's brand-
new stock symbol and price appear on the screen at the New York Stock
Exchange or NASDAQ, the company will never be the same again. Its

trajectory will be very different from what came before; so will the nature of its new employees. The company's strategy will now look as much quarter to quarter as long-term, and satisfying shareholders can become as important as satisfying customers or maintaining employee morale. Those employees, now enriched, usually say they want to stay with the company—but soon enough, the lure of their new wealth (especially when the lockout and vesting periods end and they can cash out), combined with the inevitable predictability and bureaucratization that comes with the transition from an entrepreneurial start-up to a public corporation, makes the company seem increasingly inhospitable to its founding employees.

After hundreds of Silicon Valley IPOs, many of them in the mad rush of the dot-com boom of the late 1990s, players inside the Valley and observers around the world have devised a standard model of what the process should look like. Thus the Facebook IPO of 2012 brought out legions of "expert" commentators and bloggers to predict the stock's sale price and to point out the unsettling differences between Facebook's plan to go public compared to many others. Meanwhile, despite a recent slowdown in IPOs—thanks to Sarbanes-Oxley, options pricing regulations, and other factors—the actual process of going public has become highly systematized and usually guided by consultants skilled at taking a company through all of the steps from making the decision to popping the champagne.

But in early 1971, when Intel began to ponder the notion of an IPO, tech companies seldom went public. The giants, like Motorola, IBM, and HP, had gone public many years before. However, the mother firm of the modern Valley, Fairchild, had not had the opportunity, as it was a captive division of a larger public company. And National Semiconductor had gone public years before Charlie Sporck's arrival.

So Intel's decision to go public was something rare and new in the Valley, and there were few guideposts to navigate the process successfully. The company would have to find its way on its own.

That was the bad news. The good news was that, unlike the years to come, when the Intel/Apple/Netscape/Google IPOs had made going public the goal of every budding entrepreneur and the Valley's most popular spectator sport, a 1971 Intel IPO would be of interest mostly to company employees, investors, and Wall Street brokerage houses. Besides that, compared to the multibillion-dollar IPOs to come, the money to be raised by the Intel IPO was comparatively small, even in the dollars of the era. Finally, unlike the insane IPO speculations of the dot-com age, when companies went public without ever having a profit or even a working product, Intel was a real business. In Jerry Sanders's words, "Back then, we were only allowed to go public when we were *real* companies, with real customers and real profits."[1]

The ramp-up was also neither as complex nor as exhausting as it would be in a decade, after the Apple IPO. The filings, prospectus preparation, and road show weren't the management death march they would become, but neither were they leisurely. Bob Noyce, who led the effort, spent much of the spring and summer of 1971 signing documents, writing letters to employees, investors, and potential shareholders, making presentations, and huddling with C. E. Unterberg, Towbin, the underwriters.

Twenty years before, Hewlett and Packard had set the standard for enlightened Valley leadership by instituting an innovative employee stock-purchase program. Progressive companies followed HP's lead, Intel instituting such a program right at its founding. (Clever Intel employees bought every share they could. It was estimated in the late 1990s that every dollar spent on a share of Intel founders' stock had already returned $270,000.) This fulfilled Bob Noyce's avowed goal to give everyone in the company a stake in the company's success, right down to the janitors.

But now, as going-public day approached, the Securities and Exchange Commission, the federal government's arbiter of all stock offerings, informed Intel that it could not maintain its current stock plan. To the company's credit, even as it shut down the employee stock program, it also made plans to institute a new one after the IPO. However, tellingly, Noyce, Moore, and Grove were not prepared to go all the way and award those employees actual stock options. Instead, worried that less-educated employees might not understand the volatility and high risk attached to options—not to mention the fear that employees might speculate in Intel stock—Noyce decided to institute a two-tiered program in which professionals would participate in a stock-option program while the less-educated employees would return to the old stock-purchase plan.

In a few years, the Valley would be awash in stock options, making a lot of those "uneducated" workers quite wealthy through shrewd financial management and turning the Valley into one of the world's most famous meritocracies. And Noyce's paternalistic decision would stand as one of his most unenlightened moves.

If the Intel IPO wasn't quite the era-defining event of the comparable tech IPOs of the decades that followed, it was nonetheless critically important for Intel. Simply put, the company needed cash, and a lot of it. Three major product lines, an expensive new research initiative in microprocessors, and a new headquarters that, when completed, might cost a total of $15 million to $20 million—Intel had put itself into considerable financial jeopardy that it couldn't grow out of by any level of successful sales. The 1103 was promising runaway sales now that yields were up and Noyce's penny-per-bit goal had been beaten, but without enough cash, Intel wouldn't be able to afford the success, because it wouldn't be able to pay for the materials and tools to meet that demand. The result would be catastrophic as impatient customers canceled their orders and went to the competition.

It was as stressful a time as any in Intel's history. And though Noyce, as always, presented a cool and casual image to the public, inside Intel emotions were running high. This was the interval when Bob Graham was fired, which put the marketing department in an uproar. Over in manufacturing, Les Vadász was still struggling to normalize production at a viable yield level. Faggin and his team were racing to finish the 8008 and scoping out its successor. Grove, comforted by the show of support in the Graham matter, was trying to keep the company operating at peak productivity even as he was sorting through all of the challenges presented by the move.

The "quiet period," a thirty- to ninety-day interval before the IPO, has been designated by the SEC as a time when a company should not make any uncharacteristic or outsize moves that might influence the value of the stock at opening. And yet at Intel, all of the events just described were occurring, not because of the impending offering, but because even now, several years into its existence, Intel had not yet found its normal. Little wonder, then, that in the weeks leading up to the IPO, Bob Noyce was on the phone with the SEC on an almost daily basis.

Intel went public on October 13, 1971. The offering was for three hundred thousand shares, which opened at $23.50 per share, oversubscribed, and held steady (this wasn't the 1990s). This was compared to 2.2 million shares, after splits, held by Intel's founding investors and employee shareholders. The company saw a return on its sale of stock of $7 million—in those days, enough to keep all of its programs moving forward. And those three hundred thousand shares represented only a small loss of control in the company.

Meanwhile, for the founding shareholders, as share price drifted upward with consumer demand for the stock in the following days, this offering meant nearly a sevenfold return on their original $4.04-per-share investment. Not bad in a faltering economy coming off the excesses of the 1960s. With the IPO, a number of senior Intel employees became millionaires—at least on paper—at a time when a Valley mansion cost $500,000 and a Ferrari $15,000. Many other employees—as Noyce had hoped, including some hourly workers—became wealthy beyond their wildest dreams. As for Noyce and Moore (and to a lesser degree Art Rock), the IPO meant that both men, the son of a preacher and the son of a sheriff, were now multimillionaires, among the wealthiest individuals in Silicon Valley. That was a pretty good return on five years of hard work.

A few months later, on the fiftieth anniversary of Betty Noyce's parents, Bob and Betty invited them and their extended family out to California. The Noyces chartered a bus to take them all around the Bay Area on tours. And it was on one of these trips that Bob called for the attention of the group and then held up a three-inch-diameter silicon wafer printed with 44014 logic chips. "Everybody," he announced, "I want you to see this." When he had their attention, Noyce went into a simplified version of the same pitch he was making to potential corporate customers around the world: "This is going to change the world. It's going to revolutionize your home. In your own home you'll all have computers. You will have access to all sorts of information. You won't need money any more. Everything will happen electronically."[2]

It was probably the best prediction about the future of

microprocessors made in the first five years of that new technology. And it was met with the same skepticism that most of Noyce's business speeches were being received by the world's biggest corporations. His family members smiled and thought that Uncle Bob was just being hyperbolic. Top scientists asked snarky questions like, What happens when I drop my computer-on-a-chip on the floor and it slips into a crack in the linoleum?

But Bob Noyce, showing the kind of vision normally associated with Gordon Moore, was trying to tell both groups something important. Noyce was warning the potential corporate customers that a massive paradigm shift was under way that was about to change the very nature of their business and that they had better be prepared for it. For his family, the message was more subtle, and the first clue should have been that rented bus. Implicit in his message, delivered as his executive stock lockout was ending and a fortune was about to be at Bob's fingertips, was the idea that this new invention was going to change not only the world but also *them*.

And it did. One of the least-noted features of the Intel IPO was how little it diverted the behavior of most of its biggest beneficiaries. Study the aftermath of the Microsoft IPO, and you'll see the stories of hundreds of employees, now multimillionaires, fleeing the company in the years that followed and their options were vested, to pursue their own dreams in philanthropy, entrepreneurship, and lifestyle. The result was that Microsoft suffered a devastating loss of intellectual capital from which, it can be argued, the company has never fully recovered. To a lesser degree, this happened at Apple Computer and Google, too.

But at Intel, with several notable exceptions (as we shall see), there was little loss of talent resulting from the IPO. Part of this can be credited to the times (putting in twenty or thirty years at one employer was still the standard), the culture (Silicon Valley wasn't yet prepared to deal with a vast new population of entrepreneurs, which is what most new IPO tycoons become), and the nature of semiconductor (versus computer and software) engineers.

But just as important was the emergent Intel corporate culture: Intel employees, even with money in the bank, wanted to stay. The previous few years may have been tough, and the company's long-term success—even with this welcome infusion of capital—was still not assured, but already, in a remarkably short time, Intel had fulfilled its promise and become the most innovative company in chips, perhaps even in all of high tech. At the release of those groundbreaking memory chips, the electronics world had watched Intel in amazement, and now with the first microprocessors, amazement had turned to awe. With great products, a more stable financial platform, and three of the most respected leaders in the tech world, there was no reason now not to believe that Intel could become the most successful chip company on the planet. Who wouldn't want to stick around and be part of that?

That's why, just three years after most of them had been part of the most crazed and volatile company in Silicon Valley history, Intel's employees met the announcement of the IPO calmly. Had this occurred at Fairchild in 1967, they would have burned the Valley to the ground; in 1971 at Intel, they celebrated for a few hours and went back to work.

There was only one unrepentant ex-Fairchilder, now an Intel employee, who never seemed to adopt the new, more sober and responsible style. A great scientist, but with the soul of an artist, he was the one person in the company who could declare himself immune from the tough, disciplined style that Andy Grove was enforcing throughout the organization: Bob Noyce.

There had been a second subtext in Noyce's message to the family that day on the bus. It was that this revolution is also going to change *him*.

From the day of Intel's founding, Noyce had announced that he and Gordon were building a company that would not suck dry the souls of its employees with the overwork that had already come to define the Valley. It would be a company that balanced employees' work and social lives. As we've seen, it didn't work out that way, not when product delivery dates slipped, yield rates fell, competition grew more intense, and the economy started to slide. Still, from the beginning, this philosophy meant that Gordon could take off on his occasional fishing trips and Bob could leave early one day each week to be part of his beloved madrigal singing. And they kept this up—mostly because Andy couldn't stop them—even as the rest of Intel began putting in twelve-hour days and attending weekend meetings.

But Bob's indifference to Intel's unwritten rules—or perhaps more precisely, his continued adherence to his old rules at Fairchild—extended even further across the boundaries of work and play, private and public. As early as 1969, Noyce had embarked on an affair with Barbara Maness, the twenty-eight-year-old woman (Intel employee number 43) who had been a pioneer mask designer. With Betty Noyce and the kids gone off on their usual summer in Maine, Bob had taken advantage of the moment and asked the attractive, and very bright, younger woman to dinner.

In the years that followed, Bob grew increasingly brazen about the affair, taking Maness on skiing and sailing trips, teaching her how to fly (Noyce, the enthusiastic pilot, could now start collecting planes), and even arranging to meet her at company events and off-sites. Needless to say, the affair soon became an open secret at Intel—that Bob seemed to do little to hide. Indeed, even when Noyce went back to Manhattan to be on the spot at the opening bell of Intel's IPO, Maness was with him, and they spent a torrid time in the city, unconcerned about being seen on the street together.[3]

This was the seventies, and divorce rates in Silicon Valley were skyrocketing. One of the most popular topics was the notion of the male midlife crisis, especially among men seeing real wealth for the first time in their lives. Bob Noyce, with his kamikaze skiing style and new airplanes (he began training for a jet-pilot license almost the instant he returned from the IPO), seemed to personify the image.

Anyone who dealt with the Noyces during this era quickly understood that their marriage was rocky. Betty Noyce was too intelligent and competent to be merely an appendage of the Great Man; unable (and unwilling) to escape that role in California, she spent more and more of the year in Maine, a shift made possible because one after another, the Noyce kids were heading off to college. When the couple was together in public, they seemed to rarely speak but instead would sit and stonily smoke their way through packs of cigarettes.

Bob wanted the marriage dead—and just in case Betty didn't get the message, he grew even more brazen in his behavior. One of his daughters called it "marital suicide." Betty soon found items of Barbara's left behind in the house when she returned from Maine. Noyce even convinced Barbara to sneak in through the window into his bedroom . . . where she was discovered by one of Bob's children.

Faced with this evidence of the affair, Betty demanded that he end it—which Bob reluctantly did. Barbara Maness wasn't surprised. She sensed that Noyce had indulged in the increasingly risky behavior because he was growing bored with her and was looking to get caught. As had often been the case in his life, Noyce seemed to be focused on maintaining what was perceived as proper while at the same time sabotaging the status quo and ultimately forcing others to make painful decisions in his stead.

Betty Noyce knew the game. She visited a divorce attorney, who suggested they should attempt a reconciliation. But there was no fixing this marriage. Bob wouldn't even talk about the situation with a marriage counselor. As Betty wrote to her mother in a letter she never sent, "[Bob] refuses to discuss his relationship with Barbara because he wants to cherish his reminiscences and keep her image inviolate, and because he feels it's nobody's business anyway. [It's Bob's] same old 'Let's not talk about it because I don't want to' lordly attitude, often expressed by his saying, 'it's all in the past let's forget it'—a statement he makes in almost the same breath in which he belabors me for having been short-tempered or remiss in the culinary department, or with some such historical gripe."[4]

The couple tried one last time to make it work in the summer of 1974. Bob joined Betty and the family at their vacation home in Maine. It was a disaster. One visitor described it as "open warfare," with Betty verbally pounding on Bob as he sat silently, his head bowed. And so one morning, in an uncharacteristic move that stunned even Betty, Bob announced that he wanted a divorce.

And that was it. In keeping with California's new divorce laws, the Noyces' assets were divided fifty-fifty. Simple in theory, but complicated in practice—especially when dealing with Bob and Betty Noyce. The problem was that Bob had never kept a precise record of his investments or his wealth. This posed a real danger, especially for Bob, who risked being accused of fraud if some of these assets popped up after the divorce.

Shrewdly, Noyce turned to a former assistant at Fairchild, Paul Hwoschinsky, a skilled financial expert who would go on to become a noted author of self-help wealth management books. According to Noyce biographer Leslie Berlin, Bob may have chosen Hwoschinsky precisely because, like Andy Grove and a few others, the young man was resistant to Bob's charms. As he had once told Noyce, "If you walk off a cliff, everyone else will follow you, but I will not. What I'm saying to you is that your charisma is scary. Use it wisely."

Hwoschinsky made quick work of his audit of the Noyce's most obvious assets—the million shares of Intel stock worth about $50 million, the houses in California and Maine, Bob's growing armada of airplanes, et cetera. So far, the process proved to be straightforward. But then, an embarrassed Bob Noyce showed up with a shoebox filled with IOUs. Then another. It turned out, unbeknownst of almost everyone but the recipients, Noyce had long been investing in tech start-up companies around Silicon Valley and elsewhere. Indeed, without any fanfare, he had become the Valley's first great "angel investor."

Les Vadász would tell Berlin, "He was a very generous person. He did not have money worries and . . . even if his heart wasn't in [an investment] he probably felt, 'Well, so what? There's some upside.' He didn't take it so seriously."

That September, Bob and Betty Noyce officially split. She left for Maine, never to return. In the years to come, her stock holdings in Intel would continue to grow, ultimately making her one of the wealthiest individuals in that state . . . and because of her good works, perhaps the greatest philanthropist in the state's history.

As for Bob, he was consumed with guilt over the entire affair, especially its impact on his children. As he told his parents, "Nothing else I've done matters, because I've failed as a parent."

Why had he put himself, his family, and his company through all of this? His daughter Penny thought that it was the times, saying, "I think my father really lost his compass [in the 1970s]. It was a time of such change everywhere, such liberalization, such a relaxing of rules." Certainly the photographs of Noyce taken during this period show a man dressed in the wide-lapel trendy suits of the era, a handlebar mustache (copied, as Andy Grove would soon do as well, from Ed Gelbach), and most hauntingly, a distant and angry look in his eyes. He had never looked quite that way before, nor would he in the years after.

Years later, when his life was more stable, at least as stable as Noyce would ever allow it to be, Bob would look back and say, "I didn't like myself the way I was." Even the mustache, he admitted, had been an attempt to change his identity, but "it didn't help at all," he admitted.

As for Intel during these years, the impact of Noyce's affair is incalculable. On the one hand, the legacy of the Fairchild days still lingered, so few were shocked by the boss's behavior. And it was the seventies. Perhaps most of all, everyone was so busy trying to make the company work that they didn't have time to ponder all of the implications of Noyce's behavior. But at the same time, this was a generation of men who had been raised on euphemistic aphorisms warning them against "dipping your pen in the company inkwell" and not "getting your meat where you get your potatoes." And now the head boss, the figure most associated with the credibility of the company, had done just that. Among the female employees, even though Maness was in the privileged position of mask designer, there was certainly resentment that she among all of them was singled out for special treatment. Many of the women were ex-Fairchilders as well, now older and married, and having one of their peers sleeping with the CEO did not sit well at all.

The board of directors? They probably didn't know much and would have only been concerned if it became a scandal. But Moore and Grove were a different matter. Neither man was ever so reckless in his behavior. By all appearances, Andy Grove was a happily married family man and would remain so for the rest of his life. The same with Moore, whose uxorious relationship with his Betty was the stuff of Silicon Valley legend. Their marriage was so close that it sometimes seemed they didn't even need the outside world. Moore's reaction to the Noyce affair seems to have been pretty much a shrug, the sympathy of a man who will never face such a dilemma, and his standard default attitude: "Well, that's Bob."

For Andy, who always judged all by their contribution made to Intel's success, Noyce's affair was just the latest—and not necessarily the worst—betrayal by Saint Bob, more proof that the man didn't take his company, his career, or even his life seriously. It was all a game to Bob, who was intent on following his own star no matter who was hurt—including now his family. In the end, Grove knew, to his anger and dismay, that Noyce would just charm his way out of it. The question, which he had asked himself ever since he had first joined the company—*Can I trust my fate and Intel's to Bob Noyce?*—only grew in Andy Grove's mind.

In the end, it all turned out as predicted. Betty Noyce left for Maine, Intel went on to new initiatives and distractions, Bob Noyce's reputation quickly recovered, and he embarked on a new affair with another woman at Intel (who handled the situation far more responsibly)—while Andy Grove continued to seethe.

CHAPTER 23

Consumer Fantasies

I ncreasingly schizophrenic, Intel was racing forward in two different directions. The memory side had ironed out most of the early wrinkles in design and yield. Intel was now recognized as the premier company in the business, especially in MOS memory. As the midseventies approached, the big mainframe computer companies, notably IBM, had come around to the power, price, and performance advantages of semiconductor memory. The landmark IBM 370 series, the last great mainframe and the generational successor to the Series 360, the most influential mainframe computer system of all, had been introduced in 1970, and its main memory consisted of memory chips.

IBM built its own proprietary chips, but now every company that wanted to compete against Big Blue was running to semiconductor companies to nail down its own supply of ROM and RAM chips. Meanwhile, the rise of minicomputers was in full flood. Digital Equipment, whose PDP-11 had been Hoff's model for the 4004's architecture, was now working on a small single-board version of that machine *and* a new 32-bit virtual memory minicomputer to be called the VAX. At Data General (founded by DEC refugees), teams were racing to build two comparable machines, the SuperNova and the MX Eagle (the latter project documented in Tracy Kidder's *The Soul of a New Machine*).

Hewlett-Packard had its own bid for minicomputer world dominance with the HP 3000, first introduced in 1972. So important did HP consider minicomputers to its future that when initial versions of the machine failed at customer sites, the company recalled them and reintroduced the computer two years later. This time it worked, and the HP 3000 became the most enduring minicomputer line of all, not retiring for thirty years, well into the Internet age, when it had been repurposed as a Web server.

Other ambitious young computer companies—including Apollo, Wang, Prime, and Computervision—raced into the minicomputer market with their own offerings . . . and a hunger for millions of ROMs and tens of millions of RAMs. A glorious future for Intel in the memory-chip business seemed assured.

Andy Grove, pondering the achievements of one of those memory chips, would write with uncharacteristic ebullience: "Making the 1103 work at the technology level, at the device level, and at the systems level and successfully introducing it into high volume manufacturing required, if I may flirt with immodesty for a moment, a fair measure of orchestrated brilliance. Everybody from technologists to designers to reliability experts had to work on the same schedule toward a different aspect of the same goal, interfacing simultaneously at all levels."[1]

But the emerging reality was much different—and that fact wasn't lost on the trio running Intel Corporation. The truth was that while the market for memory chips might be growing, so was the competition. Not only were American and European semiconductor companies racing into the memory business, but so too—unexpectedly, as noted above, because until now they had been considered merely manufacturers of cheap electronic novelties and calculators—were the giant Japanese electronics companies, such as NEC, Fujitsu, and Seiko (whose digital workings were designed by Intel's neighbor Intersil).

In 1972, these competitive threats in the memory business seemed small and distant. On the contrary, the memory business was growing so quickly that it seemed there might be room for everybody, with demand to spare. Two years later, the marketplace had changed radically—and not to Intel's advantage.

The first thing that happened was the arrival of all of those new competitors. Intel, along with a handful of other firms, may have enjoyed the advantage of pioneering MOS in the late 1960s, but now all of the company's serious competitors had it as well. And slipstreaming in the wake of the leader, the time it took many competitors to catch up was a fraction of what Intel had needed to achieve its market dominance. Moreover, many of those competitors, such as Motorola and TI, were bigger; others, such as the Japanese companies, were *much* bigger; and they had the growth capital that Intel lacked, even after the IPO.

Meanwhile, Intel did little to help its case. Already stretched thin pursuing both memories and microprocessors, Intel now decided to pursue the most dangerous tech strategy of all: the consumer business. In its confidence—one might say arrogance—Intel believed that it had all of the pieces in place to attack and capture the hot new world of digital watches.

As Grove recalled, "We went into the business because we thought we had a unique combination of capabilities: the CMOS chip [complimentary metal on silicon, a very-low-power chip invented at Fairchild during the Noyce era], the liquid crystal display, and assembly facilities."[2]

If that wasn't enough, Intel only had to look at the old watch maker, Hamilton, whose gold Pulsar P1 limited edition of 1972 was priced at $11,000 in today's dollars. Just think of the potential profit margins when you manufactured not only the watch, but all of its components!

The engineering imperative that defined Silicon Valley was, "If we can build a product at the leading edge of innovation, smart customers will recognize that advantage and pay us a premium price for it." As many electronics companies would learn to their dismay in the years to come, that advice founders in the face of all the subjective and irrational factors that go into an average consumer purchase.

Intel thought that if it just used good engineering and manufacturing and built a solid, reliable digital watch, the world would beat a path to its door, buying not only millions of watches, but also the Intel chips within them. The company established a subsidiary—called Microma—just for this purpose, spent millions on facilities, parts, and tooling, and set it up in offices in nearby Cupertino on Bubb Road in a building that would one day be the first official home of Apple Computer.[3]

Intel wasn't alone in this pursuit of the riches that supposedly awaited in the consumer business. National Semiconductor and TI joined in as well. So did the Japanese, most notably Seiko and Casio. Even some old-line watch companies joined in, though carefully, using OEM electronics in their own proprietary cases.

What Intel and the other chip companies discovered, after it was too late, was that watches, even digital versions, were part of the jewelry industry. Consumers bought watches for a lot of reasons: cheap functionality, social status, precious metals, glamour, as markers of wealth and social power, et cetera. And they largely reached those decisions through the influence of advertising in such channels as glossy magazines, catalogs, and television or through point-of-sale displays and the personal advice of the local jeweler. Consumers *didn't care* about performance specifications laid out on an unembellished tech spec sheet.

That initial burst of orders for digital watches had been largely an illusion. Wealthy people bought the wildly overpriced gold and silver LED versions precisely because they were new and rare and brought the owner considerable attention. The second wave, this time of LCD watches, was bought in larger volumes by people who wanted to be part of the new digital revolution, to show their bona fides as tech geeks, or simply because they were of the personality type that liked the rigorous precision of being able to track the world in tenths of seconds—in other words, the kind of people who actually did read spec sheets.

But this last group, though a sizable fraction of the population, also didn't have the kind of disposable income to be early adopters of $400 novelty watches. They would buy similar watches at a tenth that price, especially when their onboard electronics began to add multiple new functions, from stopwatches to crude games. But that kind of low-cost, high-volume production—driving the price down to commodity levels—was beyond the abilities and desires of US companies such as Intel. Their business model was based upon the scarcity and high perceived value of innovative products, not the long miserable slugfest of cheap consumer goods. That turf belonged to the Japanese, and it quickly became apparent that over the next few years, those companies would drive the low-end digital watch business.

That left the high end of the business. But here, Intel and the others ceded almost every advantage to the old-line watch companies, some of whom had been in business since the Renaissance. The watch companies understood their customers and how to create the right image. They knew the most influential types of advertising and store presentation. They understood the complex, inelastic world of pricing, and they had long-standing—even centuries-old—relationships with their distribution channels. All that Intel's Microma could offer in competition were components offering performance that wealthy watch consumers didn't really care about.

Thus Microma, which had been founded as a low-cost way to tap into a potentially huge and profitable new market, soon found itself crushed between the twin millstones of a high-end, high-profit watch business run by experts and a low-end, low-profit business being taken over by giant Japanese trading companies with the production capacity and operating capital to hang on until the market grew and segmented and margins again began to rise.

Microma had no place to go. It made one last Hail Mary attempt, paying for the production and placement of a commercial for the watch. The ad cost $600,000—a figure that shocked and appalled the bosses at both Microma and Intel, who were used to paying a couple grand at a time for full-page ads in the electronics trade press. And as consumer-electronics and watch-industry veterans could have told them, the commercial, because it was a one-off with no follow-up campaign to drive it home, had negligible effect. "Just one ad," Grove complained, "and poof—it was gone."[4]

In 1975, after less than three years, Intel killed Microma. It sold the technology, parts, and tooling to Timex and the name to a Swiss watch company. The unsold watches were handed out as five-year employment awards to Intel employees until they ran out in the mid-1980s. An example of the watch was eventually put in the company museum. With one famous exception, an embarrassed Intel never looked back.

Grove had his epitaph for the program: "We got out when we found out it was a consumer marketing game, something we knew nothing about." The one holdout was Gordon Moore. For many years after, Gordon continued to wear an aging Microma, which he called his "$15 million watch." When asked why, he would say, "If anyone comes to me with an idea for a consumer product, all I have to do is look at my watch to get the answer."[5]

Grove biographer Richard Tedlow has argued that the Microma debacle was one of the most influential events in Intel's history—for two reasons.

First was the way that the company dealt with the shutdown. Rather than, like many companies, operating in denial until it was too late to respond properly, Intel saw the trajectory of Microma's business and—while still hoping for the best—prepared for the worst. Pragmatic (and prophetic) as always, Grove told product director Dick Boucher to prepare an "end of business plan," the centerpiece of which would be the transfer and reintegration of Microma's employees back into the main company. This was a break from the usual cyclical pattern of hiring and firing that had always characterized the semiconductor industry.

For Tedlow, the message of this rehiring was: " 'You matter to us. If you give us your best, even in a losing cause, we are going to do what we can for you. We won't throw you out on the street if we can help it.' " This wasn't a radically new idea in Silicon Valley. Just up the road in Palo Alto, Bill Hewlett and Dave Packard were famously a quarter century into a no-layoff policy. But in the chip business in a company just seven years old, it was pretty revolutionary stuff. Tedlow argues that this show of loyalty by Intel to its employees, this creation of a whole population of employees who would "bleed" Intel blue, proved vital when Intel demanded considerable sacrifices from them in the years to come.

The second influence of the Microma affair was one that Tedlow found less positive. It was, he wrote, a lesson that Intel had "learned too well: 'We better stay away from consumer products,' the company told itself, 'They are not in our genetic code.' " He pointed to the Apple iPod—and he might well have pointed to the personal computer itself—as a product that Intel could easily have built first. He also quoted Andy Grove as saying, in 2005, that "All of our subsequent consumer products efforts were half-hearted" as evidence that Intel after Microma trapped itself in a self-fulfilling prophecy about the dangers of dabbling in the consumer electronics business.

Perhaps. But it is hard to argue with success. Intel may not have invented the Macintosh nor reached the dizzying heights of valuation that Apple did briefly just before Steve Jobs's death, when it was the most valuable company in the world (though Intel came close a decade before), but history may well consider the x86 microprocessor line a far greater achievement. Moreover, Intel never suffered a near-death experience, as Apple did in the 1990s. Also, comparing Intel to such a singular company as Apple is a bit unfair: better to judge the company against every other consumer electronics company of the 1970s, than to now. Intel, which has all but owned the core technology of the electronics industry for four decades, has no competition.

Still, there is some truth to this observation by Tedlow: "Ask people in the Valley that question [why didn't Intel build the iPod?] and you get a consistent answer. It is that they (or we, if you ask someone at Intel) don't market consumer products. If you ask the next question—'Why not?'—you get a blank stare. . . . If ever there was an illustration that a company's history matters, it is Microma."[6]

CHAPTER 24

A Thousand Fathers

Intel may have found itself overextended in 1974—racing ahead of the howling pack in memory chips and knowing that even a stumble might prove fatal, headed down an expensive dead-end street in the digital-watch bubble, all under dark economic clouds forming on the horizon. But there was still one business that was thriving, driving the technology, and leaving the competition behind . . . only it was a business in which Intel wasn't even sure it wanted to compete: microprocessors.

Federico Faggin and his team, now including new employee and old teammate Masatoshi Shima, had exploded out of the gate on the 8080 project once they got the green light in the summer of 1972. Within months, another old teammate, Stan Mazor, had jumped aboard to write the code. The first test of the new microprocessor took place in December 1973—and after a few errors were fixed, the Intel model 8080, the product of the century, was introduced to the public. It was priced at just $360, a bargain at the time. Even better, it needed only six supporting chips on a motherboard to give it full functionality, compared with twenty for the 8008. That meant that it could be put almost anywhere—even, in theory, in pocket calculators.

Remembering that day, Faggin would say, "The 8080 really created the microprocessor market. The 4004 and 8008 suggested it, but the 8080 made it real."[1] It also made it historic: "With the introduction of the 8080, it can truly be said that mankind changed. Unlike many landmark inventions, the extraordinary nature of the 8080 was recognized almost instantly by thousands of engineers throughout the world who'd been awaiting its arrival. Within a year, it had been designed into hundreds of different products. Nothing would be the same again."[2]

But even though Intel now had a world-historic product on its hands, those delays caused by executive indecision in unleashing Faggin and his team on the 8080 would now prove costly. Texas Instruments, still chastened by its failure to get a working model of its "microcomputer" to market before the Intel devices and by the subsequent loss of a major contract with Computer Terminals Corp., now went after the microprocessor market with everything it had.

It wasn't just about revenge. Even more than Intel at that moment, TI appreciated the potential vastness of the microprocessor business and its value as a prosperous port in the coming economic storm. And even if Intel had shown that it was technically superior to TI, at least when it came to the design of microprocessors, TI knew it had its own special advantage: it could sell to itself. That alone was enough to spur the company forward.

TI's first attempt, a chip three times the size of the 8080, was unimpressive. But what *was* impressive was the design and manufacturing firepower that this forty-year-old, billion-dollar firm quickly brought to bear to improve the design of this product and drive down its price. It wasn't long before TI won the CTC contract back from Intel.

That was just the warm-up. In 1974, just a few months after Faggin and team completed the 8080, TI went to market with the TMS-1000, a single-chip MOS microprocessor that was fully the equal of the 8080—and with a more than competitive price. It wasn't long before the TMS-1000 was being designed into TI's revolutionary new pocket calculators, kicking off a consumer electronics industry even more profitable and enduring than digital watches, and also into a new computer-terminal series (the Silent 700) that ran away with its market. TI was quickly becoming the world's largest captive and largest commercial microprocessor maker at the same time. Intel, with one sixth TI's annual revenues and not even fully committed to the microprocessor business, was overwhelmed. Only its technical superiority kept the company competitive.

Now Texas Instruments, using the unique advantage of a giant corporation, set out to strip Intel of even that edge. In its 1976 annual report, TI declared the TMS-1000 to be "the first microprocessor on a chip." Then the company's mighty legal department began filing patents on the microprocessor, which it now claimed to have invented.

The move caught Intel completely by surprise. That in itself was surprising. Apparently, despite its public claims that the 4004 and 8008 were revolutionary inventions, internally Intel had never really considered the devices to be more than an advancement, an amalgam, of the different types of circuits they contained. According to Ted Hoff, Intel "did not take the attitude that the microprocessor was something that you could file a patent claim on that covers everything."[3] Texas Instruments obviously thought differently—and now Intel faced the prospect of losing control over its most important invention.

Once again, Intel was lucky in its recruiting. In his twelve years at Fairchild, dating back to 1962, Roger Borovoy had earned a reputation as the best corporate patent counsel in Silicon Valley. In 1974, he joined his old workmates at Intel as vice president, general counsel, and secretary. He arrived just in time to help his new employer fight back against TI's legal assaults.

Borovoy was a small, voluble bundle of energy. At an Intel annual meeting in the 1980s, he famously got so excited by the proceedings that he pitched backward in his chair off the dais, causing a worried Gordon Moore to stop his presentation and shout, "Roger! Are you OK?" Like a jack-in-the-box, Borovoy leaped up waving his arms and yelled, "I'm here!" as the audience roared.[4]

But Borovoy was also a world-class patent attorney (and entrepreneur: he would help found Compaq Computer). He was also a bulldog when it came to defending the microprocessor for Intel Corporation, just as he had been protecting the integrated circuit at Fairchild. He immediately filed counterclaims for Intel's ownership of the invention. History was on his side, but not the evidence admissible in court.

The claims and counterclaims were confusing and contradictory, but they went something like this: "Texas Instruments officially announced its new microprocessor in early 1972. Intel didn't announce the 8008 until April. But Intel mentioned the product in its copyrighted brochure called *The Alternative* in late 1971. TI claimed it had made references to its designing a computer on a chip for Computer Terminals Corp. in June 1971. Intel claimed what counted was a working chip and the first deliveries by TI to Computer Terminals in June 1972 were com-posed of dead chips—and besides, it argued, the 4004, not the 8008, was really the first microprocessor."[5]

After two years of legal battles, in February 1978, the Patent Office awarded TI scientists Michael Cochran and Gary Boone the first patent for a microcomputer. Intel was awarded several related patents, but the damage was done. Today, unlike the shared Noyce/Kilby Fairchild/TI credit for the integrated circuit, *no one*, not even the President of the United States in awarding the Medal of Invention, today gives credit for the microprocessor to anyone but the team of Hoff, Faggin, Mazor, and Shima. Indeed, the only real dispute has been among those four over who should get the most credit.

But at the time, the patent was enough to keep Intel from full own-ership of its invention. And it only got worse, because by the time the 8080 was introduced, not just Silicon Valley and the tech industry, but the entire US economy had slipped into a full-fledged recession. The reasons were manifold—the end of the Vietnam War and the NASA Apollo program, the business cycle, the oil embargo, Nixon administra-tion price controls—but the bottom line was that this was an across-the-board crash. This time, Intel and the rest of the chip industry couldn't just shift production to another, healthier market. There were no other healthy markets. Already stretched with too many other product lines and businesses and facing ever greater competition, Intel had every rea-son to abandon the microprocessor business, or at least put it on hold and come back to it in better times.

Then came the biggest blow of all. Just months after the 8080 intro-duction, Federico Faggin, sick of Intel's indecisiveness and provisional support, tired of Andy Grove's management, bitter that his own contri-bution had been eclipsed by "real" Intel employees like Ted Hoff, and feeling that same entrepreneurial pull that had infected the previous generation of top Valley semiconductor executives and scientists, an-nounced that he was leaving Intel.

What happened next was an experience he would never forget. Faggin:

"It was early October 1974 and I had decided to start a new micro-processor company. . . . I was a department manager and Les Vadász was my boss, so I went to him first. Les was pretty much a slave driver himself, but he understood my reasons.

"Fundamentally, it was because I had worked like a dog—only to have everything I did taken for granted. I had to fight for anything I wanted to do—even though I was helping make the company very successful. I had built for Intel the fastest static RAM they'd ever had. And, of course, the microprocessors. Yet it still took nine months to convince them to let me build the 8080. But what really did it was that Intel took an invention of mine from my Fairchild days, told me that it wouldn't work—and when they discovered it did work, filed for a company patent on it. That was it: why waste my energies for a company that didn't care about me?"

Andy Grove called Faggin into his office. There, a conciliatory Grove asked him to stay at Intel. It was surreal moment for both of them, as Grove had shown little better than contempt for the microprocessor project as being hardly more than a Bob Noyce pet project and a drain on Intel's real business of making memory chips.

At the beginning of the conversation Andy was unexpectedly friendly, even diplomatic. Faggin: "Andy tried to sweet-talk me to convince me to stay. He told me my future would be great at Intel. It was just the opposite of the usual Andy. But I held my ground. I told him that I'd made my decision."

Instantly, affable Andy turned into terrifying Andy.

"When he heard that, Andy switched completely. He was almost vicious. I remember him telling me, 'You will never succeed, no matter what you are going to do. You will have nothing to tell your children and grandchildren.' Implicit in those words was that I would have no legacy in semiconductors. That I would never be given credit for what I did at Intel. It was like he was casting a curse on me.

"I remember looking at him and thinking, 'You son of a bitch,' but I didn't have the courage to say it to his face. I was still too respectful of his authority."

And so, Intel lost one of the greatest inventors of the age.[6]

Federico didn't just walk—he took Masatoshi Shima with him. He also took another young Intel phenom, Ralph Ungermann, who had joined Intel's new microprocessor development group just two years before and had already been promoted to manager of microprocessor R&D. They founded Zilog, just a few doors down from Microma in Cupertino. Soon they were building their own answer to the 8080. Because semiconductor fabrication laboratories had now become so expensive, Zilog—sometimes called the last great company of the first semiconductor age—had to survive for years on a venture investment by Exxon and didn't turn a profit for a decade. But in the meantime, it built some of the world's best microprocessors.

"It is interesting to speculate how different the semiconductor world might have been if Faggin had walked out of Intel two years earlier than he did; say, after the 8008, rather than the 8080. In a fast moving market, as microprocessors was, two years is a lifetime. Instead, Faggin not only waited until 1974, but helped give Intel its flagship product. So, his fate was to compete with himself."[7]

As for Faggin's reputation, Grove was as a good as his word. In the small and intensely competitive world of semiconductors, the two men occasionally encountered each other, but Grove never spoke to Faggin again. Meanwhile, Federico's name was stricken from all official Intel histories. For the next thirty years, all credit in Intel publications, annual reports, websites, and even the museum for the invention of the microprocessor was given to Ted Hoff. Mazor was forgotten, Shima was stricken from the records, and most shamefully, Federico Faggin, the man who ultimately saved Intel, was the subject of a disinformation campaign. To Andy Grove, quitting Intel was best forgotten, competing with Intel was all but unforgivable, but to do both, as Faggin almost alone dared to do, was to be cast down the memory hole.

Only in 2009, after Andy was out of Intel's leadership and Federico many years out of Zilog, was Faggin finally rehabilitated. The occasion was the premiere of *The Real Revolutionaries*, a documentary about Noyce and Shockley and the founding of the modern Silicon Valley. Federico, along with his wife, Elvia—who had fought for years to get her husband his due credit—were invited by Intel, which sponsored the premiere, to attend the event and even appear on stage. There, the old Intel veterans in the audience gave him a standing ovation. From that moment, Federico Faggin was restored to the Intel story.

There was one final, devastating shoe to drop. At the end of 1974, Motorola announced its own microprocessor.

Motorola was a company of legend. It had been around since 1928, specializing in car radios (its name came from *motion* and Vic*trola*). And because transistors were intrinsically less fragile in car radios than vacuum tubes, Motorola jumped early into the solid-state world, always just behind the pioneers in transistors, integrated circuits, and now microprocessors. And even more than Texas Instruments, Motorola had a wide range of giant corporate customers and thus was able to sell its chips to itself for use in various commercial products. A perfect example of this advantage came with the boom and bust of digital watches. Whereas the other chip companies either dropped out or competed in the low-profit commodity business, Moto, after an abortive attempt at its own watchmaking, made a bundle selling its components to the established watch companies.

The departure of Les Hogan to Fairchild had briefly set Motorola back in the chip business, but it soon came thundering back under the leadership of semiconductor general manager John Welty. Welty was a gray-haired Motorola veteran, but at heart he was a revolutionary. And though the rumor in the industry was that he had been put in charge of Moto's chip business merely to shut it down, Welty instead decided to save it—and in the process changed Motorola's plodding corporate culture (it was nicknamed the "ponderous pachyderm" after the elephant it used in one of its ad campaigns). Said Welty: "The semiconductor sales organization lost its sensitivity to customer needs and couldn't make speedy decisions."[8] He resolved to change that.

Welty had two great advantages. The first was that he had the full backing of Motorola's president/CEO, the pugnacious William J. Weisz. Weisz played for keeps, and he was intent on making Motorola a major player in microprocessors and using that success as a cudgel to reawaken the rest of the sleepy company. The second was that he hired, from Moto's failed watch project, R. Gary Daniels as manager of the microprocessor group. Daniels, a balding and affable engineer, looked like a tax accountant and seemed much too old for the job, but he had a powerful mind when it came to microprocessor design and marketing.

Daniels arrived in his new job in December 1974 knowing that

Motorola had just introduced its own single-chip, 8-bit NMOS micro-processor, the model 6800, but was nevertheless skeptical about his new job. But he quickly came to appreciate three things about this new product: it was a beautiful design capable of being the cornerstone of many generations of processors to come; it was the industry's first 5-volt microprocessor, which meant that it was the only direct plug-in replacement for existing computer logic chips; and thanks to the general slowdown created by the recession, Motorola hadn't really lost any time despite two years of delays.

Then in early 1975, a of couple months after Daniels's arrival, news came that was like a gift from heaven: General Motors wanted to talk microprocessors. It seems that the oil embargo had sent the automotive giant scrambling for new and appealing features on its cars that would either improve mileage or distract potential buyers from the loss of horsepower and big engine blocks. GM wanted to put its toe into the microprocessor—or more precisely, microcontroller (the industry term for noncomputing, usually analog, applications)—world with a custom chip version of the 6800 to be used in a dashboard trip computer. "It was a wise choice for a start because a possible failure was not life-threatening. GM would learn about microprocessors at a very low risk."[9]

GM's approach was both the reward for Motorola's fifty-year relationship with the automotive giant and also an incredible piece of good luck. Not only did it give Moto sole access to one of the world's most valuable manufacturers, but it also presented the company with a pathway to continue finding one application after another for GM and multiplying its sales to the automotive giant. Motorola wasn't going to blow this chance, and with Daniels now at the helm, it didn't. It kept the GM account for years, and those sales proved a valuable baseline of revenues to keep Motorola a dominant player in microprocessors for decades.

The years 1974 and 1975 saw other chip companies jump into the microprocessor world as well. Hogan's Fairchild soon had its offering, the F8. Advanced Micro Devices under Jerry Sanders, a company still struggling after its underfunded start, had the 2900. Sporck's National Semiconductor had the INS family. (The story of how Nat Semi got into the microprocessor field so quickly is still a matter of speculation and scandal. Don Hoefler always claimed that when Intel exhibited a prototype 8080 at a local Valley community college, the device turned up missing at the end of the day. National Semi, which also had an exhibit at the event, denied any wrongdoing . . . but soon after entered the microprocessor business. Such was life in the Wild West of Silicon Valley of the era. The spirit of Fairchild and the Wagon Wheel lived on.)

Next, in a staggering blow, Faggin's Zilog stunned the industry by quickly rolling out the superb Z80, generally considered the one microprocessor with better architecture than the Intel 8080.

Most important for the next phase of this story, a little company in Norristown, Pennsylvania, called MOS Technologies Inc. announced its own microprocessor, the model 6502. The 6502 had been designed by a group of seven refugee engineers from Motorola, and the chip was so similar to the Motorola 6800 that it was generally considered little more than a clone (it was then that the word became an industry term for a near-exact copy of an existing product) of the older, superior chip. The big difference was that while MOS Tech had looked to Motorola for its product design, it had looked to Texas Instruments for its Boston Consulting Group–based pricing strategy. That is, in hopes of capturing market share quickly, it initially priced the 6502 not where it would turn a profit now, but where Moore's Law said its price would be several years into the future.

"Learning curve" pricing eventually fell from favor in the electronics industry, largely because it tended to strip so much profit out of a market that the competitors rarely made it up. But in the mid-1970s, it was the hottest pricing strategy in tech. TI used it to steal most of the calculator business, only to find itself pounded in turn by the Japanese, who with their cheap manufacturing could also play TI's game . . . and outlast it. Now MOS Tech tried it in microprocessors. Though it failed as a strategy to give the company industry leadership—indeed, it remained little more than an also-ran—MOS Tech's price-bombing strategy serendipitously proved to be one of the most influential consumer marketing programs in business history.

Why? Because in spring 1975, MOS Tech set up shop in the MacArthur Suite in the St. Francis Hotel during the Wescon computer show in San Francisco, the first great gathering of the tribes of future personal computer manufacturers, software writers, and geeks. Then, on the floor of the show (although it wasn't allowed to sell there), the company announced that it was selling 6502s at an the astonishingly low "show price" of just $20, less than one tenth that of the 8080 and 6800.[10]

Needless to say, it quickly became a critical part of that year's Wescon experience to troop the mile over to Union Square and the St. Francis to reach into a fishbowl filled with hundreds of 6502s and fish out one that looked as if it might work, hand over the twenty bucks, and then get out of the suite quickly in case MOS Tech changed its mind.[11]

One of those young budding computer engineers was Steve Wozniak. He hadn't really wanted a 6502 to power the new computer he was designing to impress the Homebrew Computer Club. His heart had been set on the Intel 8080. He thought it was a more robust, capable, and elegant design. And Intel was a hometown company: Wozniak's father worked at Lockheed, just across Bayshore Freeway from Fairchild and the first Intel headquarters, and Intel employees and consultants lived in his Sunnyvale neighborhood—Ted Hoff four blocks away, Regis McKenna just two. And in an incredible historic nexus—sometimes compared to the kind of chance meetings that might have occurred in Periclean Athens and fin de siècle Dublin—at an anonymous intersection just behind Hoff's house, on weekday evenings just a few years before, Bob Noyce racing home to Los Altos, Ted Hoff who lived at the corner, and teenage Steve Wozniak returning home on his bike from swim practice at Cherry Chase Swim Club, would cross paths. Thus the inventors of the integrated circuit, the microprocessor, and the commercial personal computer—arguably the three most important inventions of the electronics age—converged.

Wozniak reached into the fishbowl for the 6502 because he didn't have much choice. In one of the most disturbing—and consciously ignored—episodes in Valley history, Wozniak had offered to help his younger friend, the charismatic but Svengali-like Steve Jobs, develop a new videogame for Jobs's employer, the game maker Atari Inc. Despite the fact that Wozniak, who had a day job as a technician at Hewlett-Packard, did all of the hard design engineering, worked all night long, and put his own young marriage under dangerous stress, Jobs still chose to lie to him about Atari's payment for the project.[12]

It was an early glimpse into the dark side of the young man who would one day become the world's most famous businessman and the iconic Silicon Valley figure. It was also a decision that changed modern business history. As has been often celebrated, Jobs sold his Volkswagen van and Wozniak his beloved HP-65 calculator to build the prototype Apple 1 and then start low-level production. The nearly $3,000 extra Wozniak would have enjoyed during those days, had Jobs as promised evenly shared the payment for the job, would have more than paid for a handful of 8080s with which to build his new computer.

Instead, in a bravura performance that astonishes engineers even today—Wozniak was, like Bob Widlar, one of Silicon Valley's few true geniuses—Wozniak not only managed to build his first computer around the limitations of the 6502, but even did so with an economy of support chips that seemed impossibly elegant.

The personal computer revolution, the industry that would bring the full world of high tech to the average person, was under way—and for the moment, Intel was locked out.

As 1975 turned into 1976, Intel could look back on the previous two years and see that it had begun with an earthshaking new product, the microprocessor, that was its alone to exploit in an increasingly interested marketplace. But instead of capturing that gigantic market—mainframes and minicomputers, cars, calculators, games, and now personal computers—Intel had prevaricated for months before unleashing the best designer in the business (and helped to turn him into a competitor in the process). Indeed, the company had remained unsure if it even *wanted* to stay in the business. Intel hadn't even been serious enough about the microprocessor to patent it.

Now, instead of being the sole owner of the hottest new industry in high tech, Intel found itself just one of a score of competitors—most of them more experienced, several of them with far greater manufacturing and marketing prowess, one already selling processors at commodity prices, and perhaps most irritating to a company that prided itself on its unequaled technology, one (run by a former Intel employee) that had a technologically superior product.

Meanwhile, the Japanese had yet to be heard from—though no one doubted they were coming.

Beyond this, 1975 had been the most miserable year to date for Intel.

Even God seemed to be laughing at the company's ambitions. The company's factory in Peking, a key part of the manufacturing, burned to the ground. As the recession deepened, the company accepted that it would have to cut to the bone to survive, so it had finished 1974 with massive layoffs, firing 30 percent of its 3,500 employees.

This, the biggest percentage layoff in the company's history, had a devastating effect. Sure, the other chip companies were doing the same thing in the face of the recession. National certainly did it, but that was expected from tough guy Charlie Sporck, the world's most successful shop foreman. At Fairchild, the even more scary Wilf Corrigan, a Liverpool dock kid who had become the most powerful of Hogan's Heroes, used the occasion to kick Les upstairs to chairman and then bring his own take-no-prisoners management style to Fairchild. In one of the most memorable issues of his newsletter, Don Hoefler described Fairchild employees being fired in alphabetical order over the loudspeaker, huddled at their desks in dread of hearing their names read and a security guard showing up with an empty cardboard box for each and orders to march them out of the building.

The men and women of Intel thought they were different. Intel was the enlightened chip company, and its employees were convinced (not without reason) that they were different, that they were the best of the best. Otherwise, why had a great company like Intel, with a legendary scientist like Gordon Moore, hired them? They were also family— that's what it meant to work in a company led by the great Bob Noyce. He took care of his employees, and he was famously reluctant to fire anyone. And perhaps most of all, this Intel family had been through a lot over the previous half decade: starting a company, pioneering new inventions and technologies, surviving tough teams, making the move to Santa Clara, the dislocations of the IPO. Surely after all that, the company wouldn't just let them go.

A miserable Noyce told a friend, "For a few goddamned points on Wall Street, we have to ruin people's lives."[13]

But Intel did. Arguably, it had no choice. But that was small consolation for those fired—especially if their stock options hadn't fully vested—and did little to ameliorate the damage to the morale of those employees who remained.

Only parents fully appreciate the truth in the phrase "This hurts me more than it hurts you." But in fact, no one seemed more distressed by the Intel layoff than Robert Noyce.

As the reader has no doubt already realized, Bob approached business differently than other great entrepreneurs. It was his biggest weakness, but it is also what made him a great man. Moore saw Intel, as he had Fairchild, as a platform for him to realize the technological imperative that perpetually filled his imagination. For Grove, Intel was a vehicle by which he defined himself, by which he gained the world's respect for his extraordinary abilities, and something to be fiercely protected. But for Bob Noyce, Intel Corporation was an extension of his self, the manifestation of his will and imagination, the measure of his ethics and morality. People wanted to be Bob Noyce, and being an employee of Intel in its first five years was at least a partial way to do just that, to share in Bob's charisma and success. And they loved him for it—it was the reason they forgave him all of his obvious flaws.

It was not a one-way relationship. Noyce, as always, acutely aware of the world's—and particular Intel employees'—judgment of him, did his very best to never disappoint. It was vital to him to be loved.

To Bob Noyce, already reeling from his official divorce a few months before, the Intel layoffs—inevitable as they might have been— were a devastating breach of faith with the dedicated employees he was committed to protect. According to chairman Art Rock, who loved and admired Noyce as much as anyone, the 1974 layoffs seemed to break Bob's spirit.[14] The man who had always lived his life and career as a kind of game wasn't having fun anymore. The classic photo of the era shows the two founders at an emotional low point. In the background, Moore, in silhouette, sits on a window sill of a conference room with his arms folded and looking into the distance. In the foreground, sitting at a table covered with papers and empty water bottles, Noyce slumps in his chair, jaw set, glaring emptily as if preparing for yet more bad news.

Rock would later say that during this period Bob Noyce seriously considered merging Intel with another company. That, after all, would save the rest of the employees' jobs and Intel's extraordinary technology, even if it likely would mean the end of Intel as a company and a name. Rock: "[Bob] really and truly did not like to tell people their shortcomings, demote them, or especially ask them to leave. That aspect of being CEO just tore Bob apart."

Nothing captures more the depths of his despair at the moment than the fact that he actually approached Charlie Sporck about selling Intel to National Semiconductor. Only afterward did he tell Gordon about the contact. The ever sane Moore would later say, "It was not unusual for him to talk to people about things without consulting me first"—an amazing testament to the trust in their relationship. Moore agreed to meet with Charlie, but in the end saved Intel from probable oblivion by taking Noyce aside and suggesting "that I might like to try running Intel for a while."

Only one person in the world could have made that suggestion. A relieved Bob Noyce readily agreed. He was forty-eight years old, and he would never again run a commercial enterprise.

The next question was how to reorganize Intel's senior leadership without Noyce. Not surprisingly, one individual already had a plan: Andy Grove. As Leslie Berlin describes it, "Noyce and Moore asked Grove to join them for lunch at an out-of-the-way restaurant in Sunnyvale. 'I don't think I can spend so much time on Intel,' Noyce said to Grove. 'How can we get you ready for more responsibility?' Forever unflappable, Grove paused only for an instant before he answered, 'You can give me the job.'"[15]

Art Rock, who had already outlasted the usual tenure of a venture capitalist–turned–chairman, was happy to go back to his real career and let Noyce assume his role as Intel chairman. And in a move that would define Intel's future, Andy Grove was promoted to chief operating officer and given a seat on the board of directors. Now at last, for the first time in his career, Andy was in a position of top authority over the day-to-day operations of his company. Nominally he still reported to Noyce, but now his real boss was Gordon Moore, a man he still hugely respected who, better yet, usually agreed with Andy's decisions about how to run Intel.

One might think the combination of confusion and delay, layoffs, departures, downturns, mounting competition, and executive shuffle would have been more than enough to convince Intel to kill its microprocessor program, still less than four years old. And yet instead the company decided to double down and attack the microprocessor business with everything it had.

Why the change of heart? After all, just a year earlier, Intel wasn't even sure it wanted to *be* in the business, much less throw its corporate weight behind it, despite the promises being made by Ed Gelbach, McKenna, and Bill Davidow. Suddenly now it was prepared to bet the entire company on microprocessors.

There are several explanations. One is that with Andy now in charge of all day-to-day operations, microprocessors, which until now had been a competitor and a distraction to his work in memory, suddenly became his baby. And as always with Andy, once something was on his side of the table, he was prepared to fight to the death to make it a success.

A second reason is legitimacy. Rarely discussed in studies of entrepreneurial start-ups is just how lonely it can be out there with a revolutionary new product, no competition, and a market that doesn't seem to get what you are doing. You can try to hide in the echo chamber of your own team, telling yourselves that what you've got is really great, but eventually you have to go outside and deal with investors, analysts, reporters, and potential customers. And when all of them are skeptical, even dismissive, about your product or service, it becomes increasingly difficult to retain the supreme confidence you need to keep going. That's why many of the great entrepreneurs are arrogant and obsessive to the point of megalomania, even solipsism: they sometimes have to be to take their solitary vision and make it real.

All that changed in 1975. Suddenly there were a half-dozen major players elbowing their way into the microprocessor game. In the trade press, magazines stopped asking whether the microprocessor was a viable technology or an expensive dead end and now began speculating on who would win this important new market. And Hoff's half-filled, skeptical audiences now turned into rabid, standing-room-only crowds.

This led to the third reason for Intel's decision to make a full commitment to microprocessors: *competition.*

For more than four years, Intel—alone, as far as it knew—had worked on inventing the microprocessor at a time when the world thought such a product either unnecessary or of limited interest. Then when the company had actually built the 4004 and showed it to the world, it had been met with a collective yawn: it was too low-powered, too unreliable, too expensive, too limited in its applications to replace the status quo. Then came the 8080, the single-chip microprocessor that answered all of those concerns—and now the world was saying that there were too many competitors in the space, that Intel's chips weren't as good as Zilog's, couldn't be manufactured in the volume of Motorola's, weren't as cheap as MOS Tech's, couldn't stand a pricing war with Texas Instruments'.

More than any company in the business, Intel now understood that the currently received view about microprocessors, whatever it was *this* week, couldn't be trusted, much less serve as the basis for a business strategy. So far the most successful way to compete in the microprocessor business was to just keep one eye on Moore's Law and never for a moment stop building the hell out of the next generation of devices.

This was the Fairchild blood in Intel, only now mixed with disciplined management. Intel had learned from the last fifteen years. Had its competitors? Moreover, the company also knew that its assets were better than most of its competitors realized. For example, even with the loss of Faggin and Shima, Intel believed it still had the best, most experienced processor design team in the industry. In Gelbach, McKenna, and Davidow—as would soon be seen with Operation Crush—it certainly had the best marketing team. And though it had yet to recognize their role in a larger picture, at this point, Intel at least appreciated that its design tools were an important added value to its microprocessor customers. Last but not least, Intel had its reputation, beginning with being led by the industry's greatest leader, its most esteemed scientist, and increasingly, its most fearsome competitor. So even if the competition had certain specific advantages on paper, Intel could see that it had considerable general advantages in actual practice.

But the fourth, and probably most decisive reason was that *microprocessors were starting to work.* Even as the rest of the electronics world was sliding into the recession, microprocessor sales were picking up. And being the first in the market, Intel remained the biggest beneficiary of this turnaround—especially when Digital Equipment decided to adopt the 8080 for use in its minicomputers (thus completing the circle from the moment, seven years before, when Ted Hoff saw the future of chips in DEC's computer architecture). If DEC, the world's largest outside purchaser of chips for computers, wanted Intel 8080s, the rest of the world did too.

Orders began flooding in. Dave House, who had just joined Intel as vice president and general manager of the Microcomputer Group, would laugh and say, "I think we paid for [the 8080's] R&D in the first five months of shipments."[16]

That success bailed out what would have otherwise have been a troubling balance sheet for Intel. The company ended 1974 with $135 million in sales, double the year before. But even the annual report warned shareholders, "These results alone do not truly reflect the course of the year." The memory business was in free-fall.

As the bottom was falling out of the memory business (and the rest of electronics), the last ray of hope was microprocessors. Intel's afterthought technology was now the one business keeping the company alive. What had two years before seemed a questionable strategy was now a no-brainer. For all of the positive reasons for Intel to stay in the microprocessor business, ultimately it came down to a negative reason: the company had no choice.

In keeping with the company's operating philosophy, especially under Andy Grove, Intel arrogantly decided that it wasn't going to be enough to just stay in the game. Instead, it was going to *win*. It was going to crush the competition and become the number one player in the microprocessor industry. To do that, Intel was going to ignore the warnings of the law discovered by its own cofounder and jump two generations into the future. Intel was going to design a new microprocessor model so advanced that it would leave its competitors in the dust—so far behind they'd never be able to catch up.

And so, just at its moment of triumph, when it merely had to advance its current technology to consolidate its industry leadership, Intel embarked on the most disastrous new-product initiative in its history: the iAPX 432.

CHAPTER 25

The Knights of Moore's Law

It is easy to tell the story of Intel as a parade of milestone products and to focus on that most visible feature of the company's approach—its overwhelming technical competence. From the day of its founding right up to the present, Intel has always believed, with considerable evidence, that because semiconductors are the engines of technology and Intel is the world's leading semiconductor company, it is the most advanced and capable technology company on the planet. This self-image—competitors call it arrogance—has from the beginning given Intel a kind of warrior culture. Google employees, because of that company's rigorous hiring process, tell themselves that they are the smartest people in tech; Intel employees *know* that they themselves are.

What's more, as the ultimate protectors of the sacred flame of Moore's Law, Intel employees are also convinced that the fate of the electronics world—and ultimately of the extraordinary progress humanity has enjoyed over the last half century—is in their hands. Over time, this daily sense of duty to a greater cause has given Intel a corporate culture like no other. One school of modern business theory has argued that great companies—Apple being the one they usually point to—don't just pursue business goals, but embark on a crusade, one that features a moral imperative to change the world in some way. True enough, and Apple learned that from Intel, or more precisely, as we shall see, Steve Jobs learned it first from Robert Noyce and then from Andy Grove. And if Intel's passionate sense of a higher calling, of an innate superiority, is less noisy and visible than Apple's "Kool-Aid," it is also older and in many ways even more zealous.

From a 1984 history of Silicon Valley: "The image of a giant research team is important to understanding the corporate philosophy Intel developed for itself. On a research team, everybody is an equal, from the project director right down to the person who cleans the floors: each contributes his or her expertise toward achieving the final goal of a finished successful product. . . .

"Intel was in many ways a camp for bright young people with unlimited energy and limited perspective," he continued. "That's one of the reasons Intel recruited most of its new hires right out of college: they didn't want the kids polluted by corporate life. . . . There was also the belief, the infinite, heartrending belief most often found in young people, that the organization to which they've attached themselves is the greatest of its kind in the world; the conviction they are part of a team of like-minded souls pushing back the powers of darkness in the name of all mankind."[1]

That's why Intel never became a country club and family fun center like Hewlett-Packard or Yahoo. The fun times in the semiconductor industry ended early, and like reformed libertines everywhere, the survivors responded by going just as far the other way into pious rectitude. Semiconductors were now a serious business. That's why Wilf Corrigan ran his reign of terror at Fairchild, Charlie Sporck had his employees laboring in crowded buildings under flickering yellow sodium arc lights to save money, and at Intel, a mistake might mean demotion, a shunning by your engineering peers, and certainly a lot of yelling. And despite that, the only people quitting were the ones intent on starting their own, equally hard-charging companies.

It is this steely conviction of its own superiority that has always made Intel such a dangerous competitor. The company simply cannot countenance the notion of *ever* falling technically behind any other company, be it Texas Instruments in the early 1970s or Samsung in the 2010s. It must be *better* than everyone else, or it stops being Intel. Competitors aren't just opponents; they are unworthy usurpers of Intel's crown. And on the rare occasion when Intel does fall behind the competition, as it was about to do in microprocessors, as it would do against the Japanese in the 1980s and in mobile electronics at the beginning of the new century, it is a time for raising your bets, for superhuman effort and endless hours, for ruthless self-criticism and purges, and when the company is at its worst, for equally ruthless business practices.

The most important and least celebrated manifestation of this doubling-down gambler mentality can be found in Intel's response to the quadrennial hard times that cyclically sweep across the semiconductor industry. Most of these downturns are small, but some, like one in 1974, are deep and devastating and are the great company-killing counterpoints to the crazy, company-creating bubbles of the boom times.

The natural, pragmatic, sound business-management response to these downturns is to hunker down, fire unnecessary employees, cut new projects out or down to the bone, hang on to your cash, and ride it out in hopes you can survive long enough to reach the distant shore of the next industry upturn. That's what companies in Silicon Valley have always done—with one stunning exception: Intel Corporation.

It comes down to a collision between the business realities of an industry downturn and the technological imperatives of Moore's Law. As noted, the business cycle in semiconductors turns about every four years, while Moore's Law clicks over every two or three years. That means, inevitably, the two are going to regularly overlap. When that happens, every company in high tech, but especially in the chip business, has to make a choice: will it tighten its belt and save itself or will it make the investments necessary to keep up with the law and thus remain competitive—but risk dying in the process?

Most companies choose the former. Intel *always* chooses the latter, even when it poses an obvious threat to the company's future. In practice that means that even as sales slump and profit margins evaporate, Intel continues to maintain a stratospheric level—often well over 10 percent—of research-and-development investment. This level of investment may hurt the company's stock price and stagger overworked company employees, but it has also consistently meant that once the recession ends, Intel comes roaring out of the blocks with new technologies and products and leaves its more careful competitors in the dust.

This unstated priority of Moore's Law over economic law also appears in Intel's recruiting and hiring. All but unnoticed to the outside world, this philosophy has proven even more important than technological innovation to Intel's long-term health.

The best example of this took place during these difficult years of the mid-1970s. It was during this period, in the midst of Intel's devastating layoffs, that the company made two hirings that would ultimately define Intel's top leadership into the second decade of the next century.

The first of these was Craig Barrett. Barrett, at thirty-five an old man by Intel recruiting standards, had been hired from Stanford University after a long and distinguished career in academia. He had earned his PhD in materials science at Stanford, was a Fulbright Fellow at the University of Denmark, then for two years was a NATO postdoctoral fellow in the United Kingdom, before returning to join the Stanford faculty in the department of materials science and engineering. There he compiled a notable academic record, publishing more than forty technical papers dealing with the influence of microstructure on the properties of materials, as well as publishing a classic textbook in the field.

Barrett remained at Stanford until 1974—long enough to seem a permanent fixture on the faculty—when he surprised everyone by jumping to Intel. Not only was it almost inconceivable that a tenured professor would choose to leave one of America's greatest universities; it also seemed the worst possible timing: Barrett arrived as most of the people around him were being fired—not exactly the warmest welcome for a new company manager.

But Barrett was made of tough stuff—in time, some would say too tough. Born in the southern San Francisco suburb of San Carlos—"I only traveled seven miles to go to college at Stanford"—he was, with Moore, Hewlett, and a few others, among the small group of early Valley pioneers who were actually born in the Bay Area.[2] And like those other men, Barrett had that classic Old Valley persona: a brilliant scientist, a hard-nosed businessman, and a passionate outdoorsman. For Moore it was fishing; for Barrett, who would one day own one of the world's most celebrated lodges (Montana's Triple Creek Ranch), it was hunting deer, elk, bighorn sheep, and mountain goats. He seemed happiest either in a college laboratory puzzling over a new manufacturing process or hiking the Rockies with a rifle on his shoulder.

Neither prepared him for what he encountered on his arrival at Intel. Ambition had led to him to jump to the commercial world at such an advanced age. Now he had to question the sanity of his decision as he filled his desk at Intel while those around him emptied theirs. On the other hand, Barrett's maturity at least enabled him to put his job in perspective. He knew that his career would be spent "in the shadow of the guys who preceded me,"[3] but he was prepared to accept that to be on the cutting edge of the commercial world of semiconductors. Little did he imagine where he'd be when he stepped out of that shadow.

The second hire in that recession year was even more far-reaching. Paul Otellini was much younger than Craig Barrett—just twenty-four years old—and other than a college summer job as a sales clerk at "a schlocky men's clothing store," Intel was his first real job.[4] Indeed, Intel would be the only employer he would ever have.

He hadn't planned it that way. In fact, Otellini—another Bay Area native, born in San Francisco—had just earned his MBA from Cal Berkeley and was hoping to find any kind of job in the area during the recession. Within days, and to his considerable relief, he had landed a job at Intel.

He arrived at the Santa Clara headquarters on a Monday morning in July 1974, knowing that on the Friday before, Intel's stock had lost a third of its value. Even worse, he walked into the building just minutes after the company announced its massive layoff. "My desk was still warm from the guy who had just vacated it," he recalled. "It was ugly."

That kind of experience might have turned off the young man forever on his new employer, but with Paul it did just the opposite. "At the same time, I was impressed that Intel was keeping its commitment to the college kids it had already made offers to. It would have been much simpler to just tell us not to show up." A few days later, he found himself at a brown-bag lunch with Robert Noyce and Gordon Moore.[5]

Otellini stayed for the next thirty-nine years, retiring as Intel CEO.

Meanwhile, not everyone walking out of Intel during this period was carrying a pink slip. There were a few people who managed to time the brief window between the vesting of their Intel stock after the IPO and Intel's off-the-cliff stock drop in the recession to cash out a small fortune and make their exit.

Easily the most notable of these early departures was Armas Clifford "Mike" Markkula Jr. Markkula was a Southern California boy, born in L.A. and a graduate of USC. Like many other early Intel employees, he was a former Fairchilder, though he had joined the company later than its more famous pioneers, and was still just a midlevel market-ing manager when he left to join Intel. He was hired by Bob Graham as product marketing manager for memory chips. And though in ret-rospect reporters and biographers would try to embellish his achieve-ments at Intel, in truth Markkula's time there left barely a ripple. Grove could only say of Mike that he "was always very nice to me, but I didn't have much use for him."

But if he was only a mediocre marketing manager at this point in his career, Mike Markkula was a brilliant investment manager. From the moment he was eligible, Markkula assiduously purchased every In-tel stock option he could get his hands on, and by the time of the IPO, he was in a position to convert those shares into more than $1 million—a considerable fortune in the early 1970s.

It is a Silicon Valley truism that IPOs change companies forever, no matter how much they try to stay the same. The same is true of employ-ees who get rich on those IPOs. No matter how loudly they proclaim their loyalty to the company, all that money makes them quick to walk away over the smallest slight or career speed bump.

That is exactly what happened with Mike Markkula. When Jack Carsten got a promotion Mike thought he deserved, the increasingly restive Markkula announced his retirement. He was just thirty-three years old.

Many Valleyites dream of striking it rich while they are still young and then walking away. But few stay away for long. What are they go-ing to do for the next half century? Being part of a Valley start-up is likely to be the most exciting and challenging thing they will ever do in their lives—and its siren call lures them back.

It wasn't long before Mike Markkula was looking for something new. His old Fairchild boss, Don Valentine, now a powerful venture capitalist at Sequoia Ventures, told him to check out a new little com-pany in Cupertino, into which Valentine had already put some of his own money. Markkula also approached Regis McKenna, with whom he had worked at Intel as a sort of Mr. Inside Marketing to Regis's Mr. Out-side Marketing. Regis had a suggestion: one of his neighbors—a couple, the Wozniaks, who had helped on Dianne McKenna's successful recent campaign for Sunnyvale City Council—had a son who was something of a computer genius. That son had teamed up with a couple of other young men to start a company to build low-cost "personal" computers. Mr. Wozniak was more than a little concerned about the manipulative personality of one of the boys, a kid named Steve Jobs, but the computer that young Steve Wozniak had built was reportedly quite amazing. Why not, Regis suggested, take a look?[6]

Both recommendations were for the same company, with the odd name of Apple Computer.

Markkula made an appointment to see the young founders in the Jobs garage. What he saw electrified him. The Apple I, he would say later, "was what I had wanted since I left high school"—even better, he admitted, than the Corvette he had bought with some of his Intel stock money. He wrote a check for $250,000—the money Apple needed to launch its new computer and get into production.

It is often overlooked in the popular history of Apple Computer—especially among young people who believe that Steve Jobs created the company solo and designed all of its famous products—that in the early years the company was almost always a trio, like Intel. At the beginning it was Wozniak, Jobs, and Bill Fernandez, another neighbor and the son of a judge. In those years, one could see combinations of two or three of this trio regularly visiting the neighborhood hobby shop and the local electronics stores for parts to build the Apple I.

The arrival of Markkula made him the new "third man" at Apple. Steve Wozniak would say that he thought Mike's contribution to Apple was even greater than his own—an extraordinary comment, given that Wozniak invented Apple's earthshaking first two products. But Wozniak understood that without Markkula's business experience, his network of Valley contacts, and frankly, just the fact that he was a responsible adult, turned Apple from a garage-based maker of semicustom hobby machines to a world-class manufacturer of the most influential consumer electronics products in history. Well into the 1980s, Markkula, almost always remaining in the shadows, mentored Jobs, tempered his extremes, quietly made peace with his victims, justified the young man's antics by noting that he was just a kid, albeit a brilliant one, and all in all made Apple seem like a mature, disciplined, and competent company long before it was one. Probably in large part due to his time at Intel, Mike Markkula played the role of Andy Grove at Apple, but with the calm, easy style of Bob Noyce.

With Markkula at Apple and with Regis adding Apple as his agency's hottest new client, the stage was now set for a production that never opened. Everything was in place for Apple to adopt the Intel 8080 and its descendants as the heart of the Apple I, II, and beyond. We know that's what Steve Wozniak wanted, and that combination would have changed everything. But Steve Jobs's betrayal of his best friend closed that door. For the next few years, Apple would be stuck with an underpowered central processor in its computers that not even Wozniak's prodigious design talent could overcome. And Intel would have to wait five years for the next great opportunity in personal computing, IBM, and would need Operation Crush to land Big Blue.

CHAPTER 26

(Over) Ambitions

In 1975, the electronics industry limped out of the recession of the previous year and began a slow march to recovery. Intel, however, because it had maintained its investment level despite its miserable financials for the year, the loss of 80 percent of its stock value, and its massive layoff, raced out into the upturn faster than almost all of its competitors. Indeed, in early 1976, before the outlines of the next boom were even clear, Intel upped its R&D investment.

The pattern would continue through one business cycle after another for the next thirty years. Intel always seemed to emerge from hard times stronger than any other company; it always seemed to have new products ready to introduce, and it consistently snatched away chunks of market share before everyone else emerged from their defensive postures. And Wall Street never failed to recognize, and reward, this gutsy behavior. Thus by the end of 1976, Intel's stock had climbed back from $21 per share at the beginning of the year to $88 by year's end—higher than it had been during the last boom.

Intel's financials were equally stunning: sales for the year were $226 million, up 65 percent from the year before, while earnings, at $25.2 million, were up 55 percent. But the most satisfying news was in employment. Before the crushing layoff, Intel had employed about 3,150 people. By the end of 1976, Intel employed about 7,350 people, thus erasing the labor losses of the recession and more than doubling the total. Many of those new hires were returning old hires, thus reinforcing the view that Intel never forgot the loyal members of its "family."

But Intel's turnaround was more than just its preparation for better times, more than the 8080 or the company's memory products, and more than just the rising tide of the economy. It was also Andy Grove, now in the pilot's seat at Intel. Good times made Andy angry, because Intel could never take as much advantage of the opportunity as he wanted it to. But bad times made him furious, because it was as if God himself were trying to block Intel's path to triumph. But Andy Grove always rose to the occasion, because if booms were like open-field running, busts were the equivalent of the hand-to-hand fighting and the blocking and tackling that always brought out his best.

During a recession, Andy didn't have to be patient, he didn't have to explain or justify; he just moved—decisively, quickly, and sometimes ruthlessly. He knew the old Valley saw that in good times everyone is a genius, while it is in bad times that real character and talent show through. Andy understood better than anyone in high tech that the best time to win was when your competitors were too afraid to match the risks you were willing to take.

But to take those kinds of risks, to keep your sails unfurled during the worst storms and then to run before the wind at breakneck speed once those skies cleared, required eternal vigilance. And that's why, when good times returned, as they did in 1976, Grove always looked like the unhappiest guy at the firm. That attitude was well captured in a memo he sent to Moore after the latter returned from a vacation in June 1976:

> Welcome home! You absolutely, literally, positively could not have chosen a better/(worse) week (depending on point of view) to be gone. Aside from the fact that May fell $1m short, Univac put us on total shipment hold, the ITT inspection team angrily walked out of the plant, one of our ex-employees was shot and killed in a robbery, the 19 year-old son of another one committed suicide, Helen Hoover had a heart attack, [Micromaʼs] Dick Boucher and I almost came to blows over the board presentation and various aspects of the plan, what was left of the week was pretty routine.[1]

Biographer Tedlow found a note from after another such Moore vacation in which Grove wrote simply, "It seems to me that Intel should contract with you not to go on vacation—the world invariably seems to turn into a piece of shit while you're away."

Note that this is Andy Grove writing to his *boss*. The reader can imagine what it was like to be a *subordinate* and earn Andy's ire. Ted Jenkins, a former Fairchilder and Intel employee number 22, who spent a number of years reporting to Grove (and being Paul Otellini's boss) recalled one vivid exchange with Andy: "I remember having to do a monthly progress report when I worked under Andy. I used the word 'corroborate' and he sent me a note, saying there's no such word. 'You mean "collaborate,"' he wrote. I responded with my own note and told him, ' "Corroborate" is a legitimate word.'

"He sent back one final note that said, ' "Bastard" is a legitimate word, too.' "[2]

But it worked. It always worked for Andrew Grove. And if his wrath was terrifying, it at least had the value of being fair by being universal—everyone at Intel eventually endured Andy's anger. In his old age, Grove would say that while he may have risen to this anger quickly, he also cooled down just as quickly—and that he never tried to be personal or cruel to the people who worked for him. As for competitors—and anyone, from industry analysts to reporters, who doubted that Intel would triumph—that was a different story.

In early 1976, as Intel returned to financial health and began to accelerate back to prosperity and industry leadership, Andy showed as much optimism in another note to Moore as he could probably ever muster: "Welcome back! This time I think we are going to depart from the pattern of a disaster letter greeting you. With the stock flirting with the $100 level, how could anything be wrong? In fact, things are in reasonable shape even beyond that."[3]

But it wasn't long before Andy had a whole new reason to revert to his usual impatience and anger.

The cause was the iAPX 432, the new superchip microprocessor. It wasn't designed to be just the follow-up to the Intel 8080; it was supposed to leave the rest of the industry in the dust with a technological leap so great that it would take years—even as much as a decade—for any other company to catch up. It was as if Intel were so annoyed that it had lost its sole ownership of the processor business that it was going to pull out all the stops to gain that position again.

The iAPX 432 wasn't planned to be just a breakthrough new 16-bit microprocessor hardware architecture; it was also to be a revolutionary new *system* architecture—which meant that it would also featuring radically new operating software. Either one would have been a massive undertaking. Together, they were guaranteed to be a major strain on the entire company—and would have almost no margin for error. One misstep and the iAPX 432 would not be ready by the time Moore's Law said the 8080 would need to be replaced.

In later years, Gordon Moore would say that one of the least understood lessons of his law was that the only thing more dangerous than being behind the law is being ahead of it.[4] If you fell behind the pace of technological innovation, you might still be able to make a nice little business in a niche market or find downscale applications for obsolete and commoditized chips. Thus licensees still sell millions of those first-generation 8080s, 6502s, and 6800s, no longer for computers and sophisticated digital instruments, but for cameras, toys, and cheap consumer products.

On the other hand, to try to leapfrog the competition by getting ahead of Moore's Law is always hugely expensive (costly equipment, low yield) and almost always disastrous. You either pulled it off against very long odds, or you were wiped out. There have been almost no examples of the former (Apple came close when it replaced the tiny disk drive in the iPod with solid-state flash memory chips), but there is a long list of the latter.

Perhaps the most famous—or notorious—example of a company with the hubris to try to get ahead of the law was Trilogy Systems.[5] It had the right pedigree. It was founded by the legendary computer scientist Gene Amdahl, who at IBM had played a crucial role in the design of the Model 360, the most successful computer in history. He then left to found Amdahl Corporation, located just blocks from Intel's headquarters, to build a brilliant new "plug-compatible" mainframe computer that could replace the main processor unit of the IBM 370 and still run all of its peripheral hardware and software programs. It was a roaring success.

By now the gentlemanly Amdahl, who spoke so fast that he sometimes seemed a computer himself, was generally considered to be the finest computer scientist in the world. And so, with Trilogy, he embarked in August 1979 on a project commensurate with that reputation. Raising $200 million (an astonishingly large investment for the era and a measure of the value of his name), Amdahl founded Trilogy with the express goal of designing and building processor chips so powerful and highly integrated—that is, so far ahead of where Moore's Law said chips should be at the time—that they would make possible a new generation of mainframe computers far cheaper than anything IBM had in the works.

With his pile of money, Amdahl set up a factory in Cupertino, just around the corner from where HP had built its first calculators, and set to work building prototypes of his new chips.

It was a heroic, and mad, quest. And the resulting chips were small miracles of design: huge integrated circuits almost two inches on a side, with a metal cylinder heat sink rising like a tower more than an inch high off its surface. This tower, unique to the Trilogy chip and symbolic of the whole project, was easily the most distinctive and unforgettable design feature of any chip before or since. It made the Trilogy chip appear so forward-looking and sophisticated that one could easily imagine it being the building block of the next great era of computing.

But as in a Greek tragedy, the gods seemed to punish Gene Amdahl's hubris in ignoring Moore's Law by unleashing the punishment of Nemesis. When the Trilogy chips were turned on, the cooling stack proved to be inadequate, and the sheer density of all of those transistors—devoid of the design breakthroughs that would come over the next decade—produced so much heat (equivalent to a 900-watt bulb) that entire layers on the surface of the silicon burned up. The only way to make those monster chips work in the real world would be to bathe them in some sort of cooling liquid—a nest of expensive plumbing that would have undermined their very purpose as the cornerstone of small, low-cost mainframe computers.[6]

It wasn't long before Trilogy was gone, its building sitting empty for years alongside Highway 280, an enduring warning to the thousands of Silicon Valley computer workers passing by each day.

The Trilogy story has never really been forgotten in the chip business; it has become part of the Silicon Valley mythos as a warning against technological arrogance. Unfortunately, that object lesson wasn't available two years before when Intel embarked on the iAPX 432. Still, one might think that, given the presence of the lawgiver himself as company CEO, someone might have pointed out that the company was going a bridge (or two) too far.

But no one did—and so, just at the moment when it was coming out of the recession with a full head of steam and a jump on the competition, with the technical overconfidence that would always be its Achilles' heel, Intel decided to chase a pipe dream almost as impossible as Gene Amdahl's.

Recalled Justin Rattner, leader of the 432 team, "At the time, most Intel microprocessors were going into things like gas pumps and traffic lights. With the iAPX, we were aiming to replace minicomputers."[7]

Intel was right: it would be at least a decade before anyone would be able to build a chip like the iAPX 432. Unfortunately, that list included Intel itself. And a single Intel chip would indeed replace the minicomputer, but that chip was still four generations and twenty years away.

The iAPX 432 project began in late 1974 and was still unfinished five years later. The overall design of the chip was a good one, but the state of Intel's silicon fabrication technology—and at the moment, it was as advanced as that of any company in the world—was such that the chip was painfully slow; it just wasn't capable of working at the high frequencies needed to make it run at its full potential. The project dragged on with no end in sight.

Many tech companies have died trying to complete an ambitious and expensive new product design. Others have struggled for years, until the world finally gives up, and then managed to produce a camel of a product, not a thoroughbred. (The computer game Duke Nukem is a classic example: gamers waited an entire generation for the follow-up version of this best seller, only to get a mediocre offering.) Intel, in one of its shrewdest moves—the decision bears the fingerprints of Andy Grove—decided not to put all of its digital eggs in one basket. Even as the iAPX 432 was getting under way, the company also embarked on a much smaller side project to build an improved version of the 8080 as an insurance policy.

This other chip, designated the 8085, was basically a more integrated (and thus smaller and faster) 8080. Also, taking a cue from the success in computers of the Motorola 6800, the 8085 ran on 5 volts. It wasn't a great microprocessor, certainly not compared to Zilog's new 8-bit screamer, the Z80, but it was good enough to keep Intel in the front ranks. And with Grove at the whip hand, the 8085 was introduced to the market just a year later, almost record time for a new product of its type.

But what about the impending next generation of 16-bit processors? The iAPX 432 was supposed to own that new market, but it was already slipping so badly that it had to be redesigned for the 32-bit market that was expected to appear in the early 1980s. Meanwhile, the trade press was reporting rumors that Motorola, TI, Zilog, and Nat Semi already had 16-bit processor development initiatives under way. And Intel had nothing to compete against them.

Though it wasn't yet an accepted truth in the chip business, it was already becoming pretty apparent that competition in the microprocessor industry would be a no-miss proposition—that is, if you missed a single generational jump in getting your new device out the door, you would likely be out of the race *forever*. You might be able to play in the lower-grade, commodity backwash of the microprocessor business, or you might take your design and convert it into a microcontroller to manage, say, an automobile fuel injection system or a home thermostat. But your days at the cutting edge of microprocessors, where the big fame and even bigger profits could be found, were probably over.

Once again, in this crisis, Intel showed a trait shared by few other companies in high tech and almost none in the semiconductor industry: it recognized its mistake quickly, admitted it, and then made an almost superhuman commitment to rectify it. This was the crucial middle link between the company's technical arrogance (which led it to take great risks) and Andy Grove's obsession with getting the stumbling company on its feet and back into the race.

Once again, this was an example of Intel making and recovering from more mistakes than any of its competitors. And the company had just begun.

By the time the 8085 was introduced, in November 1975, it was already obvious that the iAPX 432 wouldn't be ready in time for the 16-bit era. Intel—that is, Grove, with the sign-off of Moore and the approval of Noyce and the board—decided to divide his forces once again and embark on the creation of *two* microprocessor families.

In the memory/microprocessor schism, the challenge had been to find enough money for the little company to pay for two major product lines. Now the challenge wasn't so much money (although paying for memory and *two* major microprocessor development programs wouldn't be cheap, even in good times) as assembling enough top-notch talent inside the company's microprocessor department and through new hires to produce competitive state-of-the-art designs.

The company slammed into this problem of a dearth of talent almost from the first day. Jean-Claude Cornet, the company's engineering director for microprocessors, who had been assigned to lead this new team, found himself scouring the company for anyone who had the necessary technical skills or could learn them quickly. In the end, he found twenty people—an unprecedentedly large number for the era, underscoring the shortage of wide-gauge talent in the technology. Some of these team members came from the memory side of Intel; some had never even used a microprocessor, much less designed one.

Somehow, it all worked. As Cornet would later say, "What is remarkable is that these people had no more than a year's experience, and yet they brought a very complex product to market in less than two years."[8]

That was not quite accurate; still, this motley pickup team managed in just twenty-six months to come up with a fundamentally new microprocessor design that was so adaptable and stable that it would define the whole world of computing for the next fifteen years. It still plays an important role today.

How did they do it? It began with Cornet himself. He was faced with the stark reality that he had been assigned to come up with a brand-new product that would take on some of the most sophisticated and competitive companies on the planet and very likely save Intel in the process, or at least buy a few years for the hapless iAPX 432 to finally get to market, and he had to do so with limited resources and an inexperienced, untested team. To his credit, Cornet wasn't scared off from this seemingly impossible task. Instead, he shrewdly turned weakness into virtue—and changed the entire semiconductor industry in the process.

To appreciate Cornet's achievement, it is important to understand the context of semiconductor fabrication during that era. Though the chip business had progressed well beyond the days of lab-bench, open-window chip making, it was still far from the modern hypersterile, computer-controlled wafer fab factory.

By the same token, the *design* of chips in the late 1970s was still comparatively rudimentary. Computers were now used in some of the design work, but there were no real computer-aided design tools yet for wafer mask design. The process had progressed beyond using pens and rulers on a big sheet of paper that would be photographed and turned into a transparency, but not by much. In practice, because the design teams typically remained small, the process took longer and longer as the chips grew more complex. The new chip was planned to have many times more lines and features than the 8080, posing a serious threat to the device's development timeline. There seemed no way, using traditional techniques, to get the new chip out the door before the end of the decade—and that would likely put Intel behind the competition, perhaps even out of the race.

From desperation came innovation. Cornet knew two things. First, most of his team members didn't have the skills to take on the design challenge as a whole. Second, they still had to do their work within the constraints of the overall design: wires and channels in one part of the overall geography of the chip had to connect, with an economy of space, with similar features in other regions of the chip.

In response, Cornet came up with a celebrated solution. First, years before the notion of "concurrent" engineering had spread beyond a few large manufacturing industries, such as commercial airplanes, Cornet and his top people carefully plotted out how to take the plan for the new chip—it was designated the Intel Model 8086—and break it up into multiple design strategies that could be addressed in *parallel*. This not only had the advantage of giving the designers pieces that they could actually manage even with limited experience in the field, but it divided the total time needed for the project. With this plan in place, Cornet and staff could then plot out a very precise critical path, with specific milestones, so that each team—here's where the larger group size came into play—could go off and work almost independently and yet still finish at the same moment with all of the proper interconnections in their proper locations. It was a first glimpse of the future of all semiconductor design.[9]

Cornet still had the problem of keeping these individual teams working with one eye on the larger prize: the finished, complete model 8086. In an era when even the biggest computer terminal displays were no larger than picture-tube television screens, Cornet once again went for a novel solution. He had a giant sheet of paper created—twenty feet by twenty-five feet, the size of a large suburban living room—on which each team could draw or post small design sheets on its quadrant. Now the teams could chart the progress of their own work, align features, and also, in a spirit of competitiveness, compare their work with the others'.

Cornet's strategy worked superbly. In little more than a year after this daunting "side" project had begun, the massive paper sheet was cut up and its design entered into a computer. From there it was multiplied for the total number of chips per silicon wafer, printed on Mylar transparencies for each layer of circuitry, photographically reduced to the proper size (wafers were only about four inches in diameter in those days; today they are three times that and thus can hold nine times as many chips of the same size), and then sent to fab to be "printed" on wafers.

In June 1978, Intel shocked the chip world (which had been tracking with delight the problems with the iAPX432 and barely noticed the other Intel processor project) with the announcement of the model 8086 microprocessor. Not only was it a 16-bit device, but it was also *ten* times more powerful than the 8080. In one fell swoop, Intel had gone from being a fading and troubled competitor in the microprocessor business to being universally recognized as the new industry leader.[10]

It was a turning point—maybe *the* turning point—in Intel's story. With the 4004, Intel had created the microprocessor industry. Soon, as a furious Motorola recovered and came roaring back with its 6800, Intel would have to redefine and rediscover itself with Operation Crush to regain its leadership in that industry. But with the 8086, Intel for the first time *owned* the microprocessor industry and from now on it would fight to the death to keep that ownership.

There was more. Without the 8086—and in particular its

lower-powered, lower-cost variant, the 16/8-bit 8088—Intel would not have had a platform to mount the counterassault of Crush and capture IBM's imagination. And absolutely not least, with the 8086 Intel at last had an architecture it could build upon instead of having to start over with each generation.

It was in recognition of this upward compatibility that Intel named the next three generations of its microprocessors the 80286, 80386, and 80486—the umbrella term being the x86 family. And even after that, when competition made those designations difficult, it was still recognized that subsequent chips (such as the Pentium) were still part of the x86 family. And there was good reason for that. Officially for the next three generations—and in many respects the next ten generations of Intel (and other) microprocessors right up to the present—Intel could build upon the architectural heart and instruction set of the 8086.

It was a powerful, nearly unassailable position to be in, especially as the IBM PC and its many clones—and Microsoft with its industry-standard Windows operating system—were designed to be uniquely compatible with x86 processors. This meant that if Intel could just maintain the pace of Moore's Law, a multimillion- and soon multi-billion-dollar market would always be waiting for it. Even if IBM fell behind its clone competition, those competitors would *still* clamor for Intel's newest x86 chip.

Only Apple was an outlier, but if that was a concern in the late 1970s, when that company had a 90-plus percent market share, it was far less in the years thereafter, as Apple computers embarked on their long, slow slide to single-digit market presence. Even the introduction of the Macintosh and the departure of the increasingly destructive Steve Jobs did little to stop the slide. And every percentage point Apple lost to this Microsoft Windows/Intel x86/IBM-compatible leviathan meant the sale of even more Intel chips and further consolidated its dominance over the microprocessor world.

CHAPTER 27

Beatification

I ntel entered the 1970s as a young start-up dedicated to taking leadership of the MOS memory business. Total annual revenues in 1970 were $4.2 million, losses were $970,000 million, and the company had 200 employees, nearly all of them at Intel's rented offices in Mountain View.[1] By the end of the decade, Intel was still a major player in MOS memory, but prices of those devices were plummeting, largely due to a burst of new domestic competitors and a growing threat from abroad. However, the company was now the leading innovator and manufacturer in the hottest new industry in electronics, microprocessors. Though Motorola was coming on fast, in the last days of 1979, the Crush team would find a novel way to defeat that assault.

Not surprisingly, there were a growing number of employees inside and industry watchers outside the company who were beginning to ask why Intel was still in the memory-chip business at all.

Meanwhile, Intel had grown into a global company, with five factories in the United States and Singapore, and eighty-seven sales offices in seventeen countries, and it was now about to open a huge new fabrication factory near Portland, Oregon. The company's annual revenues in 1979 were $633 million, profits were $78 million, and it had 14,300 employees, a majority of them working at Intel's big new headquarters campus in Santa Clara. It had, against all odds, carried the torch of Moore's Law and had been rewarded with a growth over the decade that almost tracked the law's exponential curve.

During this period, as Don Hoefler's newsletter had faded, another, more analytical, newsletter had risen to take its place. It was the product of an already fabled figure, Benjamin Rosen, an elegantly spoken New Orleans boy who had most recently risen to the vice presidency of Morgan Stanley. Rosen was so intrigued by the tech revolution that he retired and started his newsletter (and later a hugely influential industry conference), which gave him access to the boardrooms of the Valley.

Rosen's newsletter, for its brief run, was probably the most insightful publication in Silicon Valley history—not least because he was among the first analysts to fully understand the implications of Moore's Law, the microprocessor revolution, and Operation Crush. And because of this understanding, he was the first to describe Intel as "the most important company in the world," a tagline that, because it proved increasingly accurate, stuck ever after. By 1980, Rosen was gone, to create the venture capital firm Sevin Rosen Funds, which would help found Compaq, Cypress Semiconductor, Lotus, Electronic Arts, and Silicon Graphics—some of them customers and others competitors of the company he had placed at the apex of world commerce.[2]

Meanwhile, Intel only seemed stronger than ever. It had begun the decade still in the shadow of Fairchild and largely acting in reaction to that other company. Art Rock had been chairman of the board; Bob Noyce, chief executive officer; Gordon Moore, in charge of research; and Andy Grove, a senior manager. Now Rock was gone, back to his venture-capital career. Noyce, battered by his experience keeping the company alive during its lowest point, had promoted himself off the battle line into the chairman's seat. Moore had taken his place as CEO, and Andy Grove was running the day-to-day operations of the company. It seemed a combination that would work.

But already three forces were at work that would once again change everything. In April 1979, Bob Noyce stepped down as Intel's chairman of the board. Life had been pulling him away from the company in several directions. In 1974 he had begun seeing Intel's director of personnel, Ann Bowers, and that November they were married. Given her job responsibilities, Bowers understood better than anyone the difficulty of her position—and when a position opened up as the first director of human relations at Apple Computer, she jumped to the new company. With both Noyces deeply involved at the highest levels in the electronics industry, they quickly became one of the area's power couples, in high demand at industry functions and by local nonprofits.

During this time, Noyce was also exhibiting signs of deep depression. Not only had the previous decade at Intel taken its toll on his emotional state, but even more so his divorce and its impact on his children. Two of his children had gotten involved with drugs, one was diagnosed with bipolar disorder and hospitalized, and another, one of his daughters, was hit by a car and was in a coma for six months. On the outside, Bob dealt with these events with his usual denial, but inside there was no denying that something had gone terribly wrong with his family and that he was part of the cause. As reported by his biographer Leslie Berlin, at a dinner for one of the recipients of an investment by Noyce's angel fund, Bob unburdened himself in a way he rarely allowed himself:

"After the dishes had been cleared and the children sent to bed, Noyce listened as the company founder explained that someday, if the business did well, he would like to move his family into a bigger, nicer house. Noyce looked up at him and said very quietly, 'You got a nice family. I screwed up mine. Just stay where you are.' Twenty-five years and a successful company later, the entrepreneur has not moved."[3]

Meanwhile, the world wanted a piece of the legendary inventor/entrepreneur, not least the electronics industry. As early as 1976, Regis McKenna had landed Bob on the cover of *Business Week,* playing chess against Charlie Sporck (the latter on the fold-in leaf) under the headline "New Leaders in Semiconductors—Intel's Robert N. Noyce, Masterminding a Radical Change in Technology." In the years that followed, he would be the subject of feature stories in all of the other major business publications (*Forbes, Fortune,* the *Economist,* and the *Wall Street Journal*), the great newsmagazines and newspapers (*Time, Newsweek,* the *New York Times*) and even, in a special edition devoted to Silicon Valley, *National Geographic.*

In 1979, even as Noyce was pulling away from Intel, Herb Caen, the legendary columnist for the *San Francisco Chronicle* and dean of Bay Area journalism, suddenly discovered him, admitting that he "had let another year go by without learning what a semiconductor is." His readers quickly fixed that, not just by offering definitions for the layman, but even more, suggesting that Caen learn about Bob Noyce. "Several people hasten to tell me about Robert Noyce of Los Altos, who not only pioneered the blamed thing [that is, the integrated circuit], he founded Fairchild Semiconductor and Intel Corp. ($600 million in ten years) and, to boot, is a pilot and champion skier. Not only that! He has just become one of only 130 people in US history to receive the National Science Medal. . . ."

"Sure," Caen added, as if a figure like Bob Noyce was too good to be true.[4]

In that same year, the hometown paper of Silicon Valley, the *San Jose Mercury-News,* finally recognized the magnitude of the business story on its beat and added the world's first daily high-tech reporter. He quickly ran off to cover both Noyce and Intel and explained for the first time to the general public the larger implications of Moore's Law. Less than two years later, as tech advertising began to pour into "the Merc" as a result of this new attention, the paper added a separate Monday business section. It soon was so profitable that it was not only making the Merc one of the nation's wealthiest newspapers, but also propping up the rest of Knight-Ridder's struggling papers. And scarcely a week went by when the paper wasn't covering news from or writing features about Intel and its founders—and syndicating those stories around the world.

This wave of attention and approbation crested in 1983 with the *Esquire* Noyce profile. Tom Wolfe had been out to the Bay Area on another story fifteen years before—a look at Ken Kesey and the protohippies living in the mountains above Palo Alto—and the resulting book, *The Electric Kool-Aid Acid Test,* had helped to make his reputation.

Now Wolfe came to the Valley to catch up on the other revolution that had taken place there in the 1960s—and the single most important person in that cultural transformation. "The Tinkerings of Robert Noyce: How the Sun Rose on Silicon Valley" would be Wolfe's last piece of magazine reporting for years. A decade later, he would write some of his most famous essays (notably "Sorry, Your Soul Just Died") for Valley-based *Forbes ASAP* magazine. When asked by the editor of that magazine if he planned to also expand the Noyce story into a book, Wolfe replied that, no, the technology of high tech, especially semiconductors, was just too much for him to learn at that point in his life and that he would leave that story to some future tech journalist. Before long, he was off to Atlanta to write about a place where culture wasn't interpenetrated with solid-state physics.[5]

But the *Esquire* article was enough. It placed Bob Noyce firmly among the heroes of the digital age. Because the author was Tom Wolfe, it also served notice to the world's elites that Silicon Valley was the engine at the heart of the zeitgeist. Because Wolfe unexpectedly wrote the piece comparatively straight rather than in the high rococo of his full New Journalism style, the Noyce piece also didn't frighten off other journalists with an impossibly high standard of narrative, but rather invited them to take their own shot at the Valley story. In time, particularly during the dot-com boom of the 1990s, they would set up temporary shop in the Valley and tell the story of Steve Jobs, Marc Andreessen, Mark Zuckerberg, and the other major figures of the next two generations of great tech entrepreneurs.

The apotheosis of Bob Noyce and his departure from the operations of Intel Corp. (he would always remain affiliated with the company and for many years would keep an office there) came, as always with his luck, at precisely the right moment. Almost at exactly the same time he became a free agent, a young middle manager at Hewlett-Packard named Ed Hayes was preparing a speech he was to give at an industry conference in Boston and would repeat a few weeks later in Silicon Valley. His theme was component quality, and armed with test results from HP's data products (computer) division, he knew he had shocking data on his presentation slides. But the lanky, red-haired, soft-spoken Hayes had no idea that he was about to drop a bombshell that would redirect the semiconductor industry, set off a global trade war, and ultimately restructure not just the chip business but the global economy.[6]

The audience, mostly US semiconductor industry managers, bored and waiting for cocktail hour, listened restlessly but politely to the beginning of Hayes's talk. After all, HP was a major customer for most of them. Then with a single slide, Hayes dropped the bomb. It offered a comparison between the acceptance rates by HP of chips delivered by US semiconductor makers versus Japanese makers. That's when the muttering and outbursts of disbelief began, both in Boston and in Santa Clara.

What Hayes showed them was a simple pair of graphs. The first compared the quality—that is, the number of working chips per shipment by vendor—of semiconductor orders delivered to Hewlett-Packard for use in its instruments, calculators and computers. The results were shocking. The US and Japanese vendors clustered in two groups on opposite ends of the graph, with the number of good Japanese chips above 90 percent, while American chips were down in the 60s and 70s. The second slide only drove the blade in deeper. Not only were the Japanese delivering far higher-quality chips than their American counterparts, but they were also delivering them twice as quickly.

For the more astute American chip vendors in the audience, the implications of these two charts were deadly. By themselves, the two slides meant little, other than that the Japanese chip companies were doing a surprisingly better job at quality and customer service than their US competitors. And "it was galling for the Americans to find that they had been overtaken by their protégés."[7] But what made those slides frightening was who had made them: Hewlett-Packard, the most admired technology company on the planet. Hewlett-Packard was the gold standard, and if HP was now giving quality and customer service primacy over technological innovation, then the US chip companies, which had been competing over who was able to best manifest the next jump of Moore's haw, had been directing their talent, energy, and fortunes in the *wrong direction*.

And it was even worse than that: the realization hit with a sickening thud that on this new battleground, the Japanese had better weapons. They had incorporated quality management into their operations; they had a trained workforce in lockstep with top-down management; and they had pliant banks and an aggressive government prepared to back them in predatory pricing schemes.

That's when the most visionary figures in the US semiconductor industry—Noyce, Sporck, and John Welty of Motorola, leaders of the newly created Semiconductor Industry Association—realized the worst. Their only competitive advantage against the Japanese was in chip innovation, but the Japanese chip companies had perfected their system of copying US designs, implementing them as products, manufacturing them at the highest levels of quality, and delivering them quickly so they could eat up most of the all-important high early-stage profits from these new designs. That would leave the US chip companies without enough capital to design the next chip generation—and the entire industry would soon grind to a halt, with the Japanese finally consuming it all.

They were trapped. And unlike the Japanese, the US government was unlikely to help, not just because banking, antitrust, and fair-trade laws prohibited American companies from acting like their Japanese competitors, but because the US chip companies, always fiercely independent from government interference, had few friends or even contacts in Washington.

And so at the moment when they should have been celebrating an economic upswing and the successful implementation of the next generation of microprocessors, the US semiconductor industry descended into a fatalistic gloom.

What made things worse was the realization that the US chip companies, not least Intel, had only themselves to blame. American chip companies had known for a decade that their Japanese counterparts were making a serious effort to get into the business, not just because it offered high value-added for a country with limited resources, but because semiconductors appeared to be the supreme technology of the emerging digital world. And their admiration for Bob Noyce convinced them that if the great man had staked his career on silicon chips, so should they. As one Japanese journalist lamented at the time, "We have no Dr. Noyces or Dr. Shockleys."

And yet though they knew the Japanese were coming, the US semiconductor industry still didn't take the threat seriously. This was despite the fact that more than a decade before, most of the same Japanese electronics/consumer products/trading combines—also not taken seriously—had licensed Ampex's audio and video tape-recording technology and then, through a combination of capital, superior manufacturing, and stunning innovation, had run away with the recording equipment industry.

Still, for some reason, US chip makers didn't think the Japanese could play at the silicon level. James Cunningham, an author and vice president of Advanced Micro Devices, would later write:

> Back in the 1960s, we used to laugh at the Japanese. . . .
>
> Usually several hundred technical papers were delivered [at US semiconductor industry technical conferences of the era]. Of these, the Japanese electronics companies might contribute only one or two, and even these were invariably of minimal importance. Not only was the technical content of these papers of little significance, but the limited English of the presenters made them all but unintelligible. It really didn't matter, as the papers said little of value anyway.
>
> It wasn't until years later, when our smugness gave way to fear and awe, that we realized that the Japanese hadn't come to talk, they came to listen, and to photograph. Every time a slide would go up all of the Japanese cameras in the room would go off all at once. We supercilious Americans even had a joke about it: "You know what that sound is every time a new slide goes up? It's the Japanese cameras going *crick, crick.*"
>
> That was more than a decade ago. We don't laugh anymore.[8]

The Japanese were astounded that the Americans hadn't seen them coming. One Japanese journalist wrote at the time, "Today, one tantalizing question still remains. Why on earth were American and European firms willing to aid the Japanese competitive efforts to close this technological gap?"[9]

In the intervening years, while the US chip makers were singlemindedly focused upon beating each other to a pulp, their Japanese counterparts at NEC, Fujitsu, Hitachi, and other companies had embarked on a systematic program to duplicate the US technology. More important, the Japanese had found two ancient (both were in their eighties) American gurus of product quality, William Edwards Deming and Joseph Juran, prophets long forgotten in their own land, and put their theories to work with a vengeance—easy to do with a generation born in postwar misery and willing to make almost superhuman sacrifices to achieve a promised prosperity. Soon quality was an obsession in these giant firms, from executive row right down to the individual assemblers on the line.

And the scope of this assault extended much further than the individual Japanese companies. Dominance of the world's semiconductor industry quickly became a priority of the Japanese government as well. Before long, the Japanese government was either looking the other way or actively involved in helping those same companies set up "sales offices"—in fact, technology listening posts—in Silicon Valley to hire local talent and pick their brains for state-of-the-art product developments.

Meanwhile, Japan's Ministry of International Trade and Industry, in concert with Japan's major banks (an arrangement illegal in the United States) helped the Japanese electronics companies embark on a massive program of tracking US patent filings in order to anticipate where they were going and beat them to the punch. The ministry provided subsidies (illegal under international law) to help Japanese companies sell their chips at artificially low prices in the United States while keeping prices high at home. In 1978, forty thousand Japanese citizens made technology-related visits, many of them subsidized, to the United States, while just five thousand American businesspeople traveled the other way. And that didn't count the Japanese students on full scholarships attending American technical universities—leading the famous Caltech computer science professor Carver Meade to mutter, "I don't think that one should be able to obtain a whole technology for the cost of only one man's work for a year."

There was an even darker side to this nation-on-nation economic assault. By the early 1980s, impatient with this passive information gathering, the Japanese chip and computer companies, with tacit government approval, made the fateful turn to actively stealing US technology. Everything peaked in 1984, when the FBI conducted a sting just a block from Intel headquarters and arrested Hitachi and Mitsubishi employees who had tried to pay $648,000 for IBM computer design secrets.[10]

Now it was an all-out trade war. And for US electronics companies, the recognition of this new reality had come almost too late. By then, the Japanese had taken 47 percent of the hottest and most profitable business in chips, dynamic RAMs (DRAMs); three years later, they had 85 percent, and half of the world's semiconductor market.[11]

When the US semiconductor industry at last recognized its predicament, it turned to the one man who held enough respect among all of the players, and in Washington, to organize a counterattack and lead the domestic chip industry back to global dominance: Bob Noyce.

The third force that would make Intel in the 1980s very different from that of the 1970s was still over the horizon as Intel was dealing with the implications of Crush, Hayes's speech, and the departure of Bob Noyce. It was the next turning of the industry's quadrennial business cycle.

The current boom would last until 1982. But after that, the semiconductor industry would tip over into one of the worst downturns in its history. The chip industry didn't yet fully appreciate that it was not the business that it was in the 1970s. In the earlier decade, it was a comparatively small industry. It had made major inroads into the computer business and had seen a moderate success in certain emerging consumer electronics businesses, and to a lesser degree in automotive and corners of the test and measurement instrument business.

Now at the beginning of the 1980s—thanks to the personal computer boom, the beginnings of the Internet, "smart" instruments, video games, digital watches and calculators, the migration of chips, processors, and controllers throughout cars from the engine to the dashboard, cable television, and the new digital control panels on audio equipment, televisions, and even appliances—semiconductors were now a truly global industry. It sold its products to scores of industries and thousands of companies around the world, manufacturing those chips in scores of factories in a dozen countries, and selling them from sales offices in more than one hundred nations.

In theory, this was supposed to be a good deal, as a slowdown in one particular market could be made up by added sales in another. True enough, but when the downturn was both widespread and international—as was the "W-shaped" recession that began in mid-1980, recovered in the middle of 1981, and then slumped again until late 1983—being in numerous markets and in multiple locations around the world was not an advantage, but a handicap.

President Ronald Reagan had come into office dedicated to rebooting the struggling US economy by wringing out the stagflation that had resulted from the energy crisis and price controls, and if that meant toppling the nation into a full-blown recession, he was prepared to do it, especially early in his administration. In the long run, his strategy would pay off for Silicon Valley (and the rest of the US economy), but in the short term, it was devastating—even more so because the downturn hit smack in the middle of the huge capital investments chip companies—and most of all Intel—had committed to make to stay competitive in the next generation of chips, which with awful coincidence were due to appear in 1982.

In 1980, Intel would enter the exclusive ranks of billion-dollar companies. Managing a company of that size, with tens of thousands of employees, scores of facilities, hundreds of suppliers, distributors, and retailers, and ever changing legions of shareholders, under the increasingly watchful eye of the Securities and Exchange Commission and with the growing need to deal directly with governments and politicians around the world—all required a different kind of leadership than that which had proven so successful in the company's first decade.

For all of the frustrations shared by Andy Grove and others about the man's casual attitude toward rules of business and propriety, his vulnerability to influence, his risk aversion, and his endless need to be liked, it was obvious to everyone—even Andy—that the one irreplaceable man during the early years of Intel (and indeed, the entire electronics industry) was Robert Noyce.

Only Noyce had the vision to see the future of chips, the will to build the two greatest companies of the era, the charisma to convince the money and talent to follow him to Intel, and the unbreachable self-assurance to take enormous, fateful, and potentially fatal risks that Intel needed to not only survive that touchy early passage but come out as the industry leader. Noyce had broken every rule, including hiding a major initiative from his own senior management, to make the microprocessor happen, and in doing so, he changed the world for a second time. He wasn't always a good man, but he was one of the greatest men modern American industry ever produced. And now he was embarking on one of the most influential endeavors of his career: to save the US electronics industry.

As always, his departure was graceful and perfectly timed. In fact, because he remained closely tied to Intel, to the outside world it wasn't apparent that he was no longer running the company. As for the timing, as with Fairchild, Noyce knew that he wasn't interested in leading a big company like Intel had become, and he instinctively knew that if he stayed much longer, the company would begin to stumble and that he would get the blame. Gordon Moore wasn't a big company guy either: he liked making technology realize its destiny, not dealing with angry shareholders and pushy reporters. He was the ultimate Big Picture guy, a visionary who could see into the technological future better than anyone alive. And so he would remain as CEO, even as he took the job of Intel chairman, less to actually run the company and more to keep his hand in on the company's daily operations to make sure they didn't go off the rails.

To stretch an analogy, evolving from a small private company to a large public corporation is like going from a republic to an empire. One needs a leader, a man of the people; the other requires an emperor, a Caesar, someone not afraid to make huge decisions, deal daily with great financial gains and losses, and control the lives of armies of people.

Andy Grove wanted that job. He had always wanted it. He had earned it. Now he set out to show that he was worthy of the job. For Intel, and by extension the world economy, the eighties and nineties belonged to Andy Grove. And happily, Grove would prove to be the greatest businessman of the age.

CHAPTER 28

The Heart of Andrew Grove

Andy Grove would die before he'd let Intel fail. And he would go down fighting with every ounce of his being. That's how you build the most valuable manufacturing company in history to that date, become *Time*'s Man of the Year, have even Steve Jobs court your friendship, and at least once, preserve the entire US economy from tipping into recession. And it's why even your industry peers visibly brace themselves before they walk up to you at a cocktail party to say hello. They know that, one way or another, it is an encounter, friendly or not, that they will lose.

To understand Intel, especially the mighty global business titan it became in the eighties and nineties, you most first understand Andrew S. Grove, born András ("Andris") Gróf. It is not an easy task, as Andy Grove, to paraphrase Churchill, is an enigma wrapped in pure will.

In the history of Silicon Valley, there are three great stories of leadership transformation. The first of these is called the Great Return. It took place in the 1990s, when Bill Hewlett and Dave Packard, retired as living legends, went back into a troubled Hewlett-Packard Co., shook up senior management, and set the giant back on the path of historic growth. That they did so based upon a single memo from a low-level secretary only underscored the extraordinary mutual trust at the core of the HP Way.

The second story might be called iSteve, after the prefix Steve Jobs attached to the historic string of new products introduced by Apple after his return at the turn of the twenty-first century. After having cofounded Apple Computer and led it to a near-monopoly over the computer industry, Steve Jobs slowly made himself a persona non grata at his own company through his childish petulance, megalomania, and cruelties toward his subordinates. He bought himself a few more years by brilliantly (and cynically) hijacking the Macintosh project after his own Lisa computer project—featuring ideas stolen from Xerox—failed. But in the end, despite his obvious gifts, Jobs became too much of a liability to Apple and was fired by the person he'd hired to be his pliable lieutenant. But then, after a decade in the wilderness, trying and failing to create a competing computer, a much older and wiser Jobs returned to a rapidly fading Apple. He was still a difficult human being, but now he understood what mattered most to him—Apple—and over the next decade before his death, Jobs turned the company into the most innovative and successful of the new century.

The third story can be called The Heart of Andy Grove. In the quarter century between his ascension to the presidency of Intel in 1979 to his retirement from the chairmanship of the company in 2005, Andy Grove was in the cockpit of the company at the very epicenter of the electronics industry, and upon him at critical moments the entire global economy seemed to rest. His public statements moved stock exchanges; when Intel coughed, the economy got sick. Presidents of the United States looked to him for advice. And his books became bibles for millions of budding entrepreneurs and businesspeople.

In other words, Andy Grove got everything in his career that he ever wanted, everything he ever thought he deserved. But it came at a great cost. His supreme confidence in his abilities as a technologist, as a leader, and as a businessman would each be shaken in turn. He would be publicly humiliated and embarrassed making a very controversial decision. And at the absolute nadir of his leadership of Intel, he would find himself and his company accused by the federal government of predatory business policies, and he would have to swallow his pride and grovel to save Intel Corporation.

And yet it was precisely these experiences, which would have broken many weaker executives, that forced Andy Grove to become the chief executive he needed to become to hold such a powerful and central position in the modern global economy. He grew to be self-effacing, forgiving of human weakness, beholden to his employees and shareholders, and given to looking beyond the walls of his company to its larger impact on the world, all without losing his fundamental toughness and competitiveness. The Andy Grove who inherited the leadership of Intel at the end of the seventies, for all of his brains and talents, wasn't qualified to run a company as important as Intel was about to become. By the beginning of the new century, when he stepped down, he made the job look easy.

"Only the paranoid survive" was the catchphrase that attached to Andy during his tenure at the top of Intel. It became the title of his best-known book. It was very good advice for anyone leading a company in the fast-changing high-tech world. But the real lesson of the Andy Grove era at Intel was that only those who learn from their mistakes thrive.

Intel came to rule the tech world during Grove's tenure not because, even for an instant, he relaxed his uncompromising attitude toward winning. He never did that. But he learned (sometimes by force) that there were multiple paths to victory besides his own.

CHAPTER 29

Mother and Child

There is a sweet—yet ultimately troubling—photograph of Andy Grove's mother taken in Hungary in 1944. She is a charming-looking woman, with short hair and a wide smile and a twinkle in her eyes. Her head tilts to the side as if she is being playful with eight-year-old András. Maria Gróf's smile is so dazzling that only after you study the photograph for a while do you suddenly notice with shock that at her lapel, almost hidden in the floral pattern of her dress, she has sewn a Star of David.

There is a second photograph, from a few months later, that is even more heartbreaking. It is a family photograph of Maria, her husband George ("Gyurka"), and little András in a collared shirt and what appear to be the straps of lederhosen.

As Grove's biographer Richard Tedlow has noted, the expressions and body language of the three figures in the little family "speak volumes." Maria in this photograph looks like an entirely different person. Her hair is piled atop her head without much attention and her face looks twenty years older, with bags under her eyes. Even her dress, though more stylish, is black and seems to hang from her shrunken body. George, in a striped suit, gray or blue shirt, and thickly knotted tie, has reached over and put a hand on her far shoulder. It is less an embrace than a gesture of protection, of giving his wife what little comfort he could offer. With its high widow's peak of deeply receding black hair, George's face is sharp with a long nose, and like his wife's, his lips are pursed with the tiniest of forced smiles. One can see neither parent's eyes, because both are looking down at their young son, who sits between them.

At the bottom of the photograph, the third point of the inverted triangle, sits little András, a burst of sunshine between the gray clouds. He looks so happy, so full of spirit, so lit by a different light that, if it were Hungary a decade hence, you would assume the photograph was doctored—or photoshopped in the Silicon Valley of András's future. For him this is a new adventure, to dress up and sit with his parents, and he is overjoyed to respond to the man with the camera trying to get him to smile. For his parents, the darkness in their faces hints at the fear that has begun to fill their days; the brave but sad look they give their child tells of the unceasing worry they have for his fate—and their determination to save him from it.

Maria and George's fears are not misplaced: this will be the last photograph the family will take together before they are pulled apart and their lives are shattered by the Holocaust.

It would be sixty years before the adult András, Andy Grove, would begin to talk publicly about that period of his life, first under occupation by the Nazis, then the Soviets. Until then, he seemed to have appeared, fully formed, on the day he first arrived at Fairchild Semiconductor in 1963 with his newly inked PhD in chemical engineering from UC Berkeley. The noted NPR interviewer Teri Gross tried in 1996, on the occasion of the publication of Grove's *Only the Paranoid Survive*, to get him to talk about those early years, . . . and instantly her talkative and opinionated guest was reduced to mumbles and attempts to change the subject.

Though this might seem a natural response to painful memories, it is interesting to note that, remarkably, yet another major Valley executive, Jack Tramiel of personal computer maker Commodore, was also a Holocaust survivor. Tramiel (born Jacek Trzmiel in Poland a decade before Grove) treated his past very differently. He was known to say, when others boasted of their educations and alma maters, "I went to college too: to the University of Auschwitz!"—a statement guaranteed to intimidate any room to silence.

It was only at the beginning of the new century, when Grove was preparing to step down from daily activity at Intel, that he finally told the world about his past by publishing a candid memoir—to the surprise of nearly everyone—about his early years and his eventual escape to America. It was called *Swimming Across*. Most memorably, in September 2003, he traveled to an event held in Chicago to help fund the proposed new Holocaust Museum in Skokie. To the surprise of those who had always known him as a religious nonobservant, Grove joined in the blessing over the bread. Then, following a brief speech, he opened the floor to questions from the large crowd in attendance.

A woman asked, "I understand that you have never returned to Budapest. Could you discuss your feelings about your homeland?"

Grove's reply, a cri de coeur, stands as the best glimpse into his childhood and its bitter memories:

I will, but I don't think I will succeed in conveying them to you. I have a hard time explaining what it is. My life in Hungary was—to understate it—a negative experience.

The obvious parts are, excuse me, obvious. The war part was obviously negative. Being shot at was negative. Living under a Communist regime and being told what to think and what to see and what to read and what not to think and on and on was pretty bad. Having my relatives imprisoned randomly was bad. But that's not . . . those things . . . Some are changed.

What didn't change in my gut and in my heart is being told at age six that "Jews like you killed Jesus Christ, and we're going to push all of you into the Danube." To have a good friend of mine at age eight, when I told him who I was—his father took all the particulars down just in case the Germans came back to make sure this one doesn't get away either.

And people on the ship coming across the Atlantic after the Revolution being told by their Hungarian minister that you have to leave your anti-Semitism behind. They were very upset about that.

My life was marred with personal experiences like that. I don't have any emotional energy to devote to that. There is nothing for me that justifies picking those scabs.[1]

Richard Tedlow, Grove's biographer, has noted that even if Andy never has returned to Hungary in person, he certainly has by proxy: "Intel processors power computers in Hungary just as they do throughout the world." Grove's books, including *Swimming Across*, have also been translated into Hungarian, and one can assume that thousands of business executives and entrepreneurs in Hungary have modeled their careers after this native son. And as the vivid memories in *Swimming Across* show, for Andy—like Holocaust survivors Frankl, Miłosz, Wiesel, and Levi—the old prewar Europe of the Jews is still as vivid in his memories as it was when he lived in it.

If the word is in any way appropriate, Andy Grove was "lucky" to have been born a Hungarian Jew—as opposed to, say, a Polish or a Czechoslovakian Jew. After a bout of anti-Semitism in the early 1920s, the country's leadership had once again become comparatively indifferent to this five percent of the population. Still, the country slowly slid into its own form of fascism, which, unpleasant as it was, proved an advantage with the rise of Hitler and the Nazis in Germany.

Little András was still a toddler when Germany annexed Austria and a two-year-old when Hitler took Czechoslovakia. He turned three on September 2, 1939. To celebrate, the family took a walk on the left—Pest—bank promenade of the Danube. András's parents had given him a little red-and-white pedal car shaped like his Uncle Józsi's sports roadster. Józsi, Maria's brother, was there as well and encouraged little András to pedal faster and faster, weaving in and out of the other promenaders, with Józsi in pursuit.

As Andy would later recall, his steering was mostly successful, but even when it wasn't and he bumped into adults, they barely seemed to notice. Everyone was looking up into the warm evening sky to the west, looking at the vertical white lines waving back and forth, sometimes illuminating the undersides of clouds.

The lines were searchlights. The previous morning, the German army had encroached on the Polish border. The next morning, the blitzkrieg would roll into Poland, crushing the Polish army, with its obsolete weapons and horse cavalry, in just a month. Meanwhile, on September 17, the Soviet army invaded Poland from the east, slaughtering Poland's officer corps in the Katyn Forest. Poland was torn in two, suffering predations from both invading armies, followed in the west by the construction of concentration camps for Jews, Gypsies, and other ethnic groups. It was in these camps where the Holocaust would begin, as the Nazis converted the imprisoned into slaves if young and healthy, murdered them if not. By the war's end, three million Polish Jews had been murdered.

Hungary, which had its own grievances against its neighbors for having dismembered the country after World War I, mostly cheered the destruction of Poland and joined the war on the side of Germany and the Axis powers. Hitler, happy to have a new ally and not another country to attack, awarded Hungary part of Czechoslovakia, then turned and sent the Wehrmacht into Yugoslavia. Anxious to regain even more of its old territory, Hungary sent its army to join the successful attack and occupation.

At this point, Hitler turned west and conquered Belgium, Holland, and France and embarked on the blitz of the British Isles. But soon his attention turned east again, this time for the largest invasion in history, Operation Barbarossa, the attack on the Soviet Union. By now, the Hungarian army was fully associated with the German army, and Hungary too declared war on the Soviet Union and joined in the invasion. Allying with Germany had given Hungary a certain immunity to the predations of the SS and the Gestapo. Unlike Holland and Poland, there were no general roundups of Hungarian Jews during that era. Longtime regent Miklós Horthy, though an admitted anti-Semite, felt it was beneath the dignity of the Magyars to imprison or murder their fellow citizens, so he resisted the horrors that were defining the rest of Eastern Europe.

Jews still had to register themselves with the Arrow Cross (the national fascist organization) and wear the Star of David, and they were denied access to many jobs and locations. There were still daily humiliations and terrors from their fellow Hungarians, but no one was being loaded into boxcars . . . yet.

Still, there was no hiding from the fact that the Jews of Hungary were on very thin ice. For George and Maria, as for many cosmopolitan, nonobservant Jews, this cruel treatment by their own neighbors and former friends was bewildering and difficult to accept. George was a milk contractor and processor who had been born in the country town of Bácsalmás. He often traveled to Budapest to sell his wares, and there he met Maria, a shopgirl, and married her.

George was outgoing and gregarious, Maria quiet and shy, but she was also much more cosmopolitan than her new husband, and when she became pregnant with András and realized she'd have to live in Bácsalmás, she despaired of never living in her beloved Budapest again. But when the butter-and-cheese company opened an office in the city, George jumped at the opportunity, and soon the Gróf house in Budapest was a gathering place for relatives, friends, and neighborhood children.

That soon changed. In 1942, András came down with scarlet fever. In those days, before widespread antibiotics, it was often a fatal illness for young children. András was hospitalized for six weeks, then sent home for an endless nine months of convalescence. At the peak of his illness, the boy's situation, in particular a massive ear infection, had become so acute that the doctors were forced to chisel away some of the bones behind his ears. The surgery helped save András, but it nearly cost him his hearing in the process.

During those long months of recuperation, András's two closest companions were his grandfather (his mother's father) and a hand puppet, given to him by his mother. He would write in his memoirs, "I cut a hole in its head behind its ears, then I bandaged it so that he looked like me." Biographer Tedlow would see this act as an early example of Andy Grove learning how to deal with trauma. Perhaps, but it seems even more a poignant attempt of a lonely boy to create a friend suffering just like him.

One day while András was still recuperating, a stunned George Gróf came home, as his son would describe it, looking as if "he was trying to smile, but there was something wrong with the smile." His father brought shattering news: he had been conscripted into the military. As a Jew, that meant he would be, in his son's words, forced "to serve in labor battalions clearing roads, building fortifications, and the like." George had been called up before to work on brief tasks, but this time it was different. He was going to war. To the Russian front.

A few days later, in the week of their tenth wedding anniversary, Maria and András took the train sixty miles to Nagykõrös, where the labor battalion had its staging area. The little family made its good-byes, not knowing when, or if, George Gróf would return.

A sick child, a husband off to the war, little income . . . Maria, riding the train home with András beside her, knew that hard years lay ahead. Reaching Budapest, she got a glimpse of just how hard: she arrived to learn that while she was gone, her father, the man who had been so helpful nursing grandson András during the boy's long convalescence (Andy would recall him as "the perfect playmate"), had suffered a stroke. He died a few days later.

Her father dead, her husband gone, her two brothers conscripted as well, and with a son to raise, Maria realized that she was terribly vulnerable. In the weeks and months to come, equally abandoned neighbor ladies and old people would often stop by the house to visit—but unlike the happy times, these visits bore an added seriousness. "Everyone seemed pre-occupied," Andy would remember. Sometimes, after he could go outside, Maria would take András to the city park, where he could play in the sun under the big equestrian statue of George Washington. The evenings were the loneliest times, and Maria would sit and drink by herself. She began drinking now during the day, too.

As for George Gróf, the Hungarian Second Army, three hundred thousand strong (including forty thousand laborers like George) was positioned on the left, northern, flank of the German army as it swept across Ukraine to the gates of the great southern Soviet city of Stalingrad. The Germano-Hungarian army fought its way into Voronezh (destroying more than 90 percent of the city in the process) and across the Don River. Then, leaving the Hungarians to guard the rear, the Germans moved on to attack and ultimately meet their fate in Stalingrad.

That was autumn. That winter, the German army suffered horribly from the depredations of the snow and cold (like Napoleon's before it) and from the counterattack by the Red Army in December 1942. Less known is the fact that a second Soviet counterattack was launched on January 13, 1943, against the Hungarians, who were as exhausted and even less well equipped than the Germans. In a battle near Svoboda on the Don, the Soviets encircled and destroyed the Second Army (and most of Army Group B) in one of the worst defeats in Hungarian history. Three weeks later, the entire German army under Field Marshal Paulus in Stalingrad surrendered, and the German eastern front was in full retreat.

It wasn't until that spring, long after the truth about the German defeat had filtered through the censors, that Maria received an official notice that her husband had "disappeared."

"I didn't know what that meant," Andy would write. "I didn't know how people could disappear. But I didn't dare ask my mother."[2]

Not only had András's home life irrevocably changed, but so had his schooldays. He attended a nonsectarian kindergarten, and though the students, because of the neighborhood, were all Jews, the prejudices of the larger world still crept in in the oddest ways. For example, András overheard his mother say to some other women—no doubt on rumors coming out of Warsaw—that "They will put the Jews in a ghetto." András, who didn't know what the last word meant, repeated the line to his equally confused classmates. Soon they were all announcing, "They will put the Jews in a ghetto," and for weeks it became a popular schoolyard game. Their horrified teacher ordered them to stop, but the students just moved the game out of her earshot.

For all his energy, András was a sickly boy. Because of his illness, he was still highly prone to infection, and this time it came as tonsillitis. András's tonsils were removed, and his ears had to be drained of fluid, a consequence of the scarlet fever.

But András, in a foreshadowing of sixty years hence, didn't let his illness slow him down this time. Soon he was back in school, still sitting at the front of the class where he had been placed because of his hearing, answering all of the teacher's questions, flirting with girls, . . . and earning the best grades in the class. He adjusted to the loss of his father because that's what children do.

But the larger world was deteriorating fast. Reeling from the defeat at Svoboda, the Hungarian government realized that the Nazi empire in the east was doomed. Seeing the inevitable, it desperately tried to broker a separate peace with the Soviets. It was a terrible mistake, not just because Stalin wasn't interested in anything but the destruction of his enemies, but because the German army, though in a fighting retreat, was also backing toward Hungary and needed to guard its rear.

What happened next was as terrible as it was predictable. Hitler ordered an attack on his erstwhile allies. Budapest was bombed. Andy has never forgotten the sight of a familiar apartment building, its front now sliced off as if by a knife and reduced to rubble, while in the exposed rooms the furniture, lamps, and paintings still stood intact. Soon after, the Wehrmacht invaded with eight divisions. On March 19, 1944, as seven-year-old András watched, the German army marched into the city: "There were no announcements and there was no fighting—they just came in. My mother and I stood on the sidewalk on the Ring Road watching as the cars and troop carriers filled with soldiers drove by. The German soldiers didn't look anything like the soldiers who had guarded my father's labor unit. Those soldiers slouched a bit, and their uniforms were wrinkled. The German soldiers were neat and wore shiny boots and had a self-confident air about them. They reminded me of my toy soldiers; they had the same kind of helmet, the same color uniform and the same kind of machine gun. I was impressed."[3]

The SS had arrived . . . and with them, Adolf Eichmann.

And so began a year of horror for Maria Gróf and confusion and strangeness for her son. The "luck" of Hungarian Jews had just run out. In the months that followed, a disoriented András was often without warning snatched up by his mother and moved to a different location for his safety. The first move was back to his father's dairy in Bácsalmás, where he was to stay with a gentile friend of his father named Jani.

The crudeness of country life didn't appeal to a city boy like András, but he resolved to make the best of it. Then he was snatched up again and sent back to Budapest. He learned years later that another gentile friend of his father who had some access to intelligence had learned that Eichmann's SS unit was assigned to scour the Hungarian countryside first in its roundup of the nation's Jews and only then assault Budapest.

Maria Gróf's judgment was perfect—as it would often be during this period—but more important was her timing. When the situation changed, she didn't hesitate to move . . . quickly. Her son inherited her decisiveness—and her understanding that life isn't a game but a series of decisions big and small that can prove to be matters of life and death.

Bringing András back to Budapest may have been the best decision Maria ever made. From May to early July 1944, the SS conducted a reign of terror throughout the country so complete that by July 9, Eichmann could report to his commanders that Hungary was entirely *Judenrein*—Jew-free—other than Budapest. It is estimated that during this brief period, 430,000 Hungarian Jews were transported to their deaths in the labor camps and gas chambers of Auschwitz-Birkenau. Staying in Bácsalmás would have been a death sentence for András Gróf.

In Budapest, they were still alive, but the walls were closing in. Once, in the city park, under the statue of Washington, a little girl he had been playing with suddenly announced, no doubt parroting her parents, "Jesus Christ was killed by the Jews, and because of that, all the Jews will be thrown into the Danube." András ran crying to his mother. They never went to the park again. And András got his first unforgettable lesson about how his gentile neighbors really felt about him.

The lessons only got more painful. Maria and her son were ordered to vacate their apartment and move into a Star House, so named because it had a big yellow Star of David painted on the front door. There were two thousand of these Star Houses, and they were to serve as the first step toward the gathering of Budapest's Jews into a single location for control and eventual extermination. Maria and András were also ordered to now always wear the same star sewn to their clothes over their hearts whenever they left the house.

Recalled Grove: "We didn't go out very much. There were few places we could go to, and the hours when we could be on the street were limited. Many stores could not serve people with a yellow star; besides that, it was a very strange feeling to walk on the streets wearing it. People avoided looking at us. Even people we knew wouldn't meet our eyes. It was if a barrier was growing between us and everyone else."[4]

In August, the Hungarian government tried one last time to surrender to the Soviets—and the Germans in response pulled down that government, bad as it was, and replaced it with something even worse: the Arrow Cross, a party of pro-Nazi street thugs led by the Quisling-like Ferenc Szálasi. By October, the Arrow Cross began forcing the Jews of Budapest into a ghetto in preparation for their own version of the Final Solution. In November, fulfilling the prophecy of the little girl at the park, the Arrow Cross marched two hundred Jews out onto the bridges over the Danube, handcuffed them together in pairs, and shot them—before dumping their bodies into the river.

Before they were done, the Arrow Cross, working under the distant orders of Eichmann, rounded up an estimated eighty thousand Jews in Budapest and shipped them to the labor and death camps. So great was their obsession with destroying Jewry that even as the Red Army approached the outskirts of the city, the Arrow Cross still indulged in an orgy of slaughter, killing Jews in hospitals and synagogues. It was during this slaughter that Raoul Wallenberg and others famously used their influence and diplomatic documents to save hundreds of individuals from a similar fate.

By then, Maria and András had disappeared. In mid-October, with her usual clear eye seeing what was about to descend upon them, Maria told her son, "We have to get out of here." Once again she moved swiftly: within days, she had found them a new place to hide.

More accurately, it was *places*. Mother and son were now to split up; Maria to stay with a workman from her husband's dairy, András with a gentile partner of her husband. It was, as one can imagine, a shattering parting: a mother entrusting her child to a mere acquaintance, a son who had already lost his father now about to lose his mother for, as far as he knew, forever. He was learning early that, in the end, he was on his own.

Maria managed to visit her son twice during this period. The second time, she arrived to find him sitting at a window, watching German soldiers loading neighborhood Jews, with their hands up, into trucks. He was crying.

The next time she visited, Maria brought news. They were moving again, this time together to the suburb of Kőbánya outside the city. There they would get new a new, Slavic, name—Malesevics—and new identity papers to confirm it. András was taught to say that he and his mother were refugees from Bácsalmás—the easiest story for the boy to remember.

That winter in Kőbánya was particularly grim, dark, and cold, with two families sharing the same apartment. Everyone used a communal toilet down the hall, and Maria carefully instructed her son to never urinate whenever anyone else was watching, as his circumcision would give them away.

In the early hours of the morning of December 29, mother and child "woke up to a strange sound. It sounded like someone was dropping planks of wood on top of each other." Soviet artillery. Maria Gróf had made the right snap judgment once again—and András was reminded once again that timing is everything. The Red Army had begun the siege of Budapest.

During this initial onslaught, one artillery round hit the apartment building they were living in, and Maria and András joined the other residents in the wooden cellar for several days. There, to occupy the time a few days later, the children were instructed to recite the catechism. András, using his quick mind now to survive, quickly asked to be excused to pee and ran to his mother for help. Maria, quickly appraising the situation, loudly told her son he had chores to do and sent him off—and their secret remained safe.

More than a half-million Russian and Romanian soldiers encircled fewer than 50,000 Hungarian and German soldiers left inside the city, with 800,000 civilians trapped as well. The fighting began in Pest, which was largely destroyed by the end of January, then crossed the river into Buda, where the defenders took advantage of the hillier terrain. In the end, Budapest became a second Stalingrad. By the time the battle ended on February 13, 1945, as many as 150,000 Hungarian and German soldiers were dead, wounded, or captured. The victorious Soviets took even greater losses, with more than 300,000 dead, wounded, missing, or sick. Meanwhile, 40,000 citizens of Budapest caught in the crossfire lay dead. Another 25,000 had died of starvation, disease, or other causes, and of these, 15,000 were Jews slaughtered by the Arrow Cross militia in a final orgy of killing.

Now the rape of Budapest began.

No sooner had the thugs of the Arrow Cross abandoned the city than the soldiers of the Red Army occupied it, bent on revenge—and exacting it in the form of what has been described as the greatest mass rape in human history. Meanwhile, Stalin had his own plans for Hungary's future.

It has been estimated that 50,000 women and girls in Budapest were raped (and usually robbed) by Red Army soldiers after the capture of the city, though some estimates are as high as 200,000. The soldiers soon were so out of the control of their indifferent officers (who often participated) that they even burst into the embassies of neutral countries, such as Sweden and Switzerland, and raped the women staffers.

For a time, Maria and András remained untouched by these horrors. Then in early January, the Red Army occupied Kőbánya, and a platoon of soldiers moved into the apartment building. Andy remembered that "they came in casually, but each of them carried a machine gun."

The first direct meeting between Maria and the soldiers proved surprisingly benign. A Russian sergeant spoke to her in German, and to András's surprise, his mother responded in the same tongue. "I had never heard my mother speak anything but Hungarian before, so I was impressed by how fluently she seemed to be able to talk with him." He was even more surprised when his mother called him over to talk to the soldier, whose name was Haie. She asked him to recite the Hebrew prayer he'd been taught in school, "Modim Anachnu Lach"—"We Are Grateful to You." András, who had been told to hide his Jewishness now for nearly two years, was confused and frightened, but his mother smiled and told him, "Just for now, it's okay."[5]

It turned out that Sergeant Haie was also a Jew and had lost his parents in the Holocaust.

But even this good luck couldn't last, not in a building full of soldiers when, despite the ravages of the war, Maria Gróf was still young and beautiful.

One evening, another Russian soldier, this one named Andrei, came into the bedroom András shared with his mother. Mother and child were already in bed. The soldier sat on the corner of the bed, talking to Maria in Russian and gesturing by poking her in the chest and then pointing back at himself. Eventually Maria nodded, got herself and András out of bed, then left the boy in the protection of another woman . . . and went off with the soldier.

She returned later, "very tense and angry," woke up her sleeping son, and led him back to their own bedroom. Even in a time of strange events and dangers, András knew that this time something was different: "I lay there, stunned and full of apprehension. I had no idea what was happening to my mother or what would happen to us both."[6]

Later more soldiers appeared—but Maria ran them off. "My mother yelled at them something about how all three women had already done it today."

As history has often shown, the behavior of occupiers is different from that of conquerors—even if it is the same soldiers and that difference is measured in days. Those Soviet soldiers who had been indulged in their looting and rape just a few weeks before were now expected to be respectful of both property and local women, especially if they shared the same building. Andrei was one of those fools who didn't adapt to the new orders quickly enough. Whether Maria instinctively knew this or was told by one of the other women, the next morning she gathered up her son and together they marched down to the local station of the military police and filed a complaint.

The MPs were apparently looking for examples and scapegoats under the new orders, because instead of sending Maria on her way, perhaps with a beating or worse, they met her and András at an apartment, where a group of soldiers and police were waiting. There, the police roughly organized a lineup that included both Andrei and Sergeant Haie. Grove: "My mother faced our Russian soldiers, then one after the other she looked each one in the eye and shook her head no. I was holding my breath when she faced Andrei. Andrei himself was beet red, and it looked like he wasn't breathing. After a very brief pause, my mother shook her head no. I yanked at her hand. She yanked back and said to me, 'Quiet,' in a fierce tone that forbade an answer."[7]

That night, Maria explained to her son what she had done. Sergeant Haie had informed her before she went to the apartment that she could get her revenge merely by pointing out Andrei, who would immediately and without trial be taken outside and shot. But he warned her, if she did so, Andrei's friends among the soldiers would take their own revenge by going to the cellar and killing Maria, András, and everyone else there. "So she had decided not to recognize him," wrote Grove.

By mid-January, the Germans and Arrow Cross militia had retreated back into the hills of Buda in a last-ditch defense, blowing up the bridges across the Danube as they went. They weren't coming back, . . . and that meant the citizens of Pest could return to the blasted buildings and rubble-choked streets that remained of their city. Feeling safe again for the first time in many months, Maria decided to celebrate their liberation by telling András that he could again use the name he had almost forgotten: Malesevics once again became Gróf.

For András, who already had seen enough in his eight years for one lifetime, this news was surprisingly disorienting: "I had become András Malesevics so through and through that for a moment I was confused."

Then "the significance of being free to use my real name engulfed me." He was so excited that he told a friend the truth—only to have the boy tell his father, who called András to his apartment and began peppering him with questions, taking notes of the boy's answers.

"I began trembling with fear and from the hatred welling up inside me," Andy recalled. He told his mother what happened, but she reassured him, telling him that he no longer had to worry; the Germans were gone. But the boy's reaction would prove to be an augury of the future: some Hungarians were already spying on their neighbors for their new Soviet overlords.

Maria soon announced that it was time for them to return to their old neighborhood. Mother and son packed up the few things they had and started off, through the snow, on a ten-mile hike back to their former apartment. Sixty years later, Andy Grove would still remember it as a surreal experience: not just the obliteration of entire districts of the city he once knew, but also by his own reaction to what he saw. So traumatized was András by his experiences over what was now half of his childhood, that as he trudged along behind his mother and passed one horror after another—blown-out buildings, bomb craters, the skeletons of horses that had been butchered on the street—he felt as if he were in a dream or watching a movie in which he was a character . . . and most of all, he was aware of his emotionless reaction to what he saw. "[I] was neither surprised nor unsurprised by anything we encountered." It is hard not to see in this response to disaster the adult Andy Grove's celebrated stubbornness and steely resolve during hard times at Intel.

Incredibly, the old apartment house was—like the human pair standing before it—battered but still standing. However, it was also occupied by squatters. And until that mess could be straightened out the next day by the landlord, Maria and András spent the night in a nearby "Jewish House" from the bad recent past.

Mother and son now set about restoring their old lives. Needless to say, the challenge was a lot more daunting for Maria, who now had to reconstruct an entire life for her son and herself, now a presumed widow. For András, life now seemed like a happy and extended vacation. He no longer had to worry about German soldiers or artillery shelling, he didn't have to remember a second name and a fake history, and best of all for a soon-to-be nine-year-old, his mother had decided he didn't need to start second grade until the next school year, six months away. "It was like being on a perpetual vacation." Worldly for his age, good at making friends, his days free and largely undisciplined, András (as Andy would later admit) became something of a handful.

Maria now filled her days with a perpetual search for her husband, from whom she hadn't heard for two years. She asked anyone she could find, in both official and unofficial channels, of any news of George. But there was nothing. Whenever there was a rumor of POWs being returned, she would haunt the train station hoping for a glimpse.

Her son found her behavior irritating. "It was obvious to me that she would never get a satisfactory answer. I could barely remember my father and now his memory, faded as it was, was tarnished by my mother's obsession."

CHAPTER 30

Father and Child Reunion

In September 1945, the same month as András's ninth birthday, once again the rumor circulated that Hungarian soldiers had been released from Russian POW camps and were on their way in several trains to Budapest. So each day, Maria made her way down to the train station, sometimes with a reluctant András in tow, . . . and each day she returned disappointed.

Her one hope lay in a secret whistle that she and George had created early in their marriage by which, if they were separated, they might find each other. Each time she went to the train station Maria listened for the whistle. And then one day, as the POWs disembarked in their rags and shuffled past, Maria thought she heard the secret whistle. She began frantically searching the crowd—as András, who thought she had lost her senses, complained.

Then suddenly, "An emaciated man, filthy and in a ragged soldier's uniform," appeared at the boxcar door. András didn't know how to react: "[I] was bewildered. This was supposed to be my father, but I didn't know him. I was supposed to love him, but I wasn't sure how I felt . . . I was embarrassed that I had been wrong."[1]

George Gróf had been through hell—and he looked it. Many of his fellow laborers had died, if not from combat, then as prisoners from disease and exposure. And George wasn't out of the woods yet. In April, he had written his family a note they wouldn't receive until after he returned home. "My dear ones: Now that it looks like the end would be here and the prospect of seeing you again, I have had another setback—a new disease, some skin ulcers. It's spreading from one day to the next. There is no medicine. They don't know how to treat it. . . . It looks like my struggles of the last three years were for nothing. And all I would like is to see you again, to know that you are alive. But I am destroyed. Just my love for you keeps me alive."[2]

George steeled himself for a miserable death that didn't come. Unlike thousands of others, he lived to see his family again. And like many men who had seen the full horror of war, he didn't speak about it during his long recovery and for many years afterward. Only as an old man did he tell his son of the savagery he'd experienced at the hands of the Soviets, and even worse, from Lithuanians serving as guards of his labor battalion. Wrote Andy in his memoir: "The story that was most incredible to me was how in the middle of one bitterly cold winter, my father's battalion was made to strip naked and climb trees, and guards sprayed them with water and watched and laughed as one after another fell out of the trees frozen to death."[3]

That autumn and winter, the Gróf family slowly healed and began to make plans for the future. There was one last matter to be erased, and it is remarkable that the adult Andrew Grove discusses it in his memoirs: his mother had an abortion. His parents would later say that they didn't want to bring another child into this terrible world, but it is hard for a reader of Grove's memoirs not to connect it to the terrible events of the previous winter. Andy has always believed that the aborted child was a girl. And though his life has been filled with women, not least his wife and daughters, that lost sister—his war's last victim—continues to haunt him.

It was a time for difficult news: many of their neighbors and relatives were dead. George's own mother, András's grandmother, had died in Auschwitz. A million Red Army soldiers were now occupying Hungary—one for every ten Hungarians—and though the war was over, they showed no signs of leaving.

Unlike its equally occupied neighbors, Stalin allowed Hungary to hold free elections. They were a publicity exercise. Hungary had been flattened by the war. A million citizens had been killed, half of them Jews in German death camps. The Soviets had executed most of the country's former leaders. There was little national infrastructure. Electrical power, water supplies, bridges, roads, and trains had been destroyed or crippled. Factories were gutted and, especially in Budapest, there was a general shortage of housing. Meanwhile, the provisional government, despite a paper shortage, had enough to start printing endless amounts of currency, quickly creating inflation that reached Weimar (and sixty years later, Zimbabwe) levels.

Stalin watched and, with his typically clever ruthlessness, decided to take advantage of the goodwill accruing from this fig-leaf gesture of allowing free elections, knowing the economy would soon collapse and the winners would take the blame. In the end, the voters elected a four-party coalition government, led by the blue collar/shopkeeper Independent Smallholders' Party. But everyone knew that the Hungarian Communist Party, itself little more than a front for the Soviets, held all of the power—backed by the Red Army, which ruled over the countryside and guarded every street of Budapest.

Still, it was a kind of normalcy. Maria continued working at the dairy shop, and George found work in a department store. The bridges over the Danube were replaced, and the city began to slowly rebuild. András was sent to third grade, where he did very well. And even his parents regained some of their old optimism and ambition, much of which revolved around their only child.

Before long George and Maria insisted that András take English lessons. He disliked them, and given that he still speaks with an accent, he obviously had limited aptitude. Yet they would prove crucial to his future. Maria was still thinking ahead.

The good times had an unexpected side effect: Andy began to put on weight at a surprising rate. "In the years after the war, I had gained weight. First I got pudgy, then I got even pudgier. Kids at school started calling me a variety of nicknames from Pufi (which means 'Fatso') to Rofi (the sound a pig makes). I didn't like being called these names, but the more I protested the louder the other kids shouted them at me across the schoolyard. So I resigned myself to being Pufi or Rofi. They became my names even in my own thoughts."[4]

Grove's biographer Tedlow goes to some length into the potential psychological sources of András's weight gain. But the reality may just be that his parents, given everything the boy had been through, decided to indulge him. One thing is certain; this was an interval in his life that Grove still remembers vividly—in some ways as much as parts of the war years—because of his humiliation and helplessness, something he would never allow in his life as an adult. Perhaps as a reminder, Andy even put a photo of his pudgy self in his memoirs.

Once off the schoolyard and into the classroom, András continued to thrive. Because his ears had not yet fully healed (they kept him from most athletic activities other than head-above-the-water swimming), teachers learned to put András in the front row. "My ears still drained and I didn't hear very well, but if I sat in the front row and the teacher stood right in front of me and spoke loudly enough for the whole class to hear, he was loud enough to hear him very well."[5]

Pudgy, hard-of-hearing "Pufi" was consistently at the top of his class in academics. It helped that at home, life was also restoring itself, as his parents' growing prosperity and the recovery of the neighborhood again made the Gróf apartment a social gathering place.

But in the larger world, the hall-of-mirrors world of Soviet totalitarianism was growing, slowly and inexorably snaring one civic institution after another, transforming it into an instrument of the state, then moving on to the next target. The first few years after the war, this trend had seemed relatively painless. But Stalinism was never benign—and in 1949, after the rest of Eastern Europe had been dragged behind the Iron Curtain, it finally showed its true nature in Hungary.

CHAPTER 31

Andy in Exile

In September 1949, the entire country listened to the radio as Hungary's minister of foreign affairs, László Rajk, was made to endure a show trial similar to those staged in the Soviet Union in the late 1930s during the Stalinist purges. Rajk, who had been tortured, confessed to everything. He was executed on October 15.

In a story that had been repeated many times in their neighboring countries, the response of most Hungarians at first was more confusion than fear. Rajk had confessed, hadn't he? Or had he been forced to? With each day, as access to the outside world diminished and the only news came from the state-run radio, the truth became more and more elusive.

For András Gróf, a natural empiricist and now a budding journalist, the growing discrepancy between what he saw with his own eyes and what was pronounced as official truth by the state-controlled media was becoming increasingly difficult to deal with, and as that discrepancy became a gulf, it became unbearable to the young man. Truthfulness, even at its bluntest, became central to how he defined himself.

On May 1, 1950, during the Communist Party's annual May Day celebration, András and some friends decided to walk over to Heroes' Square to watch the parade and the celebration that followed. While they were still blocks away, they could already hear the speakers intoning their remarks as the huge crowd roared. They picked up their pace, anxious to see—in their increasingly dreary totalitarian culture—some real spectacle and excitement. As they approached the square, still blocked from view by surrounding buildings, the group of high schoolers could hear the crowd roaring "Long live the communist party! Long live [Prime Minister] Mátyás Rákosi! Long live Stalin!"

Anxious to join the celebration, the little group trotted around the corner . . . and stopped in its tracks. Grove: "However, when we arrived at the square, the only people standing and waving were the Party members on the reviewing stand. None of the previous marchers had stuck around to watch. No other marchers were cheering. None of us were cheering."[1]

It was a Potemkin event. There were no cheering crowds; the cheering that András and his friends heard was piped in via loudspeakers hung from lampposts along the parade route. The newsreel cameras, filming for theaters around the country, would capture the fake cheers, the stolid and murderous men on the reviewing stand, the long lines of parading marchers as they passed in formation through the square. None would catch the otherwise empty square or the marchers, having followed their orders, walking away.

It was all a lie. And it began to dawn on András for the first time that everything in the New Hungary was a lie. Two years before, his father had joined his mother working at the dairy store—until the government nationalized the dairy industry in the name of the people, as it did all businesses with more than ten employees, resulting in the end of "the ready supply of fresh cottage cheese, butter and yogurt that we were accustomed to." As their income fell, George and Maria moved their son to a less prestigious, cheaper school. András quickly discovered that his new school was different from his old one. There was a different atmosphere, one largely created by the changing nature of the society outside its walls. Now, even as the new official curriculum promulgated by the educational commissars was drifting ever further from reality, the ability to challenge those obvious falsehoods was being suppressed as well. Grove: "Contradicting a position that was even vaguely associated with the Communist Party didn't seem like a wise thing to do."

Everywhere he looked, András saw this same yawning chasm between what even the censored data said was true and the official party version of that "truth." As a journalist, the young man found himself especially torn: "On the one hand, I felt that [the communists] had saved my mother's life and my own. I was very grateful for this, and my gratitude made me want to believe in them and what they stood for. On the other hand . . . they increasingly interfered in our daily life . . . all in the name of a political philosophy that I didn't really understand."[2]

The contradictions only grew louder in András's mind with the show trial of Rajk. Was he really a traitor? The verdict seemed predetermined; the execution too quick. András, growing in confidence as a journalist for both the school paper and increasingly for a national magazine for young people, sensed something wasn't quite right with the story.

Then came the May Day parade. It was assigned to András as a major feature for the youth magazine. But the fraudulence of the event was too great for him; András passed on the assignment, . . . and it was assigned to someone else who probably wasn't even there.

The contradictions of living under a communist regime were growing ever greater and ever harder to swallow. The radio and the newspapers told lies. The lessons being taught at school were full of half-truths, if not outright falsehoods. And the guests who still gathered at the Gróf household were becoming increasingly careful and circumspect about anything they said.

Worst of all were the antipodal responses to the new regime by András's father and uncle.

For George Gróf, the decision to leave the dairy store and take the job at a department store proved providential. He was soon tapped by the government to become a director at a state-owned livestock breeding and export company. George Orwell's *Animal Farm* was published almost at the same moment that George returned to Budapest after his imprisonment—and now he was experiencing the reality of some people being more equal than others in the new workers' paradise. He and Maria had been forced to fire the two young women who served as housekeepers, who had been thrilled to have the work even after being told by the government that they were "exploited."

But now, in his new position, George, according to his son, had "an elegant secretary" and even a chauffeur, both of whom were apparently immune from the same reactionary exploitation. George was now a member of the *nomenklatura*—and to his son's growing disgust quickly overcame any guilt and began to revel in his new power and prosperity.

Meanwhile, George's brother-in-law Sanyi—who had warned Maria and András to get out of the Jewish House because the Arrow Cross thugs were on their way—had taken a different path in postwar

Hungary. He had become a newspaper editor, and a brave one at that. Sanyi was unafraid of challenging the official lies of the government, and that made the government angry, a very dangerous thing in the Soviet empire. András admired his uncle greatly, and it was likely his desire to emulate Sanyi that first led the young man into a journalism career.

Then in early 1951, on an otherwise uneventful evening, Maria was reading the newspaper when she spotted George's name. The story announced that her husband was accused of consorting with "bourgeois elements." Terror paralyzed the family: that kind of accusation got people tortured, even killed.

But what was the source of the accusation? The answer soon came: Sanyi and his son-in-law had been arrested by the secret police. Andy recalled, "My aunt showed up at our house the next morning, frightened and utterly helpless. Nobody would say where they were taken or why. There were no charges and no one to inquire to. They were just gone."[3]

Would they be next? There was nothing they could do "except to wait and see what would happen next."

The answer came quickly. Without explanation—none was necessary—George was stripped of his job and told that he would be lucky to ever again hold a job at a quarter of his current salary. Writes Tedlow: "George was able to find employment, but the family's standard of living declined. Meat once a week. No more delicacies. Cheap seats at the opera. A long commute by tram, a big step down from the chauffeured car."

George, as was his personality, was stoic. Recalled Grove: "I never heard my father complain about the loss of his job, . . . in fact, I never heard him complain at all, but he became very quiet. He was the man who used to thrive on political discussions. Now he refused to discuss politics. And in any case, there was no one to discuss it with. Most of his friends still stayed away."[4]

George Gróf had become radioactive and very nearly (to use Orwell's term) an unperson. Even his friends and neighbors who didn't believe the news stories still thought it prudent not to be seen in his company. As for András, his father's "embrace of the regime and all of the perquisites that came with that collaboration"—and now his quiet bitterness at now having lost them—only increased the estrangement in a relationship that had never been fully restored after the war. A half century later, when both of his parents were gone, Andy Grove would choose to dedicate his memoirs solely to his mother. And like his uncle, he would choose to write those memoirs unflinchingly, putting accuracy before diplomacy.

The stigma attached to his uncle and father now reached András as well. Suddenly his story submissions, until now readily accepted, stopped being published and ended up in the slush pile. When he approached the editor about it, he only got an evasive answer. He told his mother what happened, and she agreed that it likely had to do with his uncle's arrest. "A career in journalism suddenly lost his appeal." Andy would never really trust reporters or editors again.

The change in the family's circumstances—along with the fact that the boy was now a teenager—once again required a change in schools. And despite the difficulties faced by his family, the young man, now fifteen, was coming into his own. He not only was getting top grades, but his giant personality and huge will were now starting to make their appearance. Soon his new literature teacher would tell a meeting of all of the students' parents, "Someday we will be sitting in Gróf's waiting room, waiting for him to see us."

András's favorite teacher was a Mr. Volenski, whose class on physics would have a profound effect on the rest of the boy's life. On parent-teacher night at Madách Secondary School, Volenski told George and Maria and everyone in earshot, "Life is like a big lake. All the boys get in the water at one end and start swimming. Not all of them will swim across. But one of them, I'm sure, will. That one is Gróf."

Those words would haunt Andy the rest of his life, and no doubt he called upon them during the most difficult moments of his career. And when it came time to write his memoirs, he not only chose his old teacher's comments for the title, but finished the book with these words:

As my teacher Volenski predicted, I managed to swim across the lake—not without effort, not without setback, and with a great deal of help and encouragement from others.

I am still swimming.[5]

It wasn't long before everyone at the school knew András. He seemed to make his mark in every class he took. "After the fiasco of my potential journalism career," he later wrote, "I was eager to cultivate an interest in a new profession that was less prone to subjectivity." He settled on chemistry. By his third year at Madách, András was so good at the subject and so trusted by the faculty that he was invited to demonstrate how to create nitroglycerin in front of a class of thirty second-year girls.

At seventeen a year older than the class, and now increasingly svelte and handsome, this was a big moment for András—and he played it beautifully. "Bang!" he wrote in his memoirs, still cherishing the moment. "The class broke into shrieks and excited applause and I was on top of the world!"[6] Andy Grove the self-dramatist was born.

He made an even greater mark on his English class. He may have abandoned journalism, but that hadn't kept him from writing. When the time came to submit his short story, called "Despair," András decided to leave its author anonymous.

What followed was every writer's dream moment: his literature teacher handed out copies of "Despair" to the class and asked everyone to read it. In the lively discussion that followed, almost everyone agreed it was a marvelous story done by a talented writer—and began to speculate on who in the class could have written it. The teacher himself said, "Whoever the author is, I'm sure that this not the last we have heard from him."

András watched and waited. He had a natural dramatic flair and a comic's sense of timing—though few would see it again until Intel sales meetings in the 1980s—and so he waited until the tension peaked . . . even until after the class began to agree it was another student, who vehemently denied authorship. That's when Andy stood up and announced his authorship of the story.

"Pandemonium broke out. The meeting dissolved into excited kids slapping me on the back, congratulating me, and shaking their heads in disbelief . . . [my teacher] Mr. Telgedi shook my hand and repeated that he thought the story was great. . . . This was the most exciting event of my life."[7]

In just three years, András Gróf had gone from the fat Jewish kid from a suspect family to the slim, handsome, most admired kid in the school. Grove: "One day at home, I caught a glimpse of myself in the mirror without a shirt on. Much to my delight, I noticed that I had muscles. I had finally lost my remaining chubbiness." It was as if he had stepped out from under a dark cloud and was now standing shining in the bright sunlight of acclaim. And it couldn't have come at a better time.

In March 1953, Stalin died. In his memoir, Andy Grove did as good a job as anyone explaining it: "Stalin's figure had been indelibly associated with the images of the Soviet Union in my mind. The picture of a uniformed, mustachioed man with a kindly expression had been everywhere—in offices, schools, at celebrations, hung on the sides of buildings—for, it seemed, most of my life. Even though by this time I had become deeply skeptical about the goodness of things Soviet, Stalin's death and the disappearance of that ever-present kindly face had a mixed impact on me.

"I was glad and I was sad at the same time. It was very confusing."[8]

As Stalin's potential successors maneuvered to see who would be the last man standing on the Red Square reviewing platform, Soviet satellite nations—especially Hungary—took advantage of their distraction by instituting reforms and generally loosening their grip on the citizenry.

It got even better. In June, Hungary's Communist leaders were called to Moscow. Within days, Prime Minister Rákosi stepped down. Better yet, over the next year, an estimated 750,000 political prisoners were released from Hungary's gulags under a general amnesty. Among them was András's uncle Sanyi, who was released in the spring of 1954. The family was restored, but whatever faith it had in the government or its benevolence was long gone.

Meanwhile, András, school superstar, prepared to apply to the University of Budapest, where he hoped to study the natural sciences. But between here and there lay two big obstacles.

The first was the oral finals at his school—which he passed easily. Second, and more ominous, was the fact that András was officially classified as a "class alien" because of his father's and uncle's records in the files of the secret police. That designation was the equivalent of an automatic veto to any potential university acceptance.

András turned to the one system that he knew might work: the underground. An old friend had a connection inside the government, . . . and somehow through some backdoor means, András Gróf's official status was quietly changed to "other." With that, the second great obstacle fell, and András was accepted into the University of Budapest. He had learned his first lesson in how to navigate government bureaucracy.

András entered the university in the autumn of 1955. Any student of the Cold War knows what was coming. But to András, life in Budapest at the university could not have been happier. He fit in quickly and earned top grades despite the fact that his professors were less accommodating to his enduring hearing difficulties. He was happy, not least because "I no longer had to be embarrassed about being a good student. At university, we were all there to learn and we all wanted to do well."

More than the formal education he received that first year, the real breakthrough for András was the expansion of his social circle. As a boy in the Jewish quarter of Budapest, on the run from the Nazis, attending schools filled with Jewish students, András had been around very few gentiles. He knew a few kids who weren't Jewish, but he counted none of them as real friends.

That changed in his freshman year at the university. Despite the fact that even there mixing between Christians and Jews was still rare, András made acquaintance with a gentile named Zoltán, whose brains and worldly ideas either matched András's own or piqued his curiosity: "Zoltán's caustic wit and his sharp insights impressed me, as did his interest in Western literature and music—he was an accomplished jazz pianist. His attempt to look Western, I soon realized, wasn't an act at all but was completely consistent with his interests."

Even more appealing were Zoltán's political attitudes. "[He] openly made cynical political comments to me, [and] I found myself opening up to him more and more, too." Though the two young men didn't yet know it, this same hunger for free speech was, still sub rosa, spreading throughout Hungary.

As close as the two young men became, religion divided them and restrained them from a true friendship. Finally one day, András (as was increasingly his way) decided to attack the problem head-on. He asked Zoltán, "Does it bother you that I'm a Jew?"

"Why would it bother me that you are a stinking Jew?" Zoltán replied. András was briefly shocked by the words, then grinned. "Yes, and why should it bother me that I hang out with a dumb goy?"

At age eighteen, András had his first gentile friend. The two even created a code for addressing each other: in Hungarian the initials for "stinking Jew" resemble the symbol for bismuth on the periodic table, and "dumb goy" that for mercury. From that day forward, the two called each other in public by their element code name as an inside joke.

As required by law, András spent the summer after his freshman year in the army. Being national service, it wasn't particularly onerous—especially for a brilliant college kid in good shape—but more like an extended camping trip. And András returned at summer's end healthy, tanned, and anxious to get back to university life.

But soon history once again intervened. The death of Boleslaw Bierut, the brutal Soviet puppet running Poland, unleashed a hidden desire for freedom in that country. This "Polish October" set off riots and demonstrations that soon spread into East Germany. The Soviets could not allow this kind of revolt and keep their empire, and soon the governments of both countries, under threats from Moscow, crushed the rebellions.

But in Hungary, where the rebellion jumped to next, the fire of freedom continued to grow, centered on the University of Budapest. Recalled Grove: "A buzz spread through the university about a march that was being organized to express our support for the Poles" to be held on October 23. The day came, and it was like a citywide street carnival as the marching students were joined by thousands of other city people from all walks of life. For András, it was thrilling, and the contrast with that empty square during the official parade six years earlier was particularly stark: "After all the years of sullen, silent May Day marches, there was something magical about a large spontaneous demonstration." András joined in.

Then things began to spin out of control. Suddenly, from what seemed like every window along the parade route, flags appeared. Red flags. The Communist flags that had flown over Budapest since the war. But now, the hammer-and-sickle emblems at their centers had been cut out—symbolizing Hungary's desire to be free of its Soviet overlords. An understandable sentiment, but also a provocative one. Grove: "Those flags were permanently altered. The act seemed unequivocal and destined to provoke a reaction. . . . I felt like we had crossed a line of no return. I began to feel a little nervous."[9]

He felt a lot more nervous as the day went on. By afternoon, the march had turned into a full-blown demonstration involving thousands of Hungarians. By evening and the end of shift, the factory workers arrived—some of them carrying a cutting torch. With it, they attacked a prominent bronze statue of Stalin, symbolically cut off his head, and took turns spitting on it. No one had any doubt after this direct insult to the Soviet Union that the Red Army would respond somehow, somewhere—and soon.

A nervous András went home that night, slept fitfully, and woke to the sound of gunfire. He decided it would be prudent not to go to the university that day. Keeping himself occupied, he was astonished to find that Radio Free Europe and the Voice of America, both of which were always jammed, were now coming in clear and strong. They were even giving support to the Hungarian uprising. What did that mean? Had the government fallen?

It had. The prime minister and his cabinet had run off to the countryside. Even the Red Army had evacuated the city. And both were still on the run, as the Hungarian countryside was now rising up as well. For a shining moment, it seemed as if Hungary might finally escape the Soviet grip and become a free and independent country once again.

But it was only a fantasy. On November 4, the Red Army attacked, sweeping over the countryside and, as the West was haunted by radio messages from the city begging for help, crushing Budapest. Recalled Grove in sorrow: "I had never seen such devastation, not even from the bombing during the war."

Then the killing began. The prime minister who ran was executed. Men and women who tried to fight the Soviet tanks and artillery were slaughtered. And, once the government was back in full control, the secret police were unleashed to round up the protesters. Soon individuals, including many students, began to disappear from streets and homes, never to return.

At the Gróf apartment, a desperate discussion was under way about what András should do. Lie low and hope that he wasn't on a list somewhere to be rounded up by the police, or make a run for the Austrian border? If the latter and he wasn't caught along the way, András already knew where he would head next: "Of course, it would be America. Or, as the Communist regime put it, 'imperialist, money-grubbing America.' The more scorn they heaped upon it, the more desirable America sounded. America had a mystique of wealth and modern technology; it was a place with lots of cars and plenty of Hershey bars."

It was András's Aunt Manci, with whom he'd stayed in the countryside years before, who forced the final decision. "András," she told him, "you must go. You must go, and you must go immediately." Her words carried the authority of an Auschwitz survivor. Maria agreed, and with her own characteristic decisiveness decreed that András would leave in the morning. That night, András "silently said goodbye" to the apartment, the only real home he had ever known—and prepared himself to leave this life, and perhaps even his family, forever.

"Struggling with the fact that they might never see each other again, George, Maria, and András" went for a walk the next morning "as if it were any normal morning"—like that day fifteen years before when they had walked together and watched the searchlights.[10] They parted without so much as a hug, so as not to alert anyone who might be watching. András then walked, without looking back, and made his way to a rendezvous point, where he met two other escapees, a boy and a girl he knew. The three of them then made their way to the train station, where they bought tickets and boarded a train for the 140-mile trip to the city of Szombathely, just fifteen miles from the Austrian border.

On the train, the trio met another girl with the same goal, who joined them. Now they were four, armed only with a list of villages to serve as guideposts along their way west. They had memorized the names, because such a list on a piece of paper would be incriminatory.

This passage, which took two days, was like something out of a nightmare. They hiked through woods and fields, mostly in the darkness. They ran into a hunchback in the middle of a field, who spoke only in whispers and gave them directions to a lonely farmhouse. There they knocked on the door, . . . which opened to reveal "a stunningly beautiful woman in a colorful peasant dress." The group slept there for the night.

That night, András used the outhouse. Sitting there, "I looked up at the bit of sky barely visible through the doorway and I thought, this is probably the last of me that I'll leave in Hungary."[11]

The next afternoon, a guide appeared to lead them. He proved to be like a wraith, disappearing for long periods, then suddenly reappearing to tell them to follow him. Finally, he stopped on the edge of a field and pointed into the distance: "Those lights are Austria," he whispered to them. "Head toward them and don't take your eyes off them. This is as far as I go."

The four students stumbled their way in the darkness across a plowed field. A dog started barking, then another. Suddenly, a flare hissed into the sky, exposing them in its harsh white light. The four of them dived for the ground, trying to get out of sight. The flare burned out, and they were once again in pitch darkness. Then a voice: "Who's there?" They realized that he was speaking Hungarian. Had they missed the border?

"Relax," the man shouted. "You are in Austria."

The four were then handed over to the Austrian police, who put them in an unheated schoolhouse for the rest of the night. It wasn't long before an argument broke out. The girls trusted the policemen and were grateful for their protection. But András argued that he didn't come this far to put himself into the hands of a different police force. What if they chose to send the four of them back to Hungary? The other boy agreed, and early the next morning, the two groups split; András and the other male student took off down the road, leaving the two girls behind.

For a young man who had never been more than a few miles from his hometown, young András acted like a seasoned traveler when he reached Vienna. He quickly fired off two telegrams: one to his parents to let them know he was safe, another to Manci's relatives, Lenke and Lajos, in New York City, asking if they would put him up were he to get to America. Later, he posted a longer letter to his folks, saying in part, "God knows, if I have a chance I will go as far as I can."

In between was a mad dash around the city, getting the permits and supplies he needed to get out of Austria and make the trip to the United States. As Andy later described it, he was "like a madman"—and in perhaps the first glimpse of the business titan to come, he admits that he didn't hesitate to "brush past the person whose turn it was supposed to be" in a line of fellow refugees at a government office. Yet even in the midst of all of this frenetic activity, András—already desperate to participate in the cultural life of the West—managed to score a cheap ticket to the famous Vienna State Opera. His critical appraisal: overrated compared to the Hungarian Opera.

The trick now was to keep moving, and that meant sponsorship by a nonprofit refugee-placement group. The best candidate was the International Rescue Committee. He rushed over to its Vienna offices and submitted his application. He then sat down for an interview.

It began well. The interviewer was pleased when Andy spoke in English, rather than through a translator in Hungarian. But then the conversation took a turn for the worse: when asked if he had fought against the Russians—an advantage in getting a privileged refugee status—András replied honestly that no, he had not, but merely participated in several demonstrations.

His response caught his interrogator by surprise. Apparently in hopes of gaining goodwill, almost *every* Hungarian refugee who had arrived in recent days had claimed to be a brave freedom fighter who had taken arms against the Red Army. Andy later recalled, "A sarcastic thought came to mind: If all these people had fought, we would have won and I wouldn't be here." Even though he knew it would be to his advantage, he was "reluctant to fabricate a story for the occasion."

He paid the price for his honesty. The next day, he learned to his dismay that he was not on the IRC list to be taken to the United States. "I felt as if someone had socked me in the stomach, then my heart started beating so hard that I could hardly breathe."[12] But he refused to give up. He at once forced his way past the line at the IRC and confronted its representatives. To no one's surprise who knew him later, through a combination of pleading and sheer force of personality, András argued his case, refusing to take no for an answer. To his amazement, the IRC officials listened, shrugged, and agreed. "I was speechless," Andy would later write.

He was going to America.

CHAPTER 32

Freedom Fighter

Within days, András and his fellow refugees were loaded on a train and sent west—through the country he dreaded most in the world: Germany. Would they be stopped at the border? And how would they react to a Jew entering their country? "The tension didn't leave me until we got to the first stop in Germany, the city of Passau."

In Bremerhaven, the refugees boarded a beat-up, decommissioned World War II troopship, the *General William G. Haan*. It was not a particularly comfortable voyage. As the refugees, all 1,715 of them, grew weary of the discomforts and increasingly seasick, arguments began to break out, many of them reflecting the old animosities, not least anti-Semitism. A Hungarian American minister, who was accompanying the crowd on the voyage, gave a homily on how the old hates had to be left behind to become part of the American melting pot—and was answered by a refugee who brandished a large knife and announced, "I hate whoever I want to hate." At the moment, András realized that the New World would never be a complete escape from the problems of the Old World.

Still, despite the difficulties, András found the voyage enthralling. As the ship passed Dover, "The thought struck me: I'm looking at England. The momentousness of everything suddenly hit me: leaving Hungary, traveling through Germany, seeing the sea for the first time, seeing England." All of this had been "unthinkable just a couple of weeks ago."[1]

The troopship arrived at the Army Terminal in Brooklyn on January 7, 1957. As Grove's biographer Richard Tedlow discovered, on that day *Time* magazine announced that its Man of the Year for 1956 was the

"Hungarian freedom fighter"—forty years before Andy Grove himself was awarded that title.

Brooklyn in the dead of winter didn't exactly provide the most attractive first glimpse of America to the young refugee. Nor did the bus trip to Camp Kilmer, a former POW camp for German soldiers, in New Jersey. During the trip across the blasted, smelly, sodden landscape of the Jersey Meadowlands, one refugee shouted from the back of the bus, "This can't be true. It's got to be Communist propaganda!" Camp Kilmer was only marginally better.

But once settled in, András was allowed to call his Uncle Lajos, Aunt Lenke, and his twelve-year-old cousin, Paul, in the Bronx and announce his arrival. The next day, Andy was standing in their apartment on West 197th Street. And for the first time in months, he was able to call his parents in Budapest.

A few days later, he ventured out with his relatives to take a look at America. Uncle Lajos worked at Brooklyn College but volunteered to take Andy into Manhattan to see Aunt Lenke at her job in a department store in midtown. As might be expected, the subway ride did little to improve Andy's mixed feelings about the wealthy and shiny America of his dreams, but then they stepped out of the station into the middle of the great city:

[I] stopped cold. I was surrounded by skyscrapers. I stared up at them, speechless.

The skyscrapers looked just like pictures of America. All of a sudden, I was gripped by the stunning realization that I truly was in America. Nothing had symbolized America more to me than skyscrapers; now I was standing on a street, craning my neck to look up at them.

Which also meant that I was an incredible distance from my home—or what used to be my home.[2]

CHAPTER 33

A New Life, a New Name

O ver the next five years, András Gróf became an American.

His initial strategy was to get a job at an established company and start a career, much as his father had done in Budapest. "My aim was to acquire a profession that would enable me to become self-sufficient as soon as possible, so I could support myself and set aside enough money to help my parents get out of Hungary and join me in America."

András initially dismissed the idea of college because of the high tuition costs: "I was about to forget about college when I learned about the city colleges. Friends told me that all I needed was ability," he wrote in 1960, when he was still astonished by the opportunity. "Americans don't know how lucky they are."

He first checked out the likeliest candidate, Brooklyn College, but its chemistry program did not fit his interests. Then there was Brooklyn Polytechnic, but its $2,000 annual tuition was out of reach. "They might as well have told me it was two million."

But the folks at Brooklyn Polytechnic were kind enough to suggest that András try the City College of New York, where tuition was free. It wasn't long before he was enrolled at CCNY and receiving a stipend for books and living expenses from the World University Service.

Grove would later say that CCNY played a crucial role in making him a true American. Forty-five years later, in a speech he made after making a major donation to the university, Grove spoke from the heart: "I asked for the admissions office, and somebody sits me down, and I tell them my story. I was wondering what shoe was going to hit me in the head this time, but they accepted me with respect, without condescension. They gave me a start, and they gave it in a classy way. It's an institution that is crucial to the workings of America, and America should be proud of it. I am."[1]

Still, American university life wasn't easy at first for this young man literally just off the boat. At first, his grades suffered—not just, as one might expect, in the language and composition courses, but even in the hard sciences like physics, in which he earned an F on his first big exam. But András Gróf didn't come to America to be defeated. He bore down even harder on his studying: "I used every minute to study" and only occasionally "rewarded myself by buying a paper cup filled with Coca-Cola from the machine in the subway station for five cents. But I tried not to do that too often because that would be five cents I wouldn't have for my parents."

On the next physics test, András earned an A. He also doubled down on his work. When the stipend ran out, he took a job as a student assistant in the chemical engineering department for $1.79 per hour for a twenty-hour week. Meanwhile, he bumped up his course load to twenty-one units—far above the standard sixteen, even though these courses were in chemistry, calculus, and physics.

The course load didn't stop him but seemed to drive him on to even greater challenges. Though he had no free time and his English was still weak, András still decided to compete in a technical-paper competition sponsored by the American Institute of Chemical Engineers. Incredibly, almost impossibly given the odds against him, András won. He was as amazed as anyone: "As a student, it was the first thing I ever won."

"Immigration is a transforming experience," András would write in his memoirs. And now there was one last detail to be taken care of: his name.

Almost from the moment of his arrival in the United States, András's last name had proven to be a handicap. Professors would stumble on the single *f* and the accent over the *o*. "In Hungary, Gróf is pronounced with a long 'o'; here, everyone read my name as if it were written, 'Gruff.'" When he told his aunt and uncle about his frustration, they told him, "This is America, just change it."

So he did just that. András started playing around with different American surnames, finally settling on Grove as being closest to the Hungarian original. For his first name, he went with Andy, because that was what he was already being called, and then formalized it to Andrew. For his middle name, he chose Stephen, a direct translation from the original István. Now he was Andrew S. Grove, the name by which he will now be known for the rest of this narrative.

One of the acts that endeared CCNY most to Andy was that when he went to the registrar's office and told them of his choice of a new name, the reaction wasn't a challenge to produce the proper documentation, but a simple "OK." The clerk didn't even redo Andy's transcript but merely took a pen, crossed out Andy's old name, and wrote in the new one. Done. And it was as Andrew Stephen Grove that he became an American citizen in 1962.

There was one more step in Andy Grove's full immersion into America. In the summer of 1957, his first in the United States, he went looking for work. He spotted an ad by Maplewood, a New Hampshire resort hotel, looking for busboys. Maplewood, a big classic Victorian pile that had catered since the turn of the century to Jewish families seeking relief from hay fever, was likely looking for Jewish college boys from Brooklyn or Hoboken, not one just arrived from Budapest. So the job interview was a bit awkward, but likely through the sheer force of his personality, Andy got the job.

It must have been a disorienting experience for someone just getting used to big city life in New York to suddenly find himself at a vacation spot in a remote corner of New Hampshire, but Andy just seemed to bull his way through it all and find his place before the veterans could even react. Within a few days of his arrival on June 10, he was already dating a girl. It wasn't a good fit, but it was a hell of a statement by Andy to everyone else.

A few days after that, he found a girl much better suited to his personality and history. Her name was Eva Kastan. She was a year older than Andy, and she had been born in Vienna. Her parents had escaped the Nazis when she was three, and the family had emigrated to Bolivia, then, when Eva was eighteen, to Queens. She also had a boyfriend. But that didn't stop Andy: they were married a year, almost to the day, after he first saw her.

The couple were married in a Roman Catholic church in Eva's Queens neighborhood. One of Eva's friends had requested that they do so. Andy would later say about the odd choice of wedding site, "[I] couldn't have cared less."

One thing the newlyweds agreed upon was that neither particularly liked New York City ("cold and wet and ugly," Andy described it). On the advice of one of Andy's professors, they decided to honeymoon in California. They loved it. Andy particularly liked UC Berkeley and the surrounding neighborhood (it's interesting that he was unimpressed with Stanford). They agreed to return; Andy hoped to get work at the Chevron oil refinery in the East Bay and Eva at an area mental hospital (she was finishing her master's degree in social work at Columbia). But for now, they had to get back to Manhattan and finish their degrees.

There were three professors who played a critical role in Andy's last years at CCNY: Morris Kolodney, his freshman curriculum adviser and the faculty member who suggested he visit California; Alois X. Schmidt, whose chemistry course had terrified a generation of young chemists out of their dreams; and Harvey List, whose course on fluid dynamics deeply interested Andy and was the reason he was investigating a career at a refinery.

Of the three, Schmidt may have had the deepest impact on Andy's personality and approach to the world. Looking back, Grove would say that Schmidt taught him "toughness," that he "legitimized . . . a brusque, no-nonsense behavior which I had no trouble adopting. Polite company frowned on these traits, but Schmidt practiced them. I thought if he can do it, I can do it." He would also say, "Harry was sending me to fluid dynamics" while "Schmidt was turning me into an asshole." A lot of Intel employees and competitors would say that he learned the latter lesson far too well.

Following graduation that summer of 1960, the young couple packed up and headed west. Andy had been accepted to graduate school at Berkeley and also had a summer job waiting for him at Stauffer Chemical Company, a few miles away from campus. It turned out to be a lousy place to work—in Andy's words, "miserable" and "run-down," and the work stultifyingly boring. Even worse was the attitude of his coworkers: "One incident stuck in my mind and had a real impact on me as a future manager. People came into work on Saturday . . . and I discovered that they sat around and bull-shitted and had an eye on the window until the most senior person pulled out. Minutes later, [the] next senior person's car pulled out; and minutes later the next one, and within ten minutes after the last senior guy left the whole place was empty."[2]

No one would ever describe Intel that way. Andy made sure of that.

Classes began that fall, and Andy, who had finished as a superstar at CCNY, was stunned to find that at Cal he was once again overwhelmed by his courses and at risk of flunking out. What made it even more painful was that, even as he was trying to make sense of what the professor was saying, he would look around at his classmates and see them casually taking notes as if the material on the blackboard was entirely self-evident.

Then, when his despair was at its lowest, an amazing thing happened. Try as he might, Andy found that he couldn't follow the logic of a particular proof being written on the board, so he raised his hand and asked about it. The professor paused, looked at what he'd written, and realized he'd made a mistake. He went back and corrected it. As Andy watched, his fellow students erased and corrected their notes with the same aplomb.

It was an epiphany for Andy Grove, "because I discovered 'these toads don't know any more than I do. They just don't dare speak up. To hell with them.'"

That realization enabled Andy to turn his grades around and ultimately become a top student at Berkeley, as he had been at CCNY. But it also exposed the dark side, the intellectual arrogance, of Andrew Grove. It is almost impossible to imagine Bob Noyce, much less Gordon Moore, contemptuously dismissing his hardworking (and no doubt equally terrified) classmates as "toads." As many would note about Andy Grove, business titan, it was almost impossible for him to straight-out compliment anyone; lauds and laurels always came with a qualifier at the end.

Though it is hard to imagine the middle-aged Andy Grove allowing himself to be subordinate to anyone, once again at Berkeley, as he had at CCNY, Andy found a mentor, this time to serve as his PhD adviser. Andy Acrivos was not only a top-notch (and highly honored) academic, but also was the only professor at Cal who could actually teach Grove something about fluid dynamics. The two men became enduring friends. It was Acrivos who devised for the young scholar a curriculum tough enough to keep him challenged.

When the time came for his thesis, Grove didn't settle for a safe dissertation that would guarantee his doctorate. Rather he decided to tackle one of the most recalcitrant problems in fluid dynamics, one that had been a conundrum in the field for more than a century. Another doctoral candidate at Cal had just worked on it for five years . . . and given up.

Andy went at it fearlessly, tossing aside the past experiments as the work of dilettantes. "I like experiments," he later wrote. "I don't like tinkering—there's a difference. . . . I went in the face of the prevailing dogma on the basis of my experiment . . . and proposed a 'Gordian Knot' kind of solution that was completely against classical beliefs. I had the guts to understand . . . what the experiment was saying. I had a PhD advisor [Acrivos] who, after a fair amount of due diligence, believed my data."[3]

Grove's dissertation ultimately produced four important scientific papers, coauthored with Acrivos, two of which were published in the top journal in the field. As Grove once told an interviewer, "This is going to sound awful, [but] it is an amazing thesis."

In the interregnum between earning his PhD and figuring out what do next with his career, Andy taught a course in chemical engineering at Cal. By then, the university had radically (literally) changed since his arrival four years before. And he quickly found himself frustrated trying to teach as the campus was experiencing the birth of the Free Speech Movement. On one particular day, Andy found himself trying to get to his classroom while the university was experiencing "one huge student strike."

As he made his way across campus, Grove grew increasingly furious, deciding about his students that "if those bastards don't show up" he would give them a pop quiz—he even figured out the questions as he marched along—and then give them all zeros on it. "I went through the empty campus and people were nowhere and worked myself up into a rage. [I] got into the building where my class was, and I slammed my door open, and there was my whole class waiting in their seats."[4]

Did the students show up out of respect for Professor Grove? Out of fear? It is a glimpse into the humorous side of Andy Grove—a part of his personality that doesn't get enough credit—that he concluded that his class showed up because "engineers were always in a world of their own. They went about their education and said to hell with all of this other shit."

Despite the acclaim he was enjoying in the academic world, Andy had already decided that it offered neither the challenge nor the power to match his ambitions. Just as important, the life of a professor wouldn't give him the money he needed fast enough to live in the Bay Area, start a family, and most of all, get his parents out of Hungary. So he began to look for a job in the commercial world. As he had already discovered on his summer jobs, work in fluid dynamics was singularly uninteresting. But when a colleague suggested he might try solid-state physics, Andy jumped at the idea.

While this might seem like an unlikely switch in scientific fields, the reality—often forgotten even by Silicon Valleyites—is that semiconductors are a chemical business, the forging of elemental materials into highly complex devices. All the other stuff—electronics, packaging, software code writing, and programming—comes later. Nor did Andy consider such a shift daunting; in fact, he considered his dissertation to have been as much applied physics as it was chemistry.

Unfortunately, not all of the companies he applied to seemed to agree; and at those where he did manage to land a job interview, his aggressive personality often got in the way. "My background was not right for the job. My personality you either like or hate, so that wasn't a currency either."

In the end, he targeted five companies. Texas Instruments turned him down. So did General Electric. Years later, he was still amazed that they passed on him. Lockheed was interested, but Andy wasn't sure about the circumscribed work at a defense contractor.

That left Fairchild Semiconductor and Bell Labs. His first contact with Fairchild didn't go well; he and the recruiter basically hated each other. Bell Labs, even without Shockley, was still the hottest corporate research laboratory on the planet. And the labs put the full-court press on Andy to join it: recruiters even came out to Andy's lab at Cal, offering big money and the chance to pursue any topic of his choice.

But Bell Labs was in Morristown, New Jersey, and no way were

Andy and Eva going back to the suburbs of New York City. Even Palo Alto and Mountain View looked better than that. Still, it seemed that Bell Labs was going to be the only choice, . . . until Fairchild changed its mind and sent a new recruiter up to Berkeley to pitch Andy about taking a job: Gordon Moore.

This was the first meeting of two of the most important figures in twentieth-century business. Young Andrew Grove, PhD, arrogant, ferocious, and contemptuous of human weakness, was ready to take apart this latest emissary from across the bay . . . but found himself melting under Gordon Moore's charm. He instantly recognized in the soft-spoken, gently humorous genius the last great mentor of his life. Andy was twenty-seven; Gordon, thirty-four. Grove: "Gordon Moore asked me about my thesis, all on his own, and listened, and got it! . . . He [was] really a smart guy—very personable, no airs. Gordon was a *big* selling factor, helping me see what I wanted to be."[5]

Within hours of arriving at Fairchild, Andy knew he had made the right decision. Here's how he described his first week on the job: "When I arrived at Fairchild on Monday morning, my supervisor, who was an electrical engineer, gave me a problem. . . . It actually wasn't that complicated, but . . . [it did require] taking a physical problem and turning it into differential equation[s], solving the differential equations, doing a family of curves [and] looking at a particular parameter."[6]

As it happened, under the regime of Acrivos at Cal, Grove had been told to take some very advanced classes in calculus, so formulating and solving differentials, which might have buried any other young engineer, was a piece of cake for Andy. "How lucky can you get?" Andy said of those first days.

In the months that followed, Grove and Moore developed a superb working relationship as they drove Fairchild's R&D operations to ever more advanced new technologies and products. They made a remarkable pair. Like his sheriff father, Moore was a big man, and he gave off an almost supernatural silence and calm; he always seemed to say what needed to be said in the least number of words, yet those words were always incisive and even wise. And as soft-spoken and genial as he was—"Uncle Gordon," Andy would affectionately call him—Moore was a man of rock-hard integrity. Just as his father was utterly fearless when he was backed by the law, so too was Gordon unyielding when it came to scientific truth, not least the "law" that bore his name.

But Gordon never was a manager. As Grove would say, "Gordon was then, and continues to be, a technical leader. He is either constitutionally unable, or simply unwilling, to do what a manager has to do." Ask him to resolve a technical question, and Moore had a ready answer—almost always the right one. "But would he interfere in some conflict between X and Y and Z? Not on your life."

Sometimes, Gordon wouldn't even step up to defend his own point of view, something Andy—who asserted his own viewpoint even when it wasn't called for—found astonishing. As a result, Grove often found himself running meetings at Fairchild (and at Intel) even when Gordon was nominally in charge. Grove: "I would be [running] a meeting and people would be bashing each other's heads and all of that. . . . I look at Gordon, something is wrong. So I'd yell, 'Stop! Gordon, what's bothering you? . . . Shut up! Gordon, what's bothering you? Shut up! Gordon, tell us whatever you wanted to tell us. . . .' [S]omebody had to stop the traffic" to allow "access to Gordon's insight."

Moore seemed to understand and appreciate what Grove was doing—once even telling Andy, "You know me better than my wife." Whereas others couldn't stand to work with Andy, with his in-your-face manner and naked ambition, Gordon, who was unique among Valley leaders for having no apparent ego or ambition, was largely impervious to Andy's competitiveness. He respected Andy because the younger man was a brilliant scientist and he got things done.

Gordon needed that kind of help, because just as he rarely controlled a meeting, he also almost never asserted discipline over the lab, first at Fairchild and later at Intel. Grove would later describe Fairchild's R&D department under Moore (and Fairchild under Noyce) thus: "There was absolutely no discipline. There was no internal discipline to the place; and there was no external discipline or expectations that were put on the lab and . . . on the manufacturing organization to support the lab."

As already noted, when this chaos in the lab, over which Grove had only marginal control, began to worsen Fairchild's overall financial health, no one was going to blame Gordon Moore, whose reputation almost matched Noyce's, but rather his high-profile executive officer, Andy Grove.

Andy could live with that. Gordon was his mentor, his friend, and the man he perhaps most admired in the world. He would willingly take a bullet for Dr. Moore.

But with Bob Noyce, it was a different story. Andy Grove never approved of Noyce as a leader, neither at Fairchild nor at Intel. Yet Grove's stated objection to Noyce's insufficiently serious attitude, fear of confrontation, and desire to be liked largely fall apart in the face of the fact that Andy was more than willing to accept most of the same characteristics in Gordon Moore. In fact, it went much deeper than that.

Part of it, no doubt, was that in Andy's mind, Moore had far greater moral standing than Noyce to demand this level of loyalty and support. But that appears to have been only part of it. Robert Noyce was one of nature's thoroughbreds; he made it all look effortless. Catnip to women, the envy of most men, handsome, athletic, successful in anything he did, charismatic, a born leader, and not least, increasingly rich—Noyce could afford to be playful with life and gracious in his victories. By the time Andy met him, Bob was already a living legend.

Almost every part of this rankled Grove. There was envy, to be sure: by the time Andy arrived at Fairchild, Noyce already was famous for the integrated circuit and for the founding of Fairchild itself, while Grove was still trying to convince the world of his talent. But even more, this refugee from Stalin and Hitler, who had survived both the Holocaust and life behind the Iron Curtain, knew just how serious and brutal life could be, how quickly happy times could turn tragic. He knew the need to always be looking over your shoulder, because indeed *only the paranoid survive.* Bob Noyce had never looked back in his life. Why should he? All he would see was a glittering wake. Bob liked to skate on thin ice; Andy only felt it cracking under his feet.

But most of all, what drove Andy crazy was that *he was at the mercy of Robert Noyce.* He could manage Gordon, but Bob, by temperament and position, could never be managed. That was why Andy hesitated to follow Gordon to Intel, something he would have done in a heartbeat otherwise. Having worked with Noyce then for six years, he was now signing up for another tour of unknown duration. His only hope was that Noyce would fail and leave the company to him and Gordon—and if that happened, he knew he could handle Gordon and make Intel a success.

Instead, against all odds, despite (or as others might say, because of) what Andy saw as a long chain of high-risk, intemperate, ill-considered decisions by Noyce, Intel had not only survived but become one of the most exciting and successful companies on the planet. Even more, the company had found that success not with its core competency in memory chips, but in a renegade skunkworks program, microprocessors, that Andy had tried to stop. Sure, Intel had been badly hit by recessions, but so had every other semiconductor company, most of them suffering far worse, and Noyce had finally resigned from the day-to-day operations of the company. But that was just Noyce once again not getting his hands dirty, skipping out once again just in time not to catch any blame . . . and once again dumping it on Andy.

In an interview in 2004 with Arnold Thackray and David Brock of the Chemical Heritage Foundation, Grove would profess to being a voracious reader of management books, in part because, given his job, it was the one part of his education that he lacked. Of all of the books he'd read over the years, Andy said that his favorite was a little book written in 1954 by the dean of management theory, Peter Drucker, called *The Practice of Management*. In that book, Drucker takes on the question of what makes an ideal chief executive officer. He says that such an individual is really a tripartite character or, as Andy read it, applying its message to Intel, *three* people: "an outside man, a man of thought, and a man of action." To Andy, who sent copies of this chapter to his two partners, Bob Noyce was Mr. Outside, Gordon Moore was Mr. Thought, and Andy was Mr. Action.

There is no record of what the other two thought of this missive, other than perhaps, *There goes Andy again*. But there was considerable wisdom in Drucker's analysis of the perfect executive and Grove's adaptation of that model to the trio at the top of Intel. This book is called *The Intel Trinity* in part in a play on Noyce, Grove, and Moore (and his law) as Father, Son, and Holy Spirit, but also because somehow these very different characters, two of whom at various times didn't even like each other, somehow came together in one of the most powerful and successful management partnerships of all time.

That they did so despite all of the friction between them is a reminder that the nature of teams is far more complex and counterintuitive than we typically imagine it to be. It is easy to assume that great partnerships are always composed of two or three naturally compatible people or if we are feeling expansive, to throw in the possibility that opposites attract in rare odd-couple pairings. Certainly Silicon Valley has lived with and been distorted by the founding partnership of Bill Hewlett and Dave Packard, two men who had a professional partnership and private friendship so perfect—they seemed to never have had an argument in seventy years—that it was almost inhuman.

By comparison, the Noyce-Grove-Moore troika was ultimately just as successful (Intel would at one point have a market capitalization even greater than HP), and yet at various times, at least one member loathed another, and resentments flowed in various directions. Such teams are a whole different area of study, one that deserves more attention. The Intel triple alliance worked better than almost any traditionally compatible partnership in high-tech history.

Not that it was easy—especially for Andy Grove. Noyce and Moore already had their fame and glory. Andy was still clawing his way to become their equal (in his heart knowing he was at least that). The two founders could dangle the rope ladder down to him—as they did throughout the seventies—then conclude he wasn't yet ready and snatch it back up. And the frustration nearly drove Andy Grove mad. Once, Regis McKenna walked into Grove's office to find a furious Andy standing in the middle of the room and holding the latest copy of *Fortune* magazine in his hands. As Regis well knew, since he had pitched the story, the cover showed Bob Noyce and Gordon Moore . . . but no Andy Grove.

"That's it. I'm done!" shouted Grove, and slammed the magazine to the carpet. "I'm through making those two famous!"[7]

But as often as Andy threatened to quit, he knew he wasn't going anywhere. Intel was the greatest playhouse in American business. And if he wasn't leading the company, he was certainly running it. And finally, as Andy himself had admitted, his personality was such that most people either liked him or hated him. Where would he ever again find an opportunity and a set of understanding partners as perfect as he had now?

No, he would swallow hard, stay at Intel, and wait for his moment to come.

As perverse as it sounds, beyond all of the competitive strengths and innovative achievements of Intel Corporation, the single most important factor in its success may be the one thing about which it has always been ashamed and embarrassed: the fact that two of the founders didn't get along, that Andy Grove didn't like Bob Noyce. For employees, it was like watching a perpetual fight between Dad and your older brother, something you didn't want the outside world to know about.

Intel PR kept Andy's antipathy toward Bob largely under wraps for all of the years Noyce was at the company—and even beyond, until his death. After that, thanks to occasional comments by Grove and others, it remained something of a rumor for another decade, until Berlin's and Tedlow's biographies brought it to the surface. In his later years (as in the PBS documentary), even Grove began to talk freely about his legendary predecessor.

So there is no doubt nor lack of evidence that this contempt was real. Yet there has also always been something showy about this feud, at least from Grove's end; Noyce, for his part, never seemed to even acknowledge any tension with Intel's "first employee." Indeed, that little dig, a perpetual reminder that the company had *two* founders, not three, always came from the unimpeachable Gordon Moore.

Speaking of Gordon Moore, he held Noyce in the highest possible respect and esteem, to the point that for years after the latter's death, Moore could barely talk about him without tears. If Grove worshipped Moore as deeply as he said he did, why did Andy not trust Gordon's judgment on the man?

Part of this is certainly Grove's pride and contrariness. After all, if everyone else in the chip business admired and even worshipped Saint Bob, imagine the cachet conferred on the one man clever enough to see through all of the bullshit and pretense surrounding the great man.

Moreover, Grove most certainly chafed at being subordinated, year in and year out, one company to the next, to a man he considered his inferior as a businessman, as a manager, and perhaps even as a scientist. Of course it didn't help that Noyce joined Hewlett and Packard in the pantheon of Silicon Valley gods, that Noyce was regularly called upon to be the industry voice in Washington, and that talk began to grow about Noyce and Kilby sharing the Nobel Prize in Physics. That kind of envy, once established, doesn't evaporate even when the subject has left this world, even when *Time* names you Man of the Year and puts you on its cover.

It shouldn't be forgotten that Noyce gave Andy two jobs, the presidency of his company, and set him on the path to huge wealth. One interesting line of speculation about Grove's strange attitude toward him comes from Grove's biographer Richard Tedlow. He, too, was perplexed by Grove's behavior.

Tedlow's answer may be surprising to those who know Andy Grove only as a business icon, but it seems pretty accurate to anyone who knew both men or followed the trajectory of Grove's attitudes over the course of his long career and retirement. It is that Grove despised those parts of Noyce that he saw and disliked in *himself*.

Those commonalities were almost invisible in those early days, both at Fairchild and the early years of Intel. Noyce seemed to revel in being a public figure; he lived fast and played hard; he inspired legions of admirers; he was earthy, self-effacing, and humorous; and he was a risk taker endlessly brimming with new ideas and schemes. By comparison, the young Grove was as serious as a heart attack, hard-edged and rigid, obsessed with rational decision making and precision. Most of all, this refugee from the darkness of totalitarianism deeply resented anyone who took a casual approach to the enormous opportunity of pursuing success in all of its forms in a free society.

It was this perceived fundamental lack of seriousness in Bob Noyce that seemed to bother Andy Grove the most. Early Valley journalists and historians, if they even knew of this one-sided feud, typically dismissed it as yet another example of Grove's arrogance and competitiveness— that he actually thought he was smarter and more capable than the great Noyce and wanted his job.

The reality was a lot more complicated. For example, Grove recognized from the start that Noyce was "a very smart guy," even going so far as to add that Noyce had "lots of ideas, some of which are brilliant, most of which are useless."[8] A backhanded compliment, certainly, but Andy almost never called an industry counterpart "brilliant." Those words speak volumes about Grove's attitude toward Noyce in those days: that the latter, for all of his gifts, simply didn't take life seriously enough, that he treated his career as a game, that it was more important for Noyce to be liked than to be a success, and most devastating of all, that he would accept failure—at Fairchild and now at Intel—rather than abandon this approach to life.

There were elements of truth in this harsh judgment, but ultimately it was both unfair and wrong. Bob Noyce's greatest gift, even more than his talent as a technical visionary, was his ability to inspire people to believe in his dreams, in their own abilities, and to follow him on the greatest adventure of their professional lives. That required a style that exhibited confidence, playfulness, and a willingness to swing for the fences (and sometimes strike out). This kind of charismatic leadership came naturally to Noyce. (Tellingly, David Packard saw a kindred spirit in Bob almost from the beginning; Steve Jobs, as we shall see, actually tried to become him.) If Noyce was undisciplined and even irrational in his management style, it was because those traits had always worked for him.

Gordon Moore saw this in Noyce and loved him for it. But Noyce had already made Moore wealthy, and Gordon had already found a measure of professional fame from his law, so, while both men (especially Noyce) wanted Intel to be successful, neither man feared its failure. Moore: "I wasn't very concerned about it. . . . Changing jobs in our industry is fairly common, and I was sure that if this didn't work out I could find something else to do, so I didn't consider it much of a risk."[9]

But Andy, who had no cushion of fame, spent every day of those early years at Intel terrified of failure and obsessed with gaining enough power in the organization to keep that disaster from happening.

At Fairchild, Grove had been further down the organization chart from Noyce than he was at Intel, where he was essentially a peer, the junior member of the company's leadership troika. He dealt with Noyce rarely in those early years—and what he saw didn't impress him. It is interesting to note that Charlie Sporck, who arguably was tougher than Grove and was in a relationship to Bob at Fairchild comparable to Grove's at Intel, admired Noyce and respected the man's judgment.

Andy Grove never did. Instead, he had pushed and prodded, complained, scrambled for power—and made Intel a better company in the process. And all the while, he exhibited a patience he almost never showed in any other part of his life. Andy waited, knowing that someday his moment would come.

And now, at the dawn of the 1980s, with Noyce at last gone, that moment had arrived.

CHAPTER 34

Bylined

O ne last point about Andy Grove before returning to Intel in the early 1980s and his assumption of its leadership. Often remarked upon at the time—sometimes with dismay by shareholders and analysts who thought he should focus his time entirely upon running Intel—was Andy's extraordinary achievement in not only running a gigantic company in a brutal market, but also somehow finding the time to write books in the thick of the battle. And even if he got help with some of these books, they still bore the heavy stamp of Grove's style and personality. And beyond their estimable quality as books in themselves, these volumes are also a reminder of the depth of Andy Grove's ambition and his drive to find immortality some way and somehow.

This "side" career in writing began in 1967, when Andy was still at Fairchild and increasingly frustrated with the company's leadership. He had considered quitting, but instead sublimated his anger and energy into writing *Physics and Technology of Semiconductor Devices*, a college textbook. All but unknown to the general public, the book became something of a classic, serving as a core text to several generations of college engineering students, and it still survives as a basic reference book today. Gordon Moore never wrote a textbook; nor did Bob Noyce, and to Andy this was a way to distinguish himself from the others, a pedestal on which he could stand beside the other two. In the years to come, *Physics and Technology of Semiconductor Devices* also proved to be a great recruiter for Intel. Young engineers wanted to work for the man who had written the classic text of the field—once again aligning Andy Grove's own ambition with those he had for the company.

In 1983, Andy wrote *High-Output Management*, once again an

astonishing feat, given the incredible demands placed upon him by the industry recession and recovery during that period. Still considered perhaps the best book on being a successful manager in the high-tech industry, it remains a bible for business-school students. In it readers find an Andy Grove who is opinionated but thoughtful . . . and, surprising to almost anyone who ever worked with him, even charming and self-effacing.

A few years later, to leverage the success of *High-Output Management*, Andy took a step that was seen as bizarre then and even more so today, signing on with the *San Jose Mercury-News* as a nationally syndicated Dear Abby–type advice columnist about business. These columns were ultimately collected as *One-on-One with Andy Grove: How to Manage Your Boss, Yourself, and Your Coworkers* (1988). If *Mercury-News* readers, many of whom had worked for Andy, found the "Agony Uncle" of the column to be a lot more sympathetic than the guy running Intel, most also admitted that he gave very good advice. And yet it was during this period that an enterprising reporter who had pried a little too far into Intel's upcoming products was told by Andy that, had she been a man, he'd "break her legs." He was joking, but with Andy Grove, no one could be too sure, because no one knew quite where his boundaries lay.

Finally, in 1996, Grove published *Only the Paranoid Survive*, whose title not only brilliantly encapsulated Andy's business philosophy, but perhaps unwittingly, the entire life of the man who had survived two of the most murderous empires in history thanks to his mother's perpetual vigilance and lightning decisiveness in the face of mortal threats to herself and her son. The book, which came with encomiums from the likes of Steve Jobs and even Peter Drucker, contained one key idea: that of the *strategic inflection point*, that moment when some combination of technological innovation, market evolution, and customer perception requires a company to make a radical shift or die.

Most companies, Grove argues in the book, do the latter because they never see this shift coming and fail to recognize it when it arrives. The only way to survive one of these points of inflection, Grove argues, is to be ever vigilant and ever prepared to move quickly and decisively—to be paranoid.

Only the Paranoid Survive was a huge best seller, and it is considered, with Alfred Sloan's thirties classic *My Years with General Motors*, as among the best business books ever written by the leader of a major corporation. In the years since, the notion of the strategic inflection point has become part of the core philosophy of modern business practice.

Still, while he was at Intel Corp., his books were seen as a sideline to Andy's real job of running the world's most important company. But in the years after his retirement and as illness forced him to increasingly retire from the public eye, those three books (he also wrote his memoir, *Swimming Across*, after he retired) have loomed ever larger in his legacy. An entire generation of business executives (and Intel employees) has grown up never having known Andy Grove as an active business titan, now seeing him as merely a historic Valley figure. But the books and their messages survive and continue to resonate with young people about to enter the business world. And thus the greatest irony of Andy Grove's career may one day be that the young man who dreamed of becoming a journalist but then abandoned that career for a life of science and business may ultimately be remembered longest as a writer.

CHAPTER 35

Riding a Rocket

The hardest thing for a young technology company to do is survive. The hardest thing for a mature tech company to do is to change fast enough to keep up with the evolving technology.

In 1980, Intel was an $855 million manufacturer of primarily memory chips. Ten years later, it was a nearly $4 billion microprocessor manufacturer. The difference between these two dates, a decade apart, was so great that Intel could have been two different companies. And that captures only part of it, because despite this rocketing growth, the company's path between the two points was circuitous and fraught with considerable peril.

To the public and in Intel's official history, the company never seemed to break a sweat. In reality, during those years Intel could have died on several occasions, or at least made the wrong strategic decision that would have left it stuck in some slow-growing, backwater market, as happened to most of its competitors. Instead, it emerged at the dawn of the nineties poised to dominate the electronics world.

Intel accomplished this extraordinary feat, against all odds, because of its tradition of technological excellence, because of the dedication and pride of its employees and the culture that buoyed them, and ultimately because of the courage and toughness of Andrew S. Grove. It wasn't pretty, and Intel made some of the biggest mistakes in its history, but in the end, it won.

As the decade began, as if knowing what was coming, the company reorganized its senior management. Gordon Moore remained chairman, Andy Grove was named CEO, and now, appearing for the first time in the public eye, Craig Barrett was named president. This was the team that drove Intel through its most important and riskiest period of growth.

In 1980, Intel set out to do what seemed impossible: to be the industry's leading supplier and leading innovator at the same time. In pursuit of this wildly ambitious goal, the company took advantage of the good times to embark on a frenzy of activity. In that year alone, the company opened a Hong Kong subsidiary and a fab plant in Puerto Rico, even as it closed its original Mountain View fab. The company also installed a massive new order-processing system to deal with the exploding demand for its products.

That year Intel also challenged Motorola's lock on the automotive market by entering into a joint venture with the Ford Motor Corporation to develop the 8061 and 8361 two-chip electronic engine-control system. This was a major step by Intel into the emerging world of microcontrollers, single-chip processors designed not for computers but to serve as brains for analog systems, from automobile engines to home thermostats to military missile systems.

Intel also that year introduced the world's first math coprocessor chip—a support device for its microprocessors to add high-speed computational capabilities. And the company teamed up with two giants of the era, Digital Equipment Corp. and Xerox, to develop what would soon be the world's dominant personal computer networking technology, Ethernet.

Meanwhile, the larger world was beginning to recognize the important of the computer chip and Intel's historic connection to it. In April 1980, Bob Noyce was elected to the National Academy of Sciences—an honor to match the National Medal of Science that had been given to him by President Jimmy Carter five months before.

Finally, to close out the year in style, *Dun's Review* named Intel one of the best-managed companies in America.

It wouldn't get this good again for a long time.

There is a phenomenon almost unique to the semiconductor business (and the industries that supply it). It can be described as a series of brief industry bubbles that are produced by a combination of the unpredictable impact of new technologies crossed with the desperation of customers to keep up with the pace of Moore's Law.

It works like this: the latest tick forward of Moore's Law creates a whole new body of opportunities for everyone from inventors to entrepreneurs to established companies; it acts as a reset switch for the entire digital world. This reset not only allows competitors to redesign existing products in hopes of getting a jump on their competitors—a classic example being IBM leaping into the market with its Intel 8088-powered PC, which officially took place in July 1981 and completely rejiggered the personal-computer business—but it also occasionally results in a radically new product that creates a giant new industry, which is what happened in 1972 when Nolan Bushnell installed the first Pong arcade game in Andy Capp's Pub in Sunnyvale . . . and created Atari Inc. and today's $65 billion computer and video game industry.

By the 1960s, the semiconductor industry was pretty adept at predicting the demand from existing industries for each new generation of chips and built fabs and ramped up production accordingly. What it could never do was predict the unpredictable, and the explosive demand that would come from a brand-new market for its chips. Nor could it risk the giant expense of putting online new capacity for new technologies that might not appear this time around.

Because these added bursts of demand almost always occurred during booms—indeed, they helped create those booms—chip companies inevitably found themselves with undercapacity, unable to keep up with demand.

In some industries, this might be good news, as manufacturers could then jack up prices and enjoy the added profits. But the whole point of getting design wins in the chip business is to capture and then hang on to major vendors from one product generation to another. That's how you guarantee that you always will have the profits you need to build future generations of products. So pissing off those customers for the short-term expedient of raising prices, even if it is only to sort out and dampen demand because of underproduction, is just bad business.

Instead, what companies such as Intel did in response to these periods of undersupply and overdemand was to create customer-priority lists. Thus, for example, the best and/or biggest Intel memory-chip customers were put near the top of the list, while smaller, newer, or less important customers—or customers who divided their orders among multiple suppliers and were thus perceived as disloyal—were put lower on the list. In downtimes, these lists were unnecessary, and during most typical booms, they did a good job of sorting demand and retaining the most important corporate clients.

But the early 1980s weren't normal times. The combination of multiple explosive new digital industries—minicomputers, video games, home game consoles, scientific and programmable calculators, and most of all, personal computers—had created a titanic demand for memory chips, in particular, 16K DRAMs. And as demand from all of these manufacturers, not just in the United States but also Europe and Asia, began to swamp the production capacity of memory-chip makers, those manufacturers, their assembly lines beginning to slow for lack of DRAMs, began to panic.

This panic in turn created two phenomena that peaked in early 1982 and (luckily) have never been quite matched in Silicon Valley in the years since. The first was double- and triple-ordering. Desperate manufacturers would place their first order with, say, Intel, and then as delivery dates began to drag out, they would place a second, even larger order in hopes of getting attention. When this didn't work, the customer would then sometimes place an identical order for plug-compatible chips from one or two other manufacturers.

The result, as one might imagine, was a sudden ballooning of demand beyond anything the chip industry had ever seen before, and though the semiconductor industry suspected that some of this demand was artificial, it had no way of knowing just how much.

As the months passed and the situation continued to deteriorate, increasingly desperate customers, many of them facing a threat to their very survival, began to look for alternative sources of memory chips, and the semiconductor gray market, which had always existed on the fringes of the industry, rose to meet them. This was the second phenomenon, one that caught the traditionally crime-free tech industry completely off guard.

The official gray market specialized in dealing in obsolete chips and consisted of small warehouses scattered around the Valley run by men willing to hold on to their inventories, usually bought from semiconductor companies when they retired product lines. As the demand for

16K DRAMs grew, these brokers found themselves visited by big companies willing to pay almost any amount to keep their production lines moving. Seeing big profits and having unique access to the back doors of the big chip companies, some of these gray marketers crossed over into more questionable practices.

The most legal version was to buy scrap DRAMs—that is, those that didn't meet their official performance specs and were trashed—then test them and sell the best of them (sometimes illegally described as fully working chips). Most customers didn't care, as long as the chips worked well enough for the applications in which they were used.

But soon, this source of supply began to dry up as well—not least because chip companies like Intel found themselves besieged by consumers angry about the poor performance of the devices they had purchased. They began destroying the chips—or at least thought they did, because the gray market had now turned into a full-blown black market. Chips being sent to scrap dealers ostensibly to extract the gold off their leads (pins) were harvested first for even the most poorly performing live chips. Meanwhile, all sorts of criminal enterprises began to insert themselves into the DRAM delivery system: small-time hustlers who didn't know what an integrated-circuit chip was other than it was worth its weight in gold; organized crime groups, including local Vietnamese gangs; espionage agents from other countries, both enemies, like the Soviet Union, and erstwhile allies, like France and Israel; and unlikeliest, radical political groups like Oakland's violent Black Guerrilla Family, which looked to the black market in chips as a way to finance its political activism.

It wasn't long before scrap memory chips were being stolen off the Valley's loading docks (sometimes with the help of hookers to create a distraction). Employees with gambling debts were being blackmailed to leave back doors unlocked during night shifts, and various other stratagems were used to separate chip companies from their production. At least one local low-grade criminal was murdered over a busted chip deal.

Meanwhile, the memory-chip drought began to take its toll in the Far East, where the young consumer-electronics business was a huge consumer of memory chips. So it wasn't long before Asian businessmen carrying briefcases full of cash began arriving at San Francisco Airport with orders to find DRAMs at any price. This led to some odd moments, such as the time when a chip seller and buyer, having taken adjoining rooms, opened the adjoining doors and tried to hand the money and goods to each other, only neither was willing to let go first. The result was slapstick in which the two parties, each gripping both briefcases through the partially opened doors, played a desperate tug-of-war while trying to hide their faces from each other.

Tech booms, with their prospect of overnight riches, almost always produce a certain amount of criminality and fraud. Witness the dot-com bubble of the late 1990s. But it has never been as bad as it was during the boom between 1979 and 1981. Not only was there the gray/black market surrounding DRAMs, but the speeded-up production and increased quotas led to an epidemic of amphetamine abuse as fab and assembly-line workers tried to keep up with the amped-up pace. Drug dealers openly worked out of the trunks of their cars in the vast National Semiconductor parking lot (and took orders from Intel employees, among others). Meanwhile, the Japanese had their listening posts and were ramping up to all-out industrial espionage. And a half-dozen nations had spies in the Valley—some of them (like the Russians) hoping to steal the technology they would put in the nose cones of ICBMs aimed right back at Silicon Valley.

Even legitimate companies were cutting corners in the face of this unprecedented demand. Apple, for example, in an early program it doesn't like to discuss, tried to keep up with demand for the Apple II by hiring local immigrant Asian women jammed into apartments in Sunnyvale and elsewhere in the Valley to stuff chips onto motherboards and get paid piece rate—Valley sweatshops.

By this point, Silicon Valley, and particularly the semiconductor industry, was screaming along at very near its breaking point. Intel and its competitors were racing to bring online new capacity as fast as they could. Customers were at wit's end, not just dabbling in the gray market, but even threatening their legitimate suppliers.

Then in the second half of 1981, as it was destined to, the boom collapsed. The good news was that, as the opportunities for profitable crime evaporated, the bad guys moved on—though the Japanese and the government and industrial spies remained.

The bad news was that now chip makers, including Intel, got a devastating lesson in just how artificial the stratospheric levels of orders really were. Once demand in the consumer and industrial electronics markets ground to a halt, they quickly caught up with their need for memory chips . . . and then started canceling their orders. They didn't cancel just one order, but all of the duplicate orders they had made months before in their panic for components. Thus one order cancellation might quickly turn into four different cancellations at one chip vendor and perhaps three or more elsewhere. A 10 percent drop in real customer demand instantly turned into a 30 or 40 percent drop in actual orders. In industry terms, the "bookings-to-billings" ratio, which had been well above 1.0, overnight slumped to well below that balance point.

The crash, which came sooner and was deeper than anyone had predicted, was probably the worst that has ever hit the semiconductor industry and the rest of high tech. Intel's financials slumped for the first time in its history. The company, which had averaged nearly 100 percent growth in revenues per year since its founding thirteen years before, suddenly saw them fall—from $855 million in 1981 (and that was benefiting from a good first half before the crash) to just $789 million in 1982—and it would take all of 1983 to get back up to annual revenues of two years before. This was despite the introduction—at last—of the IBM PC, which captured the world's attention and quickly gave IBM market leadership.

What made this downturn particularly devastating in high tech was that it occurred simultaneously with—and helped to accelerate—a shakeout taking place in the recently high-flying disk-drive business. In 1980, more than 150 5.25-inch Winchester-hard-disk-drive companies had burst on the scene to take advantage of the growing need for mass memory storage by PC makers. They couldn't all succeed, and when the personal-computer industry slipped into its own shakeout (the now long-forgotten first-generation PC makers like Adam and Sinclair that were crushed by Apple's success), the crash was quick and brutal. In the end, there were fewer than a dozen disk-drive makers left, serving an equal number of computer makers.

The semiconductor industry, too, was undergoing a shakeout, as the losers in the race to get the IBM PC contract (and thus all of the PC clone contracts as well) began to fall back, racing to capture other, lesser markets. Thus National Semiconductor, Texas Instruments, and Fairchild, which had been major players in microprocessors just a few years before, now all but gave up on computers and shifted into microcontrollers for automation or into lower-power processors for consumer and industrial products. Zilog, which arguably had the best processor architecture of all in the Z80, made a similar move to get out of Intel's way, by turning to its own microcontroller, the Z8.

By the time this downturn was over, most of these companies, whatever their balance sheets said, would be struggling to find their place in the semiconductor world, and some would already be sliding toward oblivion. And by the end of the decade, Zilog would be almost a forgotten player, earning its revenues licensing its chip designs for use in watches, toys, and other budget applications. Fairchild, once the mightiest company of them all, would be owned by National Semiconductor, which itself would be struggling to find a purpose. And the largest profit center at Texas Instruments would become its legal department as it sued every other semiconductor company over its IC patents.

Only one semiconductor company seemed well positioned to suffer minimal damage—perhaps even grow—during the 1981–83 crash. That was Motorola. After the debacle of losing the IBM PC contract to Intel and Operation Crush, Moto had come roaring back. And in 1979, even as Big Blue was appraising Intel's 8086/8088, Motorola was preparing to unveil one of the greatest chips ever designed: the Model 68000.

It has been speculated that if Motorola had been a little quicker with the 68000 or IBM a little slower in its due diligence, the history of the semiconductor and personal computer industries might have been a lot different. The reality is that even if IBM had seen the 68000, it probably wouldn't have adopted it for the PC. At that moment, taking a gigantic risk entering into a new industry, Big Blue wasn't interested in the coolest chip on the planet; it wanted high-quality volume production and a ton of support, and that's what Operation Crush and the 8088 gave it.

But the rest of the digital world looked at the 68000—a true 16-bit processor capable of a blistering two million calculations per second—and swooned. IBM might be designing the Intel 8088 into its new computer, but the rest of the world started designing the Motorola 68000 into almost everything else, especially automobiles. And when the crash hit in late 1981, it was those blue-chip customers that enabled Motorola to endure the downturn better than any of its rivals: indeed, the company *made* money during this period, its chip sales growing from $1.2 billion in 1980 to $1.6 billion in 1983, with *no* operating losses. This success also enabled Motorola to go far afield in search of new customers. In particular, it built its own version of the 8088, the 8/16 bit 6809, and took off in pursuit of consumer-electronics customers in the Far East. The company also began to experiment with the use of microprocessors in the telecommunications industry, a move that would have major implications in another fifteen years with the rise of the cell phone.

Intel could only look at Motorola's success and grit its teeth. It had the most important contract in the history of high tech, but it was just beginning to ramp up and wouldn't produce enough revenues to make up for the slump taking place everywhere else in its business. Processors were hurting enough, but the memory-chip business, now that the bubble had burst and DRAMs were in oversupply and almost being given away, was a disaster.

Intel had some important new processor designs in the works, but the company had invested most of its available capital in building those new fabs all over the world. Moreover, in mid-1980, there was as yet no clue just how deep this crash might go or how long it might last. So despite the gains being made by Motorola, prudence dictated that Intel should hunker down—cut overhead, lay off employees, delay new product development, and freeze all new construction.

That is what Intel had done in 1974, when it inflicted upon itself the great layoff. The company also had a fab (the Aloha 4 facility near Portland, Oregon) under construction back then, and its response that time had been to stop all construction. The result was an unfinished and empty building, with slobber- and snot-covered windows, guarded by a single Doberman and immortalized in Intel's history as the Aloha Doghouse.

But in what may have been the most important executive decision he made in his entire career at Intel, Andy decided to throw out the industry playbook and take an incredibly high-risk roll of the dice with the company he had only just inherited. This would be Andy Grove's Big Bet—and what made it even more remarkable was that Grove, new in the job, still less than twenty-five years in America, finally in the position of authority and power he'd long dreamed of and schemed for, had every reason not to make such a wager. Everything argued for him to go conservative and careful and not put at risk everything he'd gained. And yet as had been seen before and would be seen again, hard times and big threats only seemed to make him bolder. And he would never be so bold with so much at stake as at this moment.

Intel, Andy decided, would go full speed ahead through the mines and shallows: no layoffs, continued full investment in R&D and product development. Construction would continue on the new factory, Fab 7 in Albuquerque, to be completed in autumn 1982, though it would not go online until demand returned. Meanwhile, its staff would be hired and trained, but would temporarily be placed elsewhere.

Finally and most notoriously, in October 1981 Andy announced the implementation of what would be called the 125 Percent Solution. For the next six months, every employee at Intel was expected to work two extra hours each day without pay.[1]

"At a company where sixty-hour work weeks were already the norm, the 125 Percent Solution was something of a nightmare. With winter approaching, employees typically arrived in the darkness and left in the darkness. Families were strained, emotions—in a company already known for 'constructive criticism' [i.e., shouting]—were raw. But the employees also knew that the alternative might be to join the legions of their industry counterparts haunting employment lines and reading the want ads. Employees jokingly took to wearing '125% Solution' sweatbands in keeping with the company's new sweatshop image."[2]

The 125 Percent Solution wasn't easy on Andy Grove, either. The outside world questioned his judgment. During the 1974 recession, Bill Hewlett and Dave Packard had instituted the Nine-Day Fortnight, in which their employees gave up one day's pay every two weeks to keep HP from resorting to layoffs. But that was a pay cut, not a demand for more than a day's extra work per week—and HP wasn't in the chip business. Jerry Sanders of AMD, at an industry conference in New Orleans in 1980, had with typical flamboyance announced a no-layoff policy at his company and challenged his competitors to follow along in an industrywide pact.[3] But that was during good times, and AMD was a small company—and no one believed Jerry anyway. Intel, along with the others, had refused to sign the pact, and yet now here it was doing everything in its power not to lay off a single employee.

Many suspected that Grove, who already had a widespread reputation as a hard-ass, had some secret motive—publicity? a stealth project?—for this move. That mystery worked to Intel's advantage, but it also meant that if the strategy failed, Grove would be doubly blamed precisely for *not* having such a trick up his sleeve. He also looked like a hypocrite when he ordered every lobby crew to keep a list of all employees who arrived late, . . . only to be late himself on the very day that a reporter from the *New York Times* was waiting in the same lobby to do a story on the campaign.[4]

Luckily, in the end, the 125 Percent Solution seemed to work. Dick Boucher, director of marketing communications and business development, said, "We accomplished a lot," and then added the obvious: "Although it was not universally popular, it was the right thing to do."

The closing of the 125 Percent Solution on March 31, 1982, was cause for celebration, including beer parties featuring mugs 25 percent larger than normal.

"Intel counted among its gains the accelerated shipment of new microcomputer systems, a jump in microcontroller sales and a speed-up in the preparation of federal tax returns to get an earlier refund. Whether any of these accomplishments were the result of the extra two hours' work each day is debatable. But one fact remained: Intel maintained full employment in the worst economic times."[5]

CHAPTER 36

For the Cause

There was one more certain benefit of both Andy Grove's Great Bet and the 125 Percent Solution. It was that Intel, rather than backing off, forged ahead with its new-product development programs. The result, which was completed in March 1982—the last month of the Solution—was arguably the single biggest new product announcement in the company's history.

During that month, Intel announced *four* important new processors:

- *The 80186 and 80188.* This was a pair of 16-bit microcontrollers designed to be embedded in instruments, computer peripherals, and other electronic devices. This was Intel's response to Motorola's growing dominance in the controller industry. Long obsolete, these two devices are largely forgotten today, except that the 80186 was the first Intel chip to use the 80x86 (usually abbreviated to x86) nomenclature that would become synonymous with future generations of Intel microprocessors.
- *The 82586 coprocessor.* This device was designed to work alongside the main Intel microprocessors to power connections to local area networks. This chip and its descendants would help power the modern PC-oriented, networked organization, and in time the birth of the Internet.

Most important of all:

- *The 80286 16-bit microprocessor.* This was first "modern" microprocessor for personal computers, and the first true member of the 80x86 family architecture that would define the microprocessor to the world into the next century and to the present. The 8086/8088 had powered the IBM PC, but it was the 80286 that would power all of the clone PCs now being created by hundreds of companies around the world, beginning with Compaq Computer, that would adopt Wintel (Microsoft Windows operating system/Intel x86 CPU) in the years to come. In the process, this army of PC clone makers—which soon included giants like Dell, HP, Sony, Toshiba, Acer, and Hitachi—would drive the price of a personal computer down from thousands of dollars to hundreds, sell millions of machines, . . . and set off the truly personal computer era that would make possible the World Wide Web and create the modern economy.

The 80286 was not the stunning breakthrough design of say, the Motorola 68000 or the Zilog Z80, but it wasn't meant to be. Intel had a lesson from both the computer and software industries: *upward compatibility*. Andy was playing the long game now: the core architecture of the 286 was designed to survive almost intact through multiple new generations of chips in the x86 family over the years to come. That meant that programs designed for 286-based personal computers would also run on future hardware upgrades of those machines run by descendants of the 286. And that in turn meant that the current library of application programs—from productivity tools to games— didn't have to be rewritten but merely had to be upgraded as well. The 286 could even run all of the software that had been designed for the 8086/8088.

The impact of this Wintel combination of an open IBM hardware platform universal Windows operating system, and x86 processor CPU was quickly apparent at the retail level. Whereas just a few years before, the personal computing world had been little more than Apple Computer and also-rans, now the dominant shelf (and soon aisles, then store) space was devoted to Wintel machines and Windows software, with Apple's (closed) hardware and software reduced to an ever smaller footprint and eventually just a few shelves. Apple's fortunes during this era followed the same trajectory: the company that had once owned 95 percent of the PC business now saw its market share in free fall.

Meanwhile, by 1988, six years after the 80286's introduction, fifteen million 286 PCs had been sold around the world.

The 80286's advances in technology over its predecessors were as profound as its impact on the marketplace. The device was originally released in 6 MHz and 8 MHz versions, which Intel eventually scaled up to 12.5 MHz, making it much faster than the 8086—and some of Intel's second sources (notably AMD and Harris) created superfast versions that could run up to 25 MHz, or almost five million instructions per second. Ultimately the capability of the 80286 was judged to be about twice that of the 8086, perhaps the largest intergenerational jump between processors ever accomplished by Intel.

The 80286 featured 134,000 transistors (compared to 29,000 in the 8086) and some added functionality, including multitasking (the ability to perform more than one operation at a time), built-in memory protection (until now found only in big computers), and real-time process control. Unfortunately, many of these special features, notably multitasking, were specifically coded for IBM applications, not Windows. This led a furious Bill Gates to declare the 80286 a "brain dead chip." The love-hate relationship between these erstwhile allies would remain until IBM left the PC business . . . and then continue between Intel and Microsoft until the twilight of the PC era a decade into the twenty-first century.

The fact that Intel could introduce a new microprocessor model in the middle of the worst downturn in semiconductor industry history was a testament to the commitment of Intel employees during the 125 Percent Solution and to the fearlessness of Andy Grove in charging ahead even as Intel's competitors reined in. And when Compaq adopted that product, the 80286, as the engine of its pioneering Wintel clone personal computer, the game was over. Intel had won the microprocessor wars for at least the next quarter century . . . if it could just live up to the responsibilities of that leadership, and with Andy Grove in charge, there was little doubt that Intel could do so.

Only Motorola remained in the arena. The rest of the chip industry either faded away into oblivion or moved away in search of some other profitable product.

Intel's remarkable collection of product announcements in March

1982 did a lot to enhance the company's reputation as an innovator and a fierce competitor. But for now, at least, in the thick of a devastating and attenuated recession, it did little for Intel's bottom line.

By midsummer 1982, just a few months after the celebrations over the completion of the 125 Percent Solution, Intel top management realized that the recession wasn't ending, the company's financials weren't ramping back up, and what had appeared to be new growth in chip orders proved instead to be just a temporary inventory adjustment by Intel's biggest customers. The company would have to tighten its belt once again.

Semiconductor industry downturns weren't supposed to last this long, and in implementing its austerity program, Intel had assumed that the market would once again be gaining steam by mid-1982. But it wasn't happening; the downturn seemed to stretch on into the indefinite future, and Grove and his team, to their horror, now realized that their recent hiring of three thousand employees (taking advantage of the glut of available, newly laid-off top talent in the industry) had been dangerously premature. Intel couldn't afford them, and the company's projections showed that if 1983 was as bad as it appeared, it would also have to lay off a number of company veterans.

Once again Grove, with Moore's backing, moved quickly. In November, Intel announced an across-the-board pay cut of up to 10 percent, combined with a pay freeze through 1983. This one hurt: "This time the thrill was gone. Employees had signed on to the 125 Percent Solution with a certain esprit de corps. After all, that was only work time, the least expensive commodity in Silicon Valley. But this time it was money [much scarcer in this, one of the most expensive communities on earth]. And though there were a few brave 'Thank God We Still Have Jobs' parties, the reality was that this cutback was felt strongest by the people Intel wanted to keep most: the bright, fast-track types who measured their success by the growth of their salaries."[1]

Then, a month later, Intel dropped another bomb on its employees. It announced that it had sold a 12 percent share ownership in the company to IBM Corp. for $250 million. It was a win-win for the two corporations. A quarter-billion dollars was chump change for mighty Big Blue, and it was willing to pay an inflated price—that $250 million should have bought as much as a 25 percent share in Intel—to have a deeper engagement with, and more influence over, its most important supplier. Intel gave up only a nondecisive share of the company in exchange for more than a quarter of its annual revenues, enough to keep the company financially strong through the end of the downturn and beyond. IBM shareholders got a stake in one of the world's hottest companies, while their counterparts at Intel got a nice price bump.

The only players feeling left out in all of this were Intel employees, who asked themselves why their employer, now that it had another quarter-billion dollars in the vault, continued the pay cut and freeze. Because no one was going to blame their beloved Uncle Gordon, most of the anger (probably accurately) shifted to Andy Grove, as people forgot that just a few months before the same ruthless boss had gone to extraordinary lengths, even risking the future of the company, to save them in their jobs.

Either way, an estrangement was growing between Intel senior management and the rank and file that would never be fully restored to status quo ante. Every large and maturing company has this moment, when it stops being family and becomes simply an employer. This was Intel's moment, nearly fifteen years after its founding; the happy times had lasted longer than at most companies. Although Intel would continue to enjoy some of the most loyal employees in tech and certainly in the hard-boiled semiconductor industry, this was an important point of inflection in the company's history. Would it have been different if Bob Noyce was still in charge? Maybe. But without Andy Grove at the wheel, Intel might not have made it through the 1982 crash.

Still, when Grove's *High Output Management* was published the next year and quickly climbed the best-seller charts, more than one Intel employee grumbled that the real secret to Andy's high-output philosophy was demanding more work while cutting pay.

CHAPTER 37

East of Eden

J erry Sanders had always had a special relationship with Bob Noyce. Back in the Fairchild days, when "Hollywood Jerry" was still in his twenties and trying to make his mark as a Fairchild sales rep in Southern California, he had taken a lot of flak from his fellow employees for past attempts to be a movie star, for living perpetually beyond his means, for his flamboyant clothes (though it wasn't true, everyone at Fairchild told a story of Jerry and pink pants at IBM), and most of all for his clever and boastful talk. It was the last that got him in the most trouble, not least because, as many people agreed, behind the pose, Sanders was one of the brightest people in the Valley.

The one person who always defended and respected Jerry Sanders was Bob Noyce. It is a measure of Noyce's insight as a leader that he saw greatness in the loudmouthed kid long before anyone else did and consistently promoted him up through the organization. Noyce may have also seen something in Sanders that he knew was in himself—not least an eye for the ladies. Noyce treated Jerry like a wild but beloved son, and Sanders, having spent his life in search of a father figure, found him in Bob Noyce.

It is not surprising, then, that when Sanders finally left Fairchild and was trying to start a company that might ultimately be a competitor to Intel, Noyce helped him from the start. Needless to say, this didn't go down very well at Intel, especially with Andy Grove.

What emerged in the years that followed was an almost biblical story that would profoundly shape Silicon Valley's history. Andy Grove served the great man, the Adam of the semiconductor industry, as his chief lieutenant. Yet as we've seen, Noyce suffered Grove for his brains, his talent, and his toughness, and Grove, though he served his boss with everything he had, was largely contemptuous of Noyce for his lack of seriousness and his need to be loved. By comparison, Jerry Sanders, who had built a company largely to attack Intel, loved Noyce, honored him, and never said a negative word about the man in his life. Noyce reciprocated by standing by Jerry when no one else would; he enjoyed his company, and was willing to help the younger man any way he could.

Whether Andy Grove was jealous of this relationship or thought it was bad for Intel has never been entirely clear, but ultimately it didn't matter, because for Andy the personal was Intel—and he was against the deal. And the fact that he didn't figure out a way to quash it then would haunt him for the rest of his career.

Needless to say, this was a three-way relationship that was fraught with peril. And when in 1976, the still-struggling seven-year-old Advanced Micro Devices designed a reverse-engineered clone of Intel's hot new 8086, Intel was ready to sue AMD into the ground, as it would do to every other chip company that tried the same thing in the decades that followed. But that's when Noyce intervened. Why not, he argued, let AMD build the clones? After all, the company is so small that it poses no threat to us. Better yet, he argued, why don't we just license the 8086 to AMD as a second source? We know them, they don't have the manufacturing capacity to challenge our dominance, . . . and we can handle Jerry. After all, he's family.

Andy Grove didn't agree, but what did that matter? Bob Noyce was boss. And besides, "second sourcing" was a long-established practice in the semiconductor industry. Giant manufacturers, from IBM to General Electric to Boeing to the Detroit automakers, didn't like the idea of having their product manufacturing at the mercy of output of some tiny sliver of glass being built by a bunch of flaky young engineers in Santa Clara or Phoenix. The risk was just too great. So they demanded that their semiconductor suppliers license their technologies to fellow chip companies to guarantee a continuous supply.

Needless to say, the chip companies grumbled at this arrangement, which meant they had to share their precious intellectual property with competitors. And of course, these companies didn't go out and sign second-source deals with companies that might actually propose a true competitive threat if the originating company stumbled. And they also filed as many patents as they could to minimize their loss of ownership. But in the end, they made the second-source deals not just because they had to, but also because the contracts they got in return were hugely profitable.

So in many ways, the AMD second-source deal on the 8086 was a good business strategy, as even Andy Grove—whatever his feelings about the Noyce-Sanders relationship—had to agree. And yet at the same time, his concern was well placed, because Hollywood Jerry, with his convertible Bentley, his tailored suits, his pomaded coxcomb of blond hair, and most famously, his spats, was, beneath that flamboyant exterior, one of the cleverest and most relentless competitors in the history of high tech.

According to its agreement with Intel, AMD had the right to "copy micro-codes contained in Intel microcomputers [i.e., microprocessors and microcontrollers] and peripheral products sold by Intel."[1] It was such a minor story that no newspaper and only a few trade magazines even carried the news.

Six years later, there were no stories at all when the two companies renegotiated the deal. But this time, the story behind the deal had changed completely. During the intervening years, AMD had done its job well and lived up to Intel's expectations—it was a reliable backup but no real threat. But now with the recession on and Intel with overcapacity in its fabs, Andy was willing to risk going it alone. That it also removed Jerry Sanders, whom Grove had always detested as having even more of the qualities that he didn't like in Bob Noyce, was icing on the cake.

But now IBM intervened. The same wariness that had led Big Blue to inch into the PC business with the lower-powered Intel 8088 as the CPU instead of the 8086 carried over into a similar attitude when it came to chip deliveries. IBM wanted a second source for the Intel 80286, which was planned to be the brain at the heart of its new PC AT, and since Advanced Micro Devices had done a solid job with the 8088, IBM wanted that company back in on the new deal. Intel could only gulp and agree; it didn't want to fight IBM over this and potentially lose the most important contract in tech, and it also did not want to go out and find yet another second source . . . and risk having two companies cloning its flagship product.

But unlike the original contract, this second AMD second-source deal had two gigantic downside risks. The first was that AMD, largely thanks to the 8086/88 deal, was no longer a struggling start-up but a healthy midsize semiconductor company full of a lot of talent and run by a brilliant, risk-taking CEO. It could very well gobble up a lot of the 80286's sales, especially if demand jumped after the recession and Intel struggled to keep up. The second was that, in sharing the design and coding for the 80286, Intel was also surrendering part of its destiny. For all of its revolutionary impact, the 8086 was essentially a one-off product; Intel could get away with licensing it in the knowledge that the device had only a limited lifespan. But the 80286 was, by its very design, the first in a family of microprocessors that would share a common architectural core and would be upward compatible for generations to come. Thus in giving AMD the license to the 80286, Intel potentially opened the door to AMD cloning all of its future chips for decades to come.

In the end, Intel's—and Andy Grove's—worst nightmare came true and both things happened. Many people have called Jerry Sanders, the "Clown Prince" of Silicon Valley, a fool, but no one has ever called him foolish. He knew that the Intel x86 family was going to be the dominant technology in the entire electronics industry for many years to come. Once he got his hands on the second-source agreement for the 80286 and saw its wording, he knew exactly what he had, . . . and he had no intention of ever letting it go. Bob Noyce had given Abel the greatest gift in the chip industry, while Cain could only fume and plot revenge.

The Traitorous Eight of Shockley Transistor at the birth of Fairchild
Semiconductor, 1960. (Wayne Miller/Magnum Photos)

Intel's first
employees, circa
1968: Robert Noyce
and Gordon Moore
are in the front. Ted
Hoff and Andy Grove
are the two men with
glasses behind
Moore, with Grove on
the right. Les Vadász
is the man half-
turned directly behind
Noyce.

Dr. Gordon Moore, cofounder of Intel Corporation. This photograph was taken at Fairchild Semiconductor, circa 1965, at about the time he formulated Moore's Law.

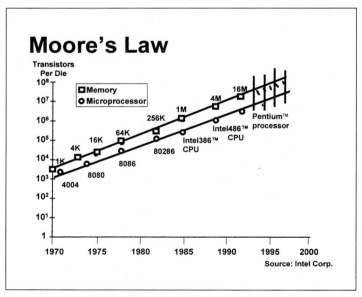

Moore's Law. Intel's processors are the dots.

Dr. Robert Noyce, circa 1970, not long after the
founding of Intel Corporation.

The Mayor of Silicon Valley: Bob Noyce, late 1980s.

Andy Grove, almost unrecognizable in his glasses, in the early days of Intel, circa 1969.

Intel's cofounders in battle, circa 1975.

The nearly unique photograph of Intel's Trinity together, circa 1975.

The microprocessor era begins, 1976.

Federico Faggin, the creator, at Intel, circa 1970, during the development of the first microprocessors.

Ted Hoff, the visionary behind the architecture of the microprocessor, late 1970s.

Stan Mazor, the programmer, circa 1975.

The expensive foray into consumer electronics: Gordon Moore wearing his Intel Microma watch, circa 1980.

**Future Intel CEO Craig Barrett soon after joining
the company, circa 1980.**

Paul Otellini, the lifelong Intel employee, circa 2004.

CHAPTER 38

Turn at the Tiller

I n 1983, though its employees were still under the recessionary pay freeze/salary cuts regime, Intel managed to post its first $1 billion revenue year: $1.122 billion, up 25 percent from $900 million the year before, with much of that growth in the last quarter.

Clearly the recession was finally coming to an end, and Intel had positioned itself beautifully to come roaring out into the new boom and dominate it. This was a credit to Andy Grove's leadership: for all of the dire predictions made at the beginning of the crash that Intel was going to spend itself into bankruptcy, the company had clearly not only survived, but (other than Motorola) it was in better shape than any of its competitors.

And if Andy Grove was at his best during tough times, during the next few years, he showed that he could deal with the very different demands of good times as well.

For Intel, 1984 was a year of consolidation. That August, IBM was slated to introduce the PC AT, and in preparation for the expected demand for 80286s, the company ramped up its production, bringing Albuquerque up to full operation and opening another fab in Singapore.

This was also the year that two other senior Intel executives made an enduring mark on the company.

Les Vadász had played a vital role at Intel from the moment he was hired as the company's first official employee. He had been the critical figure in overcoming the fatally low yield rates and getting the company's first memory chips to market. Then as head of Intel's MOS group, he had been the top supervisor in the development of the 4004 and thus led Intel into the microprocessor business. His strengths were quiet competence, discipline, and consistency. Everyone liked Les as a person, and he created a stable environment that enabled his people to do their best. It sounded simple, but especially when maintained over dozens of years, such a management style was rare in Silicon Valley and almost unprecedented in the semiconductor business.

Now in 1984, as Intel was coming out of the blocks into the next boom, Les was given the assignment to create a new initiative: the Intel Development Operation (IDO). Vadász had been an advocate of this program, saying: "One of Intel's strengths is its ability to focus precisely on a defined technology direction. However, when you are so focused, it's harder to tap other business opportunities that are relevant and worthwhile but don't happen to fit the current corporate blueprint." With IDO, he explained, "The idea was to fund good ideas, build a fire wall around them and see what developed—a kind of internal venture capital operation."[1]

In other words, without Bob Noyce around to identify important new projects, circumvent the rules, and nurture and defend those projects until they were strong enough to stand on their own, Intel had to find another, more formal way to preserve that process. That Vadász would understand this need for an Andy Grove work-around and then would successfully lobby to make it happen captures more than any other act in his forty-year career at the company why Les Vadász was a vital and beloved figure at Intel.

As for IDO, in assuming the role of an intrapreneurial venture capital firm, it could be described as a moderate success. In the end, it lacked only one thing: Bob Noyce, with his genius, his risk taking, and his charismatic leadership. Still, it did have some interesting victories. In 1984, computer visionary Justin Rattner approached the IDO board with the notion that Intel should use its processors in large clusters to create a new kind of "parallel-processing" supercomputers. IDO agreed to fund him, and over the next thirty years, Intel, using each new generation of its microprocessors, built some of the world's most powerful computers—and earned Intel a total of $5 million in the process. Rattner told the writers of Intel's annual report in 1992, "Parallel processing was a very risky idea; it required both seed capital and a significant period to develop a business." Intel gave Rattner both.

That same year, IDO also backed an internal venture that would eventually become Intel's Personal Computer Enhancement Division (PCED). This start-up set out to determine if Intel could sell its new math coprocessor chips and add-in memory boards directly to PC owners via the growing number of computer retailers specializing in customers who wanted to upgrade their current machines instead of buying new ones. Jim Johnson, who started PCED with fellow Intel employee Rich Baker, later said, "Marketing to PC consumers is in Intel's blood now, but at the time it was almost unthinkable."

Intel IDO invested in a number of other new initiatives in the years that followed. None proved as big as those first two projects. Nevertheless, the bang for the comparatively small number of bucks invested by Intel through IDO was considerable. But even more, the value of such an operation, with the mind of Gordon Moore and the heart of Bob Noyce, serving as an outlet for the very best of Intel employee creative energies, was immeasurable. It was an archipelago of the old Intel inside the new. Les Vadász said it best: "There is more to the technology business than accounting for the money. You nurture a sense of what is possible: it's never what drives you; it's what *could be*."[2]

The other figure who rose into the limelight at Intel during this period was Craig Barrett, the old Stanford professor who'd made a stunning life change at the peak of his academic career.

As already noted, while still in academia, Barrett had earned a considerable reputation in materials science. Now he turned that talent to understanding the nature of semiconductor chip yields. Thanks to more powerful and precise new fabrication equipment (aligners, steppers, etc.) chip yield rates had improved greatly since the early years of Intel and dramatically since the early days of the chip industry. But they were still lousy compared to most other industries; companies still threw out 50 percent or more of new chip models in their early days of production— and were lucky to get to 80 percent yield even on mature chips that had already been produced in the millions of units. This poor quality was matched by compromises throughout the rest of the manufacturing process. Some companies (though there is no record of Intel doing this) actually sent dead (or compromised) chips from their early production runs to customers as a sort of placeholder until they could send live ones.

Yield was always a problem in the chip business, but in the early 1980s, it took on a whole new level of importance. One reason was the recession. A company like Intel, trying to maintain its high level of investment even during hard times, desperately needed profits . . . but realized that most of the investment capital was going out the back door (often to criminals) as scrap. The company realized that if it could just increase its yield rates by 10 percent, it would have the profits it needed to reinvest in new product development.

The second reason was the Japanese. HP's announcement a few years before about the comparative quality of US versus Japanese chips had been devastating to American semiconductor companies. The US companies had convinced their competitors—and themselves—that miserable yield rates were intrinsic to chip fabrication. The success of their Japanese counterparts in blowing apart that myth had been especially humiliating. Now if they were going to restore their image and catch up, they were going to have to solve the yield puzzle for themselves.

Craig Barrett had seen the specs that compared Intel's manufacturing quality with those of other world-class manufacturers, including some of Intel's Japanese competitors, and he was as dispirited as everyone else. "Basically," he would recall, "all our results—yields, throughput time, capital utilization—were pretty abysmal."[3]

The difference was that Barrett thought he knew the path to a solution. Intel needed to go back to first principles and start from there. "We set our expectations higher. We trained our engineering staff in statistical process control. We gave more attention to equipment selection and management. We pushed our technology development."

Barrett's biggest breakthrough was Copy Exactly, the absolutely exact reproduction of successful existing practices and facilities in other locations. "I got the idea from McDonald's," he recalled. "I asked myself why McDonald's french fries tasted the same wherever I went. That's what I told my guys, 'We're going to be the McDonald's of semiconductors.'"[4]

The idea didn't go over well at first. In fact, it resulted in a contentious meeting in which Barrett outvoted his managers. Copy Exactly it would be—and when testing the process proved too much of an interference to Intel's working fabs, Barrett and his team cleared out a fab (a big financial bet) and turned it into manufacturing-process test site.

Recalled senior vice president Gerry Parker, "We had always been proud that we developed our manufacturing processes in the factory. However, it was clear that interrupting production to tweak our processes was too disruptive. So we turned Fab 5 [in 1985] over to technology development. When we got the yields up on a new process, the manufacturing teams [throughout the world] had to replicate the process exactly."[5]

At the heart of this story about Intel is the message that the company succeeded and reached the pinnacle of the modern economy not because its leadership was so brilliant (though it was) nor because its employees were so bright (though they were), nor that it had the best products (sometime yes, most times no), nor that it made it made fewer mistakes than its competitors (completely wrong), but because Intel, more than any company America has ever known, had the ability to *learn*—from its successes and even more from its failures. Most companies achieve an initial success and then just struggle to hang on to their gains, never getting much smarter as they age. But for thirty-five years, through the Noyce and Grove eras, and always under the watchful eye of Gordon Moore, Intel consistently got *better*—not only bigger, richer, and more capable, but *wiser*. And that is why during that era no company in the world could keep up.

Intel also became better because, somehow, under the leadership of a dreamer and then a taskmaster, it remained open to those lessons wherever they came from and didn't just follow orders from above. Art Rock, Les Vadász, Ted Hoff, Ed Gelbach, Regis McKenna, Bill Davidow, and many others at different times moved Intel in a different direction, imposed their vision on the company, or just told the company it was making a mistake and offered a likely solution. What made Intel a great company was that it not only allowed such individuals to complain, but listened and reacted when they did.

No story better captures all of this than Craig Barrett's yield project at Intel. The notion of Copy Exactly is so primitive, so cargo cultish—if something yields good results just repeat it mindlessly and keep doing so forever—that it couldn't help but clash with the refined empirical and scientific view of Intel's manufacturing engineers. Their training and their sensibility told them that they needed to know the *why* of reduced yields. Barrett instead told them not to worry about cause, only effect: see what works and just keep doing that. Intel had to learn to be dumb and simple—the most difficult thing imaginable to a company named after intelligence and building the most complicated devices in the world. It was a crucial lesson.

Best of all, it worked. Within a few years, yield rates rose to 80 percent at the company, with equipment utilization rising from just 20 percent to an astonishing 60 percent.[6] Other than the microprocessor, it can be said that no single idea from a single employee ever made a bigger financial contribution to Intel than Barrett's Copy Exactly. And that contribution went further than the bottom line. When the Japanese made their full-on assault on the US electronics industry, American chip companies—long humiliated by HP's quality comparisons in the late seventies—no longer had to apologize for low yield rates. Following Intel's example, they could now show product delivery quality comparable to the Japanese. Thus, as much as anyone, Craig Barrett saved the US semiconductor industry in the 1980s.

None of this was lost on Andy Grove, who began to move Barrett up through the senior ranks. Barrett quickly became the man to watch at Intel, Andy's heir apparent. The question now became whether, with such a boss, he could survive the wait.

CHAPTER 39

Eminence

In 1982, the *Wall Street Transcript* named Gordon Moore the outstanding semiconductor industry CEO for the second year in a row. In 1983, Tom Wolfe wrote his famous profile of Noyce in *Esquire* magazine. A year later, Gordon Moore and Bob Noyce were named to the Institute of Electrical and Electronics Engineers (IEEE) Hall of Fame. In 1985, Noyce was inducted into the National Inventors Hall of Fame. During those years, Intel was consistently named one of the best companies to work for in America. And in 1984, *Fortune* declared it one of eight "Masters of Innovation."

Inside Intel and indeed throughout most of the electronics industry, it was generally known that the real person who had driven Intel to its current success was Andy Grove. Anyone who knew Gordon Moore recognized that he was not the kind of man who could crack the whip hard enough to have made Intel such a disciplined, hard-driving success story. But in the larger world, Gordon, because he was such a famous scientist (Moore's Law was now being applied to all of the electronics industry) and because he was officially chief executive officer of the company, was given almost all of the credit. That he was so humble and self-effacing about his honors made his achievement seem even more remarkable. Meanwhile, Noyce, now a living legend, was still being given credit for all of Intel's success—despite the fact that his connections to the company (he was now in the placeholder position of vice chairman) were now tenuous.

Andy could only grit his teeth and stick to his job. As much as it was a statement of his own hard-earned knowledge as a manager, 1985's *High Output Management* was also an announcement to the world that if it wanted to learn about the *real* story behind Intel's success, it should forget those famous guys and talk to Andrew Stephen Grove. But despite all of his efforts and as the episode with Regis in Andy's office underscores, it was Noyce and Moore who got all of the honors during these years.

It is ironic that as much as Grove felt eclipsed by Noyce and Moore in the first two decades of Intel, he has equally overshadowed them in the years since. And if it is generally recognized that the early, diminished appraisal of Andy was unfair, less appreciated is that the current perspective on the three men may be equally unfair. Thanks to Andy's historic achievements in his years as CEO as well as the influence of his best-selling books, it is he who has now largely eclipsed the two founders in the Intel story. Noyce and Moore would never entirely disappear, of course, both because they were the founders and because of, respectively, the integrated circuit and the law.

But as the years passed and new generations rose to rule Silicon Valley, the two men slowly slid into Andy's shadow. Of the two, Gordon probably got the worse deal, despite Moore's Law, the fact that he was likely the most respected figure in the Valley, and along with being one of the richest men in America, was one of its leading philanthropists as well. But the general assumption—even, in time, within Intel—was that he had been more of a titular, and comparatively ineffectual, CEO for almost a decade, while it had been Andy who had always run the show, even as he waited all those years to finally come out behind the curtains in 1987 as the official (rather than de facto) chief executive.

The truth is very different. Lost in this celebration of Andy Grove's achievements is the fact that he had one boss from the day he arrived at Fairchild Semiconductor in 1962 until he stepped up to become chairman of Intel in 1997. That's thirty-five years in which one man, Gordon Moore, served as manager, mentor, protector, promoter, and defender of one of the most complicated, mercurial, and difficult employees in modern business history. Thirty-five years: as even Andy would agree, most bosses wouldn't have been able to handle him for thirty-five minutes. If Andy Grove pulled off one of the greatest business careers of the twentieth century, it was because Gordon Moore built and maintained the stage on which he was able to perform. He brought out the best in

Grove—not just because Andy trusted him completely, but because he trusted Gordon's *judgment* above anyone's. Gordon Moore didn't have to be a great manager of Intel; he only had to be a great manager of Andy Grove. And he was that in spades.

The second undervalued contribution that Gordon Moore made to Intel was the law—not Moore's Law the equation, but Moore's Law as an operating principle. Gordon more than anyone else understood that it wasn't really a law in the sense that its fulfillment over the years was inevitable, but rather that it was a unique cultural contract made between the semiconductor industry and the rest of the world to double chip performance every couple of years and thus usher in an era of continuous, rapid technological innovation and the life-changing products that innovation produced.

That much was understood pretty quickly by everyone in the electronics industry, and it wasn't long before most tech companies were designing their future products in anticipation of the future chip generations promised by Moore's Law. But what Gordon Moore understood before and better than anyone was that his law was also an incredibly powerful business strategy. As long as Intel made the law the heart of its business model, as long as it made the predictions of the law its polestar, and as long as it never, ever let itself fall behind the pace of the law, the company would be unstoppable. As Gordon would have put it, Moore's Law was like the speed of light. It was an upper boundary. If you tried to exceed its pace, as Gene Amdahl did at Trilogy, your wings would fall off. Conversely, if you fell off the law's pace, you quickly drew a swarm of competitors. But if you could stay in the groove, as Intel did for forty years, you were uncatchable.

That Intel not only found that groove and rode it, but soon became synonymous with the law itself can be credited only to Gordon Moore. Andy made it happen, but if he had deviated from Gordon's core strategy, his tenure at Intel might have been much shorter. One can even see Moore's hand in that all-important decision by Intel to maintain its onerous R&D budget even during hard times like the mid-1970s and early 1980s. For Gordon Moore, short-term financial worries could not be allowed to deflect Intel from its perpetual breakneck pursuit of the law.

CHAPTER 40

Crash

Intel closed out 1984 with breathtaking revenues of $1.6 billion—a nearly 50 percent jump from the year before. The sacrifices of the previous two years had paid off even more than anyone had imagined.

But the business cycle that had always defined the semiconductor industry (and by extension, the rest of tech) was still in effect, and so if the 1981 recession had been attenuated, it just meant that the upturn would be shorter this time. And that proved to be the case. By the second half of 1984, chip orders were already slipping.

This time, it was the Japanese chip companies that tumbled the tech world into a recession that was steeper than expected. Anticipating huge demand—not least by taking orders away from US chip makers— the big Japanese companies had invested heavily in fabs to build both DRAM and EPROM chips. They misread the true demand for these chips—and the amount of similar investment by their counterparts— and created a massive overcapacity in their production. Before long there was a glut of memory chips on the world market. And that led once again to a free fall in the market price.

But there was also a second, hidden, factor in this price drop: the Japanese semiconductor industry had decided to use this downturn to mask a carefully constructed attack designed to wreck the US semiconductor industry. This strategy, known as two-tiered pricing but more commonly known by the colloquial term *dumping*, was simple. Japanese chip companies, with support from the Japanese government (in particular the Ministry of International Trade and Industry, MITI), held its prices for memory chips artificially high at home—for which their Japanese industrial customers were subsidized in their purchases with investment capital, while prices for those same chips sold overseas, especially to the United States, were allowed to fall impossibly low levels.

How low? At one point, Japanese EPROMs were selling in the United States at one half their cost to manufacture. There was simply no way for the US semiconductor industry to compete with these prices; even matching them would be corporate suicide. And once the US companies realized that they weren't up against just their Japanese corporate counterparts, but against the Japanese government itself in an undeclared trade war, they knew that their only recourse was their own federal government.

This was not an easy decision. Silicon Valley had consciously built itself to have as few ties to Washington as possible. Sure, it had found much of its seed money in the fifties and sixties from defense and aerospace contracts, but as soon as the commercial, industrial, and consumer electronics industries had opened up, the Valley had turned away from military contracts and never looked back. Even David Packard, the patriarch of the Valley, returned in 1971 from a stint as deputy secretary of defense and quickly dropped nearly all of Hewlett-Packard's custom defense contracts.

By 1985, the thought of going hat-in-hand to Washington, DC, and begging for support from Congress in fighting Japanese business predations was anathema to the boardrooms of Silicon Valley. Moreover, unlike more mature industries, in which the competitors have divided up the turf and settled on a wary peace, the semiconductor industry was still caught up in internecine warfare. Many of the feuds dating back to the old Fairchild had not yet been settled; indeed, you could still see an aging Bill Shockley sitting in a local Valley watering hole complaining about the Traitorous Eight.

The real question was not whether the chip industry should go to Washington for help. That quickly became inevitable, as the cheap Japanese chips began gobbling up market share. The question was whether it *could* go, whether the various players could actually work in concert for once to achieve a common goal. And that question quickly devolved down to whether there was one figure in the US semiconductor industry who was trusted enough and could represent the industry's needs forcefully enough that all of the many companies could agree on him and would rally around him. In the end, all agreed that there was only one such person.

Bob Noyce.

Noyce had been enjoying the near-decade since he had departed from the day-to-day leadership of Intel Corp. He spent four years as Intel's chairman of the board—and by all accounts did a superb job in a position for which was uniquely suited. After that, he happily stepped down, letting Gordon take the chairmanship and taking the undemanding role of vice chairman. He still kept an office at Intel, choosing to work out of a double cubicle, thus setting a standard for democratic leadership that Valley executives have struggled to match ever since.

When he was around, which became less and less in the years to come, Bob could be found sitting at his standard cubicle desk or at the standard meeting table a few feet away, looking at first glance like some senior marketing manager or perhaps a corporate finance guy. But any such notions soon disappeared once you entered that little open office. Not many Silicon Valley middle managers had the National Inventors medal hanging on the wall, or photographs with heads of state. And then there was Noyce himself: you didn't have to know who he was to appreciate that you were in the presence of an extraordinary, powerful, and perhaps even great, man. And even when, in his deep and cigarette-gravelly voice, he joked and chatted and peppered his sentences with the occasional expletive—all designed to put you at ease—the vibration still filled the air. It has been said that only two men in Silicon Valley history could command the attention of a room merely by entering it: David Packard and Robert Noyce.

As he became wealthier and increasingly famous, other concerns began to draw Bob away from Intel and the vagaries of corporate life. He sat on committees and boards. He and Ann Bowers bought a big house in the hills behind the Los Altos Country Club and regularly held fund-raisers there for various charities. Bob brought to all of these charities the same charisma and ingenuity that he brought to everything else in his life. Joggers on the narrow road around the golf course learned to dive out of the way when they saw Bob's silver Mercedes almost leaving the ground as he crested the last hill, late again for some fund-raiser he and Ann were sponsoring.

He also had time to be an athlete and a daredevil again, just as he had been as a young man. He bought several planes and regularly flew them. And with his new wealth and time, he also took his skiing to an even higher level: with Bill Davidow and Jim Morgan, the CEO of one of Intel's suppliers, Santa Clara neighbor Applied Materials, he took up helicopter skiing in the Bugaboo Mountains of British Columbia.

All three men were expert skiers, but this was very dangerous. Eventually the growing number of avalanche fatalities forced them to quit. But before then, Jim Morgan would recall, Noyce always took greater risks and took more challenging paths down the mountains than his two companions. "He just took crazy risks," Morgan later said, "stuff Bill and I were frightened to do. We were always afraid Bob would kill himself."[1]

Despite the widespread resistance in the Valley at the time to Washington, Bob had also dabbled in politics. In 1977, he had teamed up with Sanders, Sporck, Motorola VP John Welty, and Fairchild president Wilf Corrigan to found the Semiconductor Industry Association (SIA). The SIA's "essential" task—besides putting on a wildly popular annual dinner at which it predicted the next year's chip book-to-bill ratios—was, in the words of the minutes of an early board meeting, to "slow down what the Japanese government is doing in support of its industry, and speed up what our government [is doing]."[2]

At that point, the biggest threat was acquisition. In the 1970s, nearly twenty US semiconductor companies had been bought, mostly when they were most vulnerable during the crash, by foreign interests. By the end of the decade, that trend had begun to slow, in part because of the terrible track record of those companies after acquisition. Big companies didn't know how to deal with either the culture of chip companies or the demands of Moore's Law and so fell behind and lost key talent. In addition, most of the big chip companies, notably National Semiconductor, adopted "poison pill" stock plans to head off unfriendly takeovers.

As this record suggests, even at this early date, Noyce was already worrying about the Japanese threat. Of all of the US semiconductor leaders, he had been to Japan most often. And because of his singular position as the most famous and admired of these leaders—indeed, he was often credited with inspiring the Japanese chip industry—he had been given access to the highest reaches of the Japanese electronics giants. He had no illusions about those companies. While some of his counterparts still dismissed them as "toy makers," Bob had met their top engineers and seen their world-class fab lines. As he would muse in 1978, someday future business historians might note that "in addition to originating and nurturing a vibrant semiconductor industry, the United States also lost it—the same way we have lost the steel industry and the TV market—to foreign competition."

Noyce was also personally angry at the Japanese—for having betrayed what he saw as American trust—and at the US chip industry (including himself) for having fallen for the ruse. As Ann Bowers would tell Noyce biographer Leslie Berlin, "Bob was a very trusting person. You had to practically punch him in the nose to have him think there was something untoward happening. You can imagine how he felt when he realized that all those years he had been hosting the Japanese guys they had been trying to get [American] secrets."[3]

In the meantime, however, another challenge presented itself, and both Noyce and the SIA turned their attention to it first.

After the explosive growth of the technology industry in the 1960s, new company creation had slowed in the 1970s. This was not because there were fewer entrepreneurs. On the contrary, there were more than ever, including the likes of Bill Gates, Steve Jobs, and Nolan Bushnell. But venture capital funds were smaller and more conservative about making investments. That was due to the high federal capital gains tax, which discouraged potential investors from taking a chance on such high-risk opportunities as new high-tech ventures. By 1978, that tax stood at 49 percent, and the comparative dearth of new companies, the lifeblood of Silicon Valley and the tech world, threatened the long-term health of the industry.

One of Bob Noyce's new post-Intel commitments was serving as the chairman of SIA's board and head of the organization's Trade Policy Committee. In the former role, in February 1978, he traveled to Washington to give testimony before the Senate Committee on Small

Business about the need to reduce the capital gains tax (or more precisely, the differential between that tax and the regular income tax). He was backed in this testimony by the American Electronics Association (AEA), which had served as his primary sponsor.

It proved to be one of the most important and far-reaching contributions Noyce ever made to the electronics industry. By the end of the year, Congress had passed and President Carter had signed a bill that reduced the capital gains tax to 28 percent (and relieved restrictions on pension fund investments in venture funds). Three years later, President Reagan would cut the tax again, to just 20 percent, the lowest since the Hoover administration. These lower rates set off the biggest entrepreneurial start-up boom in high-tech, and perhaps even US, history. As a result, the 1980s and 1990s would see almost continuous growth in high tech, with the traditional downturns of the business cycle shorter and shallower, the coffers of many venture capital funds filled to bursting (jumping from $50 million to $1 billion in just eighteen months), and the creation of new companies at many times the rate of just a few years before. These new companies survived and thrived in—as well as contributed greatly to—what would be largely credited to both the "Reagan revolution" and the Clinton administration as the longest period of economic expansion in American history.

Many people, from economists to tax-reduction activists to politicians to business executives, had made this prosperity possible. But the pivot point, the event that set the conversation going and made the subject legitimate, was Bob Noyce's testimony before the US Senate.

Now, in the early eighties, the Japanese semiconductor challenge was back with a vengeance. Every US semiconductor company suffered. Intel, as noted, had suffered layoffs—in the end they would total two thousand employees—and still had to sell a minority ownership in the company to IBM. AMD lost two thirds of its net income between 1981 and 1982. National Semi saw a $63 million reversal in profits (from $52 million to an $11 million loss) in the same period. It was with rueful humor during those dark days that Noyce proposed to start an SIA meeting by offering to lead the members in a group prayer.

By 1984, as the Japanese chip companies began to crush the prices on memory devices and gobble up ever more market share, the situation became critical. Within another year, Japan's share of the world semi-conductor market had surpassed that of the United States. Japan was now number one.

Interviewed by the *San Jose Mercury-News*, Noyce was filled with despair. Everywhere he looked, he said, he saw signs of "the decline of the empire. . . . Can you name a field in which the US is not falling behind now, one in which the US is increasing its market share? We're in a death spiral." America, he feared, might become "a wasteland," adding, "What do you call Detroit? We could easily become that." Worst of all, he said, as if warning himself, the growing despair that was beginning to replace the Valley's essential optimism was its own vicious circle: ". . . optimism is an essential ingredient for innovation. How else can the individual welcome change over security, adventure over staying in a safe place?"[4]

Noyce knew what he had to do. He threw himself into the fray, later estimating that he spent half of the decade in Washington, DC. (Moore: "Dealing with Washington was something I was very happy to have him do because then I didn't have to go back there.")[5] And even when he was back in the Valley, he was stumping for the SIA's strategy against Japan, including giving testimony before Congressional committees that had decamped to San Jose to hold hearings on the topic.

One again, his appearances were electrifying. As always, Bob seemed calm, assured, and measured in his argument, the very embodiment of all that was good about American business and entrepreneurship. For many Valleyites and government leaders, it was the first time they had seen the legend—resonant voice, incomparable legacy, and native charisma—in the flesh. It had its desired effect.

The reality was that Bob's appearances and testimonies were both carefully rehearsed and aligned to his long-term strategy for winning the trade war. Noyce worked constantly with Intel's future VP of legal and government affairs, Jim Jarrett, then a newly hired publicist, to craft speeches that positioned him not as a spokesperson just for Intel or even the semiconductor business, but for the entire US electronics industry. In time, the two men developed a collection of speech "modules" that could be mixed and matched for different audiences and occasions, leaving room for Bob's usual improvisations.

As Jarrett told Noyce's biographer Leslie Berlin, "He wasn't a great speaker from an oratorical style standpoint, but he had so much charisma that it really over-rode his moderate style as an orator. . . . His presence was really the thing that mattered. He did not have a Baptist-preacher approach. [His] was a thoughtful, kind of modestly presented style. He was just being himself."

At the heart of his speeches were a series of claims that, though a bit suspect in their accuracy, nevertheless had a powerful impact. The US semiconductor industry, Noyce told his audiences, provided the "crude oil," the core components that made the rest of the electronics industry (now the nation's largest manufacturing employer) and thus the US economy run. The chip industry's sales were the real index of the nation's health, and over the previous decade, that industry had added more than $2 billion to the US economy just in corporate taxes and directly or indirectly provided livelihoods to millions of Americans. It was the first time that the semiconductor industry had ever made the case for its primacy to the American public.

Whether they were fellow chip industry managers, engaged Valley citizens, or members of Congress, audiences ate it up. And Noyce never once exposed the fact that he was performing a tightrope act.

He had been put in this delicate situation because the electronics industry had broken up into two camps, each with a different notion of how to fight back. On one side were those who believed that government intervention was the only solution. It was their belief that the only way was to fight fire with fire. If Japan's MITI was going to help its domestic industry with trade protectionism, capital investments, and violation of international trade law, then the US Department of Commerce—or Defense—should play the same game, targeting the Japanese semiconductor industry (with tariffs, even embargos) in a high-stakes game of reciprocity.

This plan had the support of a number of liberals in Congress, the so-called Atari Democrats, including Al Gore, Paul Tsongas, Patrick Leahy, and Gary Hart.[6] On the other side were the advocates of fair trade, including President Reagan, who believed that a combination of enforcement of international law with an even greater unleashing of American competitiveness would ultimately outstrip the Japanese. The

Atari Democrats, backed by a number of the Valley's more progressive CEOs, also wanted to focus the fight on semiconductors, or at most, electronics. The free-traders argued for an across-the-board fight to drop the barriers to free trade.

Noyce, who always sided with free competition and entrepreneurship, naturally gravitated to the Republican side. But shrewdly, he diplomatically maneuvered between both camps, alternating between his role representing the high-tech trade associations and his personal role as an influential industry leader. Thus he politely refused an offer to join an industrial policy group led by California governor Jerry Brown, which included friends like Charlie Sporck and Steve Jobs, yet when the time came for the SIA and AEA to petition the US Trade Representative for relief against Japanese violations under the Trade Act of 1974, Noyce put all of his reputation and energy behind the move.

He did much more than that. Noyce also helped convince Intel to join other electronics companies in funding a political action committee (PAC) to promote the industry's views inside the Beltway. He also helped support the creation of the bipartisan group of Congressional representatives who supported the US chip industry, the Congressional Semiconductor Support Group, which petitioned the White House and cabinet officials to increase their help.

Most influential of all, though, was the small army of tech CEOs who volunteered to constantly travel to Washington and try to influence lawmakers face-to-face. Sporck would later say that this group was *"the* secret to the SIA's success."[7]

Clyde Prestowitz, the administration's counselor for Japan affairs, would later write that "appealing to Washington was not easy for these men and the others like them in the industry who embodied the ideals of the American dream. Coming from modest, even poor backgrounds, they had succeeded through initiative, inspiration, and perspiration, in founding an industry widely seen as the key to the twenty-first century. They had done it on their own as lone riders without government help—indeed, sometimes in the face of government harassment."[8]

No one was more important in this campaign than Robert Noyce.

Wrote the *Harvard Business Review*, "He is something of a legend in the electronics world. . . . The Washington establishment wanted to get to know him as much as he wanted to develop political contacts."[9] Prestowitz called him the one irreplaceable figure in the campaign.

Noyce downplayed his importance, crediting more to his wealth than his prestige. "I found that money gave you power, that your opinion was more highly valued if you were rich."[10] But more than anyone, he knew the value of being Dr. Robert Noyce, inventor of the integrated circuit and founder of Fairchild Semiconductor and Intel.

By the late 1980s, the Japanese semiconductor threat had receded, and by the 1990s, the Japanese chip industry had returned to its position as a respectable but mostly unthreatening number two. There were several explanations for this. One was the pressure placed upon the Japanese government by the US government, including $300 million in fines and the passage of the Semiconductor Chip Protection Act. Japan eventually caved and agreed to operate under international trade and US copyright laws.

Another was that the Japanese were playing under a very short time clock: high-quality manufacturing and price-bombing worked only with *existing* technologies. Once the US chip companies came up with the next generation of chips, the Japanese could copy them and then build them faster, better, and cheaper. The Americans eventually responded (for example, Barrett at Intel) by approaching the Japanese in speed and quality, yet still remained the industry innovators. That meant that in time (and a very short time, given Moore's Law) the American chip companies accelerated away, funded by the profits they now retained in a fair marketplace.

Finally the Japanese were also racing under another, bigger clock: demographics. The war generation was still working in the 1980s and still maintained its long-established habits of hard work, discipline, and thrift. Ultimately, those traits were just as important as anything taught to Japanese workers by Deming and Juran. But this generation was now growing old and facing retirement, and the next generation of Japanese, raised on the fruits of the sacrifices of their parents, had much less interest in devoting their lives to either the nation or their descendants. In other words, the Japanese chip industry had to win the semiconductor wars in the early 1980s, or it would likely never have the chance again. And in the subsequent decades, that would prove to be the case.

Bob Noyce emerged from the US-Japanese chip trade war a national figure. In Washington, he represented the very best of American industry. To his peers in tech, Bob was a hero. And to the world at large, the name Bob Noyce now entered the public consciousness as the embodiment of the age of the microchip.

But it was in Silicon Valley where Bob's achievement met the strongest response. Though rarely discussed, even corporate CEOs keep in the back of their minds a secret career strategy. The same ambition and hard work that got them to the top only makes them want to climb even higher. And in the Valley, the ultimate career standard was set by David Packard: start a company in a garage, grow it into the leading innovator in its field, then take it public, then take it into the Fortune 500 (or better yet, the Fortune 50), then become the spokesman for the industry, then go to Washington, and then become an historic global figure. Only Packard had accomplished all of this; he had set the bar, . . . and the Valley had honored his achievement by making him the unofficial "mayor" of Silicon Valley, the second person to hold this title after Packard's own teacher and mentor, Stanford professor Frederick Terman. Now, when the Queen of England came to the Valley, Packard was her host; when any global organization needed the validation of a Valley connection, it was Packard's name they pursued for their letterhead.

Fred Terman had made Silicon Valley possible through his pioneering electronics course and his creation of the Stanford Industrial Park. Dave Packard had built the Valley through Hewlett-Packard and set the highest possible standard of corporate culture with the HP Way. Now, with Terman gone and Packard growing old, it was Bob Noyce's time. After all, he had not only invented and husbanded the technology that now underpinned the tech world but (thanks to Moore's Law) guaranteed its importance to the global economy for many years to come. And now he had saved the US electronics industry from the greatest threat it had ever known. Most of all, he had matched Packard's career achievements—something no one had done before. And despite some near misses (Steve Jobs, Eric Schmidt, Larry Ellison, Mark Zuckerberg), no one has done it since.

Still in his midfifties, Bob Noyce promised to be around for decades to come, and the Valley felt secure in having him as its figurehead. From now on, he would speak for Silicon Valley on the world stage, and Bob Noyce felt comfortable assuming that exalted role.

CHAPTER 41

Memory Loss

There were many business casualties in the chip trade war between the United States and Japan, companies that died or were battered out of competition or were sold off. Surprisingly, among the biggest companies in the war, the one most changed by it was the one generally considered to be the greatest victor: Intel Corp.

Intel had been built as a memory-chip maker and had been very good at it. At various times during its first fifteen years, the company had not only been the industry leader in memory-chip sales, but also the leading innovator. Memory had put Intel on the map, and it had funded the invention of the microprocessor.

But the arrival of Japanese competition in the industry and its subsequent price manipulation had been devastating to Intel's financials. By the mid-1980s, it was becoming increasingly apparent inside the company that while Intel could defend its microprocessor business almost indefinitely, it could no longer guard its memory business. There were now just too many competitors, and pricing was likely to be depressed far into the future. Even if Intel could claw its way back in a couple of chip generations to be the obvious industry innovator, there was no guarantee that even that position would produce enough profits to keep the company prosperous.

All of this had been apparent to many of Intel's senior managers as early as 1982, and they had pressed Grove and Moore to get out of the memory business. Then as the market crashed, their calls became even louder. But still the two remaining founders hesitated; after all, the Intel they had built was a *memory* company. Grove: "Then in the fall of 1984 [everything] changed. Business slowed down. It seemed that nobody wanted to buy chips anymore. Our order backlog evaporated like spring snow. After a period of disbelief, we started cutting back production. But after the long period of buildup, we couldn't wind down fast enough to match the market slide. We were still building inventory even as our business headed south."[1]

Moore: "We had been in the memory business from the beginning, and it was the first product area we pursued. We'd been quite successful in it, generally. But we had not done a good job for two generations and we'd lost our leadership position. We had made the R&D investment to jump ahead of the industry once again at the one megabit level—a million bits of memory in one chip. We developed the technology. We had the product. But now we were faced with the decision of investing $400 million in facilities to become a major player again. It was at a time when the memory industry was losing a significant amount of money. It looked like it was going to be overcapacity forever and, well, it was emotionally difficult. The numbers just stared us in the face. The chance of getting a return didn't look very good."[2]

Grove: "We had been losing money on memories for quite some time while trying to compete with the Japanese producers' high-quality, low-priced, mass-produced parts. But because business had been so good, we just kept at it, looking for the magical answer that would give us a premium price. . . . However, once business slowed down across the board and our other products couldn't take up the slack, the losses really started to hurt. The need for a different memory strategy, one that would stop the hemorrhage, was growing urgent."

It was becoming increasingly apparent that Grove and Moore were the last two leaders left at Intel who were still holding out for the company remaining in the memory business—but in the end, they were the only two who mattered because they had final sign-off. And the topic was so painful for them, so fraught with emotion, that as Grove later admitted, he could barely bring himself to speak about it.

Grove: "We had meetings and more meetings, bickerings and arguments, resulting in nothing but conflicting proposals. . . . Meanwhile, as the debates raged, we just went on losing more and more money [1984 revenues were $1.63 billion, 1985 revenues fell to $1.36 billion, and 1986 revenues fell again, to $1.26 billion]. It was a grim and frustrating year.

During that time we worked hard without a clear notion of how things were ever going to get better. We had lost our bearings. We were wandering in the valley of death."[3]

Given Andy Grove's history, those were strong words. But they reflected the attitude of much of the rest of the US electronics industry. Remember Noyce's comments from the same time about the end of the American civilization.

What was holding Intel back? According to Grove, there were two reasons, both of them based on the role of memory chips as the technological heart of Intel: "One was that memories were our 'technology drivers.' What this phrase meant was that we always developed and refined our technologies on our memory products first because they were easiest to test. Once the technology had been debugged on memories, we would apply it to microprocessors and other products. The other belief was the 'full-product-line' dogma. According to this, our salesmen needed a full product line to do a good job in front of our customers; if they didn't have a full product line, our customers would prefer to do business with our competitors who did."[4]

But in the end, Grove began to realize that only he and Gordon really believed those two reasons. And as Andy found himself being grilled even during visits to outlying divisions, it was slowly becoming apparent that there was a third reason for their resistance to the move: pride. Neither man wanted to admit that they had been beaten in the company's core business under their watch—and face shareholders and the public with that failure.

Grove: "I remember a time in the middle of 1985, after this aimless wandering had been going on for almost a year. I was in my office with . . . Gordon Moore, and we were discussing our quandary. Our mood was downbeat. I looked out the window at the Ferris wheel of the Great America amusement park revolving in the distance, then I turned back to Gordon and I asked, 'If we got kicked out and the board brought in a new CEO, what do you think he would do?' Gordon answered without hesitation, 'He would get us out of memories.' I stared at him, numb, then said, 'Why shouldn't you and I walk out the door, then come back and do it ourselves?'"[5]

Metaphorically, that's exactly what they did. To a general sigh of relief from their management team, Moore and Grove made the decision to take Intel out of the DRAM business. They then took that decision to the board. Art Rock, who still sat on the board, later called the vote to abandon the memory business "the most gut-wrenching decision I've ever made as a board member." By comparison, Noyce voted yes without any hesitation. Ann Bowers would tell biographer Berlin, "He thought the Japanese were already beating the heck out of memories. He hoped microprocessors would offer a way out." The motion carried.

If the decision was met with cheers by Intel's management, it wasn't so happily received by the rank and file. Many of them had spent their careers at Intel in the memory business, and they believed in Intel's memory devices at least as much as they believed in the company itself. Now they feared—not without reason—that they were about to be jettisoned, as had the Microma employees less than a decade before.

They weren't far off. Moore: "Now, deciding to get out of a major business like that means a whole lot of things had to happen internally. There were literally thousands of people that had to be re-employed doing something else or let go. We hope we could re-employ them."[6]

But in the end, for most of Intel's memory division employees, that would prove to be impossible. In 1982, Intel had been the second largest independent semiconductor company in the world; in 1986, it was number six. As Moore would say that November, "Intel introduced the DRAM. We used to have 100 percent share of the market. And we went from 100 percent to zero percent over time." And despite selling the memory business to Zitel (thirty Intel employees went with it), Intel lost $173 million in 1986, the company's first year in the red since it went public fifteen years before.

In the end, Intel cut 7,200 jobs, a third of its employees, many of them former members of the memory division. It was "a truly awful process," recalled VP of corporate programs Dick Boucher. "We were not laying off people who were incompetent—many were long-tenured and had successful careers."[7]

During 1985 and 1986, the company also closed seven factories. It was not an idle decision: senior management was holding secret meetings in Bob Noyce's living room to discuss "how to shut down Intel, if it comes to that." There was little consolation in the fact that six other US memory-chip companies were quitting the business as well, leaving just two, or in the fact that Intel's biggest Valley counterparts (most of the smaller ones had already died), including National Semi, Fairchild, and AMD, were in dangerous straits as well. It didn't even matter that Intel now had the weight of a billion-dollar company. Boston's Wang Laboratories, once a dominant computer-workstation player, was showing that a multibillion-dollar electronics company could die in a matter of months.

But there were two big differences between Intel and all of the others. One was that even in one of its darkest hours, Intel still maintained R&D and capital expenditures at a stunning 30 percent of its revenues, even as losses mounted. The second was that Intel had the microprocessor contract for the IBM PC and its clones, and it had just announced, at the nexus of this investment and this contract, the next generation of that chip. And so even as it struggled to survive, Intel was still ready to make its move.

CHAPTER 42

Andy Agonistes

The Intel 80386 microprocessor was formally introduced in October 1985,the same month that Intel got out of the memory business. Large-scale deliveries took place in the third quarter of 1986. Reflecting the global nature of the personal-computer industry, the 386 was simultaneously announced in San Francisco, London, Paris, Munich, and Tokyo.

The 386 was Intel's first 32-bit microprocessor, and it featured 275,000 transistors—nearly double the 80286 total. Unlike its predecessor, it was designed to run multiple software programs at the same time. This made the 386 not only many times more powerful than the 286, but a particularly popular CPU for the new generation of computer workstations and high-end personal computers. Yet it could still execute most of the old code designed for the 286, and even the original 8086. So powerful was the 386, that tweaked up over the years to be several hundred times faster than the original, it continued in use in embedded systems and mobile phones (such as the BlackBerry 950 and Nokia 9000) well into the first decade of the twenty-first century, when Intel finally shut down production. The designer of the chip, Intel's John H. Crawford, went on to lead the design of future generations of 80x86 chips.

IBM wanted the 386, . . . just not yet. It preferred to stick with the 80286 for a couple more years. That opened the door to the cloners, and in August 1986, an aggressive new player, Compaq Computer, became the first PC cloner to introduce a 386-based machine. This announcement opened the door for other clone competitors to jump into the market. And by the end of 1987, IBM had become just another player in the industry it had created, . . . while Intel and Microsoft had a giant new industry all to themselves.

Well, not quite. By this point, Intel had several competitors for its own microprocessor design. Cyrix was a company created by disaffected Texas Instruments scientists in 1988 who decided to reverse-engineer the most popular processor on the market, the Intel 80386, in such a way as to not violate any Intel hardware or software copyrights and patents. Intel quickly sued Cyrix but largely lost the case. An out-of-court settlement cleared the way for Cyrix to make its chips in fabs that Texas Instruments had licensed from Intel, and the resulting chips were sold by TI mostly to computer enthusiasts. More a nuisance than a real threat to Intel, Cyrix was eventually bought by National Semiconductor.

Another new company, Chips and Technologies, also reverse-engineered the 386, but the resulting product had a number of flaws and didn't last long on the market. It eventually made a good business in PC chip sets and computer graphics chips . . . and was bought by Intel in 1997.

The real threat to Intel, the one that angered Andy Grove the most, was Advanced Micro Devices. In his mind, Jerry Sanders had talked his way into the 8086 second-source deal by playing on the soft-heartedness of his surrogate father, Bob Noyce, then had cemented the arrangement, over Intel's objections, when IBM had demanded that AMD serve as backup for Intel's deliverables.

But Noyce was now out of the decision process, and IBM was falling behind its own clone competitors and had lost its leverage over Intel. More than that, Intel had been able to endure AMD's presence during the good times, when neither company could keep up with demand, but now during a downturn, Andy saw no reason why his company should endure this parasite on its inventions, which was snapping up 15 percent of the market for them. Those lost millions and the profits they would have produced would have gone a long way toward improving Intel's balance sheet and saving some of those lost employees. He bet that Intel's production and reputation were both now sufficiently robust that it no longer needed a "safety" manufacturer as backup.

For his part, Jerry Sanders was a desperate man. AMD had lost

$37 million—a higher fraction of its revenues than Intel's $173 million loss—and his company was on the ropes. Intel had made a deal with him, and he was going to hold them to it. In the end, as he saw it, it all came down to that phrase—that AMD had the right to "copy micro-codes contained in Intel microcomputers and peripheral products sold by Intel." And that single word *microcomputer*. What did that mean? It had been a common-enough term in the early days of the business, but it had long been supplanted by *microprocessor* (the chip used in comput-ers) and *microcontroller* (the chip used in noncomputer systems, such as automobile fuel injections and thermostats).

For Sanders, the word *microcomputer* meant *microprocessor*, and just because Intel had upgraded the original 8086 and given it new names like 80286 and 80386, those devices still used the original core set of instructions—the microcode—and therefore, the contract still held. And unless some court told AMD otherwise, Sanders was going to con-tinue copying the latest Intel microprocessor and keep his company alive. He also believed that, for all of its huffing and puffing, if it was going to keep its giant corporate customers, Intel would have to accept the backup of AMD as a second source.

This was not a situation that was going to be settled by negotiation. No matter what the facts were, in the contract and on the ground, the lawyers for both companies soon understood that this fight was also personal. And thus when Intel introduced the 80386, Andy pointedly also announced that for the first time Intel would remain the sole pro-ducer of its leading microprocessor. "We didn't want to put the chip on a silver platter and ship it off to other companies. We wanted to make sure we would get something of value in return." Intel buttressed this decision with the argument that it was abandoning second sourcing be-cause the products produced by those contractual partners (i.e., AMD) weren't up to Intel standards. Most of the world saw this as a smoke screen. Intel just wanted to take back control over its own business and the money that went with it.

Grove: "I think AMD, especially, thought we were bluffing. They did not meet their commitments, thinking we'd need their help eventually to meet demand. It was like a game of chicken, and we didn't swerve."[1]

Two implications, both historic, emerged from this one decision.

AMD, realizing that it really was being shut out of the x86 universe, responded by going ahead and cloning the 386 anyway. Intel sued. Neither side was willing to cut a deal, and *Intel v. AMD*, which in various incarnations would last into the twenty-first century, would be the biggest and longest lawsuit between two companies in high-tech history.

At the same time, in making such a move, Intel gave itself an unprecedented challenge in the semiconductor world. It had the most popular new chip in the world, one whose total demand was, for now, incalculable. And whatever that demand turned out to be, Intel was going to have to meet it . . . alone. Suddenly Craig Barrett's work just a year before on manufacturing productivity, as well as Intel's massive commitment to capital improvements during the hard times, loomed large in importance to the company.

Grove: "We had to commit to supplying the entire needs of the industry. That motivated us to get our manufacturing performance up to snuff. We developed multiple internal sources, so several factories and several processes were making the chips simultaneously. We made major commitments to production ramps, and we didn't hedge."[2]

The good news was that, in taking this huge risk, Intel at least had—at last—an economic wind at its back. In July, the United States and Japan signed an accord to open the Japanese market to US semiconductors. The war wasn't yet over, but the end was in sight. Then in August, Intel won a dispute with Japanese giant NEC that upheld Intel's right to copyright its microcode. Besides AMD, that would spell the end of offshore cloners.

In early 1987, the US semiconductor industry saw a rate of (real) orders greater than it had seen in four years. By the end of the first quarter, Intel saw profits for the first time in a year: $25.5 million. Throughout the world, company employees celebrated with "Back in the Black" parties.

This achievement was recognized both inside the company and outside it. In April, the same month as the great financial news, the board recognized Andy's extraordinary contribution through the difficult days by naming him, on Gordon Moore's recommendation, the chief executive officer of Intel Corp. Moore, in accepting the title, remained as chairman of the board, but removed himself from the day-to-day operations of the company.

It was an important statement on the part of both the board and Dr. Moore. Since he had arrived in the electronics industry a quarter century before, Andy Grove had always been kept on a short leash—his brains and skills widely admired but his style and risk taking always suspect. Now with this promotion, his board of directors and his mentor were announcing to the world that Andy had paid his dues, that he had shown considerable wisdom and poise leading the company through the toughest time in its history, and that he now deserved the right to be the CEO of this billion-dollar corporation.

In a final touch, the board also voted to buy back, three years later, IBM's 12 percent stake in Intel. When that investment had been made in 1983, it was seen as an unwelcome intrusion by a big customer on a much smaller supplier that had to agree to the deal if it wanted to keep the big PC contract. But in the disastrous years that followed, the IBM investment quickly came to be seen as a timely lifeline that had helped to save Intel.

The electronics industry appreciated Grove's achievement as well. Now Andy was being truly ranked as an equal to the founders. Thus it was fitting that that June, when Bob Noyce received the National Medal of Technology from President Reagan, Andy received the 1987 Engineering Leadership Award from the IEEE.

Andy Grove now had everything he had ever dreamed of: a billion-dollar company, the hottest chip on the market, full ownership of the company's core technology, the economy heading into the greatest boom in American history, and the biggest new technology ever, waiting just around the corner. He was in the cockpit of the most important company in the world, with a supportive board and, as long as the stock held strong, no one to answer to but himself.

Equally ambitious men have been broken by such an opportunity—by hubris, by betrayal, by even cleverer competitors, and by their own flaws. Andrew Grove, formerly András Gróf, was fifty-one years old. He had been in the United States for thirty-one years, and a part of the semiconductor industry for twenty-four. He had been hugely successful, but that success had largely been in the shadow of even more famous men. He enjoyed great respect from his peers throughout the electronics industry, but that respect was tinged with doubt: perhaps Andy was too hotheaded, emotional, and shortsighted to be more than a chief operating officer, to be a chief executive.

He would have fourteen years to make his own mark on the world, to prove that he was the great leader he believed himself to be. If he succeeded, Andy knew that he would make history and be recognized as one of the best business executives of the age. If he failed, he would not only be blamed for wrecking Intel, but for crippling much of the rest of the electronics industry. Most of all, he would be forever known as the man who crashed Moore's Law—the ultimate betrayal of the man who had been his friend and mentor.

If Andy Grove was afraid, he never showed it. He only grew more ferocious. And in early 1988, he unleashed Intel Corporation to take the leadership of the electronics world.

CHAPTER 43

Mentoring a Legend

For Robert Noyce, the mid-1980s were a happy time. The hard times that had hit the semiconductor industry and Intel had been worrisome, but he was no longer in the thick of that fight. He didn't spend sleepless nights worrying about how the company would meet payroll or go through the misery of conducting a major layoff under his name. As vice chairman, his sole job was to advise Gordon and Andy, upon whose shoulders the responsibility for the fate of Intel now rested.

Of course Noyce was deeply interested in the fate of the company he cofounded, and much of his wealth depended upon the value of Intel stock. But he had other interests as well, not least the fate of the US electronics industry. And there, as the leader of the US side of this trade war, he had enjoyed one victory after another and had become a business hero in the process. The most powerful members of the federal government sang his praises. And even the Japanese, who had been his foes, regarded him with the same or even greater awe.

At home, Bob's life was as busy as ever, though in a different form. The Los Altos house was the site of continuous activity: fund-raisers, cocktail parties, and dinners. As director of human relations for Apple, Ann was quickly coming into her own as a major Valley figure, and the pair of them achieved a degree of social power together that Silicon Valley had seldom known before and never has since.

A regular visitor to the Noyce household during this period was an intense and quixotic young man in the midst of a serious life crisis and desperate for help.

Steve Jobs had been the most celebrated young entrepreneur of the age just a few years before. He had made the cover of *Time* magazine before his older and more successful Silicon Valley counterparts. And with the Macintosh computer, introduced in early 1984, Jobs had become something even more: the embodiment of a new generation.

But then it had all fallen apart. In 1983, in an effort to put someone in charge of a fast-growing Apple Computer that was beyond his business skills, Jobs had recruited the president of PepsiCo, John Sculley, to serve as Apple's CEO. Jobs believed that he could control Sculley and thus remain the de facto leader of the company. But Steve Jobs had become too much a destructive force inside the company, and—to nobody's surprise but Jobs's—Sculley proved to be a far superior corporate politician and infighter. Before long, Steve Jobs was driven out of the company he had cofounded and for which he was the most visible face.

Jobs founded a competing company, NeXT, almost in a fit of pique. Its products were at least as innovative as Apple's and even more stylish (and one was used to write the code for the World Wide Web), but the company never really gained traction. Jobs was becoming a lost soul and at great risk of ending up, while still under thirty, a once famous but now largely forgotten Valley figure.

It was in this state of mind that Jobs went searching for a mentor, a father figure who not only knew what it was like to live at the top of the tech world and be under constant public scrutiny but who, unlike him, had succeeded. It wasn't long before he found himself invited to dinner with the man he admired most.

The relationship between Steve Jobs and Bob Noyce, as with most things in Jobs's life, was both curious and compelling. After all, Jobs's unforgivably selfish behavior toward his partner Steve Wozniak at the birth of Apple had meant that Apple could not use the Intel 8080 and 8086 in the Apple I and II—a combination that would have changed history. Instead, Apple (and now NeXT) and Intel had become competitors-once-removed.

Still, the ties to Apple during that company's early years were very strong. After all, the third founder, Apple chairman Mike Markkula, had been an Intel marketing executive. Regis McKenna, Intel's agency publicist and marketing guru, was now enjoying even greater fame doing the same work with Apple. And of course, Ann Bowers had grown close to Jobs during the years the two shared at Apple. Bob himself would sometimes call Apple, and as Markkula described it, "come over to Apple and just hang around. Go in the lab and talk to the guys about what they were doing."

Still Jobs was not the most welcome guest at the Noyce household. Noyce enjoyed the young man's company—perhaps because he was too smart, too old, and too famous to fall under the spell of Jobs's intensity and vaunted "reality distortion zone." Looking back, Bowers would tell Leslie Berlin that Bob treated Jobs "like a kid, but not in a patronizing way. He would let him come and go, crash in the corner. We would feed him and bring him along to events and to ski in Aspen."

Noyce even invited Jobs to join him on a flight in Bob's Seabee seaplane, a trip that almost ended in disaster when Noyce accidentally locked its wheels while landing on a lake. Later, when the pair attempted to land on a runway, the plane nearly flipped over. Only Noyce's superb piloting saved them from oblivion. "As this was happening," Jobs would later say, "I was picturing the headline: Bob Noyce and Steve Jobs Killed in Fiery Plane Crash."

But as Ann Bowers knew as much as anybody, being around Steve Jobs was thrilling but ultimately exhausting, because the young man seemed to have no boundaries. He showed up at the door unannounced, called at midnight over some notion that had just captured his fancy, and generally acted as if he had no appreciation of the personal lives of other people. Even Bob, late one night after getting off a call from Jobs, said to Ann, "If he calls late again, I'm going to kill him." But he still took the next late-night call from Steve Jobs.

What Jobs seemed to want most was not specific advice from the older figure; Bob didn't know enough about personal computers to give it, anyway. Jobs wanted a vision of how to live, how to succeed in the Valley and be beloved, not hated, to be the object of esteem, not ridicule. Jobs: "Bob was the soul of Intel. [I wanted] to smell that wonderful second era of the Valley, the semiconductor companies leading into the computer [age]."[1]

Bob Noyce now had two surrogate sons, Jerry Sanders and Steve Jobs, with whom in many ways he had a better relationship than with his own son, still bitter about his parents' divorce.

CHAPTER 44

The Man of the Hour

In 1986, in the wake of the success of the international trade bill and the retreat of the Japanese semiconductor industry, the American chip companies held a series of meetings under the auspices of the SIA and led by National Semi's Charlie Sporck. There they began to discuss the need for some sort of joint venture to develop cutting-edge design, manufacturing, and quality-control tools to make sure that the US semiconductor industry would never be caught flat-footed again as it had with the HP presentation in 1979. By May 1987, they had proposed, and the SIA had approved, a business plan for the creation of Sematech (semiconductor manufacturing technology), an industrywide consortium hoping to restore America's now-lost lead in semiconductor manufacturing tools, equipment, and technology.

Despite the fact that there was almost no precedent for a government-backed industry-run research consortium, Congress—now fired up from winning the Japanese trade war and anxious to maintain its contacts with an industry as wealthy as semiconductors—seemed eager to allocate $500 million ($100 million per year for five years, matched by the members of the SIA) to get things under way.

Still, there was some backlash to the idea. The National Science Foundation thought the chip industry had "wildly overstated" its problems, but still agreed to the consortium. The White House was equally skeptical, but its Science Council voted for it by a one-vote margin. In Congress, some representatives who had been on the side opposite to the Valley's traditional antigovernment conservatism grumbled at the sudden willingness of the same industry figures to run to Washington for handouts. "Most of them are right-wing Republicans . . . basically against any government intrusion. But when they get into trouble because they've been damn fools, then they come to us for bailout."[1] The *Washington Post* took particular interest in Noyce's constant trips to DC to pitch legislators, noting that this legendary proponent of entrepreneurship and individualism was "in Washington pleading for a handout."

These comments certainly stung—and Bob Noyce knew that he was being hypocritical in his current actions versus his past positions on the federal government. On the other hand, he was also a staunch patriot; he believed that the Japanese had not played fair and that he had to do whatever it took to keep the US semiconductor industry alive in the face of these illegal onslaughts. Moreover, Noyce was hardly doctrinaire. After all, he had started Fairchild largely with defense and aerospace contracts, and he had no illusions about the importance of the federal government to the creation of Silicon Valley. And if Intel, following HP's lead, had pulled away from government contracts in recent years to focus on the commercial and consumer markets, that was a business decision, not a philosophical one.

That said, it had to sting when members of the semiconductor industry itself began to come out publicly against Sematech. Leading this countercharge was T. J. Rodgers, the pugnacious and voluble founder of Cypress Semiconductor, headquartered just a few blocks away from Intel. TJ, whose celebrity (and notoriety) as the most outspoken figure in Silicon Valley began at this moment, represented the new generation of small chip companies, many of them operating without their own factories, specializing in custom or niche chips. TJ held press conferences and made numerous television appearances, glibly calling Sematech "an exclusive country club" for big chip companies who would never share their government-funded findings with little chip companies.[2]

TJ had two simple arguments. First, when Silicon Valley and especially the chip industry had done so well without government interference, why should it now open the doors and invite the feds to come in and interfere with that success? Second, if the federal government was so willing to put up several hundred million dollars to help semiconductor manufacturing in the United States, why not cut that money up into $10 million tranches and use it under the guidance of the SIA as venture-capital investments in start-ups targeted at the manufacturing weaknesses? Then the market would discipline those companies and select the winners—as it always had.

Twenty-five years later, as Valley residents watch Air Force One regularly landing in the Bay Area and have to make way for presidential motorcades going to the latest fund-raising dinner with Valley executives (Regis McKenna held the first of these dinners, at his Sunnyvale home, for President Clinton), it's obvious that TJ's first warning was more than accurate.

As for Rodgers's second argument, it was particularly devastating because he sounded more like Bob Noyce than Bob Noyce did. But Noyce, whatever his private thoughts, had chosen to represent not his own views, but those of the industry. In the end, Sematech was funded, and construction began in Austin, Texas. At a gain to his public popularity but at a cost to his personal credibility, Noyce had again won.

CHAPTER 45

Juggernaut

In 1988, Intel came roaring out of the gate. Revenues for 1987 had totaled $1.9 billion, a 51 percent jump from the year before. Much of that growth had come in the second half, signaling that the next boom was on.

Intel was ready for it. The 80286 chips were flooding out of the company's fabs throughout the world, feeding a seemingly insatiable demand for personal computers that was being fed by an explosion of new software applications, from word processing to spreadsheets and games, but most notably Microsoft Windows 2.0, the upgrade to its revolutionary Windows operating system introduced in 1985. Windows was already the de facto standard for the world's Wintel computers, but the second version drove that point home as huge crowds camped overnight at the world's computer shops for a chance to buy it.

With its microprocessor business roaring along, Intel now turned its attention to its microcontrollers. On a single day in April, promoted by Intel as the Big Bang, Intel rolled out a total of *sixteen* new products and development tools for embedded control applications. With this announcement, Intel was making a secondary point: most companies like to introduce one product at a time in order to wring as much publicity as they can out of each one. Sometimes if it is two lesser products or two important but closely related products that might capitalize on each other's attention, a company will "bundle" them in a combined announcement. But with Big Bang day, Intel was telling the world that it had so many innovations in its microcontroller business that it could just roll them out all at once in one big market-crushing announcement. This was confidence to the point of arrogance. This was Andy Grove's

Intel. And it worked. After the Big Bang, there was no question that Intel now owned the controller universe.

Intel made a second move in 1988 that showed its confidence. Having left the memory business, it now went back into it. Intel, in fact, hadn't really entirely left the memory business before. It had gotten out of its biggest business—DRAMs—but had still retained its smaller EPROM business, mainly because this kind of chip was used in close support to its microprocessors. Now, in a stunning turnabout from the trade war at the beginning of the decade, Intel spotted a brand-new technology being developed by Toshiba—"flash" memory—and not only made it Intel's own, but actually improved on its performance, price, and manufacturability.

The advantage of flash memory is that it combines nonvolatility—that is, it stores data even when the electrical current is turned off—with electrical erasability. Flash's initial appeal was to makers of embedded systems, because the technology made it easy to upgrade existing hardware just by changing or upgrading the onboard software. But over the next dozen years, as the density of flash memory began to race up the curve of Moore's Law, it became increasingly possible to replace magnetic memory—i.e., disk drives—with flash memory arrays that were faster and lighter, used less power, and were more rugged in real-life applications, while taking up the same or less space. This revolution in memory, in which flash slowly replaced the hard disk just as the hard disk had replaced magnetic core, would remain almost invisible to the outside world until 2005, when it burst into the public consciousness with the introduction of the Apple iPod Nano. It was flash memory combined with a new generation of mobile-oriented microprocessors and high-resolution displays that made possible the smartphone boom that followed soon after.

As early as the beginning of the 1990s, Intel could already see this future emerging. Said Dick Pashley, vice president of Intel's Semiconductor Products Group, "It looks as if solid-state technology will eventually render conventional disk drives obsolete in mobile systems. . . . Disk drive replacement was a radical idea five years ago, and people were naturally skeptical. A very dedicated and immensely creative team clung to the vision, and today people are believers."

Like DRAMs before it, flash memory proved to be Intel's hidden moneymaking engine. While the rest of the world was focused on the company's famous microprocessor business, Intel spent twenty years quietly and consistently upgrading the power of its flash memory chips, finding more and more new customers, and cranking out profits. It wasn't until the early years of the new century, when the company had become complacent about its ownership of this business, that a competitor appeared—South Korean giant Samsung—that stunned Intel out of its placidity. In the intervening years, however, flash was not only a reliable cash cow for Intel, but the stage on which the company made one of its biggest—and least celebrated—contributions to the electronics revolution and the Internet age.

CHAPTER 46

Empyrean

J ust months into its creation, with more than two hundred employees hired, Sematech was already in serious trouble. The problem was finding a chief executive with both the availability and the skills to run such a complex operation. The *New York Times* reported:

> The consortium has been trying to hire an experienced executive from one of its member companies. But the companies are reluctant to part with key personnel and have made it attractive for the executives to stay. This difficulty is viewed as an indication that the 14 member companies are less inclined to share their manpower for the common good than has been publicly proclaimed.
>
> "Sematech will suffer if we don't get someone in soon," said Robert N. Noyce, vice chairman of the Intel Corporation and a member of a three-member committee charged with finding a chief executive.
>
> The inability of the industry to find a leader for its much touted research consortium is frustrating the Department of Defense, which is underwriting half of the cost, and impeding planning for the project, government and industry officials said.[1]

The big companies of the semiconductor industry had supported the idea of an industry consortium, even if it meant getting into bed with the federal government, as long as they got some competitive advantage out of it and as long as they didn't have to sacrifice anything but money. But Washington's expectations were different: this was a grand experiment in a government-business joint venture whose image with voters depended upon success, so the government expected the semiconductor industry to put as much on the line as the government had.

As the pressure mounted, it was obvious to everyone who should run Sematech. Bob Noyce had led the campaign to make the consortium happen, he was the one individual (besides Gordon Moore) who was respected by everyone in Washington, Silicon Valley, Phoenix, and Dallas, and perhaps most important, he apparently wasn't doing anything else.

The pressure put Bob in a difficult position. On the one hand, he had taken heat for contradicting his own long-established position as a free marketer by supporting this industry consortium, and there was a very good argument for him to conclude that he had done more than enough for his industry and just quietly back away from the whole impending debacle. No one would have blamed him if the rest of the semiconductor industry couldn't find a single senior executive to volunteer for the job. He had already done his part, and brilliantly.

On the other hand, his name was now synonymous with Sematech. He had made public statements about the crucial importance of finding the right CEO for the consortium. If Sematech cratered now, Noyce would be closing out his career (he was now sixty) with his biggest failure.

In the end, it appears that all of those things slid into the background, and what remained foremost in Bob Noyce's mind was *duty*. Whatever his own views, whatever the mess he was about to dive into, Bob had made a commitment to the success of Sematech, and if no one else would take the job, then he would do it himself.

"The country had made a commitment to this concept," he told an interviewer at the time. "I felt that if I didn't follow up with a personal involvement when I was needed, it was betraying a trust to the people of America."

Silicon Valley was stunned by the announcement. The *San Jose Mercury-News* made the story its front-page banner. Bob Noyce *was* Silicon Valley. And now he was headed off to Austin to run some thankless, government-overseen project designed to help a bunch of chip companies—most of them not in the Valley—that couldn't cut it against the Japanese. It seemed a waste of time for a great man.

Needless to say, that wasn't how the rest of the country felt. If Bob Noyce had taken the job, then Sematech must be real; it must be valuable. And it would succeed, because Bob Noyce never failed. And indeed, the very news of Noyce's ascension to the CEO's position had an instant effect upon the disheartened troops at Sematech. Wrote the *New York Times*: "Hundreds of engineers who had dismissed the venture as a fringe experiment jumped at the chance to work for him."[2]

What Noyce found when he arrived in Austin was even worse than he had anticipated. "But even with Dr. Noyce as the glue, the consortium's members were hard to hold together. Cultural differences sometimes seemed insurmountable among the 250 people assigned to work at Sematech by their companies (450 others are permanent Sematech employees). I.B.M. employees, the 'White Shirts,' were put off by the more informal Californians from Advanced Micro Devices. A.T. & T. people were conspicuous for their secretiveness and reluctance to speak up. People from Intel were too blunt."[3]

This disharmony was even worse than what Noyce had seen in his final days at Fairchild. Exasperated, he called on an old acquaintance who had helped him with some of his start-ups, Bill Daniels, a meetings specialist from Marin County. In time, the combination of Noyce's presence and Daniels's group exercise—in which he reminded the Sematech employees that they were still in the economic equivalent of war against their Japanese counterparts—began to slowly transform the operation. Employees began to work together and create a distinct Sematech culture. The sense of duty and teamwork was reinforced by a new look for the facility, which was painted red, white, and blue—and which the *Times* compared to the United States Olympic Team training camp.

Noyce had succeeded once again. And indeed, Sematech survives to this day. It has never been as transformative as its early promoters claimed it would be, and yet it was anything but the failure its detractors predicted.

MIT's *Technology Review* magazine assessed Sematech in 2011: "Before Sematech, it used to take 30 percent more research and development dollars to bring about each new generation of chip miniaturization, says G. Dan Hutcheson, CEO of market researcher VLSI Research.

That increase dropped to 12.5 percent shortly after the advent of Semat-ech and has since fallen to the low single digits. Perhaps just as import-ant, Sematech set a goal in the early 1990s of compressing miniatur-ization cycles from three years to two. The industry has done just that since the mid-1990s, speeding innovation throughout the electronics industry and, consequently, the entire economy."[4]

Technology Review concluded that "Sematech has become a model for how industry and government can work together to restore manu-facturing industries—or help jump-start new ones."[5]

By the time that article ran, Sematech's biggest current endeavors had to do with improving the manufacturing processes used in solar technology. And irony of ironies, its membership now included those same Japanese companies it had been founded to defeat.

It was a great (if largely unrecognized) victory for Bob Noyce, but it came at great cost. He often spent the work week in Austin and then flew back to Los Altos to preside over various Valley events and fund-raisers. There was little time now for his usual outlets of skiing and flying his planes, though he did swim almost daily at his Austin home. Unfortu-nately, the stress of the job had brought back his old chain-smoking.

On one of those trips back to the Bay Area, in which he gave a speech on Sematech, he took the time to be interviewed on a locally produced but nationally syndicated television show. Before the cameras rolled, he sat back in his chair and told the host, whom he knew well, that he was always happy to be back in the Valley, even if it was only for a couple of days. He joked about now being a Texas cowboy, saying that he had to "scrape the shit off my boots" before landing back in the Bay Area.[6] He was proud of what he had accomplished with Sematech, he took enormous pride in what Intel had become, and he was obviously beginning to ponder the next chapter in his life. Steve Jobs heard he was in town and, anxious for Bob to meet and approve his fiancée, invited Noyce to his house for dinner. They stayed up talking until three a.m.

Like any group of employees who had ever worked with Bob Noyce, the people at Sematech adored him. And when an article appeared in the *San Jose Mercury-News* during his visit that quoted a semiconduc-tor equipment company executive as saying that Americans needed to "change their idols" and then nominated Noyce "for the pedestal," those employees decided to greet Bob on his return with a party, complete with T-shirts that featured that quote, Bob's photograph, and the line BOB NOYCE, TEEN IDOL. Sematech officially declared that day, June 1, 1990, Bob Noyce Day.

By all accounts, Noyce was surprised and flattered by the event, not least the posing for photographs with pretty female staffers.

That was Friday. On Saturday he held some business meetings at home to catch up on what he had missed. And on Sunday, he took his usual morning swim. Then, feeling tired, he went inside, lay down on a couch . . . and died.

CHAPTER 47

The Swimmer

The news, which reached Silicon Valley later that Sunday morning, hit like a body blow, one from which Silicon Valley has never really recovered. There has never been as devastating a piece of news before or since in the Valley. Even Steve Jobs's death, because it was expected, wasn't as shocking. Today, even young programmers working at Twitter or Facebook, for whom the name Robert Noyce might have no resonance, still feel the effect of his early death.

Silicon Valley had been comfortable with the idea of Bob Noyce as its spokesman. The "Mayor of Silicon Valley" was expected to serve in that unofficial role for another twenty years, until he was the Grand Old Man of Electronics, as Packard had been. But now he was gone, . . . and no one since has taken his place. Of his generation, Bob Noyce had been the only one trusted with the role. The leading figures of the next generation, like Jobs and Ellison, were simply too dysfunctional to speak for anyone but themselves. And by the time the dot-com and Web 2.0 generations came along, the Valley had become so fractured into mutually exclusive industries that they hardly knew one another, much less agreed on a leader and spokesman.

In the last decade of his life, Bob Noyce had brought the Valley together more than it had ever been, and almost from the instant of his untimely death, it again began to come apart. The Fairchildren had been Noyce's children, and without Bob there was no one to serve as the intermediary, the peacemaker—and nowhere was this more obvious than in the lawsuit between Intel and AMD, or more precisely between Andy Grove and Jerry Sanders, which only grew more bitter and unresolvable as the years passed.

The news of Bob Noyce's death struck particularly hard those who knew him—friends, old Fairchilders, competitors, Intel employees, old friends from Grinnell and his childhood, even those in Washington who had known him only briefly but had felt the touch of his personality. It seemed impossible that Bob, still healthy and vital, busy planning his next ski trip, already preparing for the next great venture in his life, could die from a heart attack. Jerry Sanders wept, even tough old Charlie Sporck was in shock and could hardly speak. Steve Jobs looked heartbroken, abandoned, and lost. Even a year later, when Gordon Moore, Bob's partner for more than twenty years and compatriot for more than thirty, tried to speak about his friend, tears would fill his eyes and he would be momentarily too stricken to talk.

On Monday morning, Intel's employees around the world arrived at a different company than the one they had left the Friday before. They clustered in groups to talk. Many cried.

A week later, more than a thousand people attended a memorial service for Noyce in Austin. Hundreds more gathered for a ceremony in Japan.

In San Jose, on an official Bob Noyce Day, the largest theater space in the South Bay, the Civic Auditorium, was filled with two thousand Valleyites for a memorial ceremony presided over by Bob's brother Gaylord. Afterward, the tearful crowd made its way outside to the steps and walkway. As they watched, hundreds of red and white balloons were released into the air. Then with an enormous roar, just one hundred feet above their heads, Bob's new Cessna Citation jet, which he had never had the chance to pilot, flew past in his honor. For many in the crowd, this moment was the most memorable and uplifting of all.

President George H. W. Bush called Ann Bowers to offer his condolences, as if Bob was a head of state. Memorials were placed in the *Congressional Record* by more than a score of senators and representatives. Defense secretary and future vice president Dick Cheney called Noyce "a national treasure." Newspapers around the world prepared extended obituaries for many readers who had first heard the man's name just a few years before. In fact, one wire service carried the wrong photo to accompany its article. And that evening, to the surprise of millions of viewers who had never heard of Noyce, Peter Jennings, anchor of the

ABC evening news and a closet technology fan, devoted a short segment to Bob Noyce. Earlier that year he had named Noyce his Person of the Week and had celebrated his life. Now he announced his death.

The *San Jose Mercury-News* created a four-page special supplement about Bob Noyce, printing it in time to hand it out at the funeral. It was filled with encomiums from people high and low—from his old professor Grant Gale and Gordon Moore to a bank teller who had handled the check when he was buying one of his planes.

Apple Computer's official comment, no doubt influenced by Steve Jobs, may have been the most powerful of all. "He was one of the giants in this valley who provided the model and inspiration for everything we wanted to become. He was the ultimate inventor. The ultimate rebel. The ultimate entrepreneur."

Upside magazine, the Valley's most popular (and cheekiest) business magazine, set aside its usual style:

> He filled any room he entered with a bewildering impression of power and command. He appeared neither humble—he well knew his place in history—nor arrogant. He was just his own man; slightly reserved, ultimately unknowable to everyone but those very close to him. The only real clue was when he was angry or intellectually engaged. This his eyes would seem to go black, burning a hole through your forehead to pin you to the chair behind.

> But perhaps the best way to describe Bob Noyce to those who will never meet him is that before and after everything else, he was a swimmer. And that fact, more than any of his famous achievements, may be the most telling characteristic of the man. In swimming there is the seamlessness of action, the balanced use of every part of oneself, the pursuit of perfect motion, the essential solitude.

> Dr. Noyce dived competitively at Grinnell, scuba dived throughout his life, and turned the backyard of his Los Altos home into a swimmer's paradise. Even his voracious winter pursuit of skiing throughout the world seems consonant with his love of water and fluidity of motion.

> Bob Noyce spent most of the few spare moments of his life swimming, and with a kind of noble symmetry, his fatal heart

attack came after a morning swim. One imagines him on those last laps of the pool, even after a lifetime patiently working to refine his stroke, to perfect his movement through the water until he left just the slightest of wakes. And, one imagines on that fateful morning, as always, Bob Noyce made it look effortless.[1]

PART V The Price of Success (2000–2014)

CHAPTER 48

Family Matters

A nd then there were two. The trinity that had founded and led Intel to the top of the semiconductor industry was now down to a duo. And with Gordon Moore now removed to the chairmanship, Intel increasingly became a solo performance by Andy Grove.

And of course, he wouldn't have had it any other way.

The nineties would prove to be the strangest, unlikeliest, and most exhilarating decade in the history of the electronics industry. Many had dreamed of escaping the torture of the industry's four-year business cycle, a chance to just keep moving forward without the pain of cutbacks and layoffs. Their prayers were answered. But the nineties also proved an enduring lesson in what happens when a tech-industry bubble just keeps inflating, giving its participants no time to catch their breath, regroup, or pause to rebuild. The greatest boom of them all was just around the corner, and as Intel mourned the death of Bob Noyce and the end of an era, it had no idea that what was coming, just months away, would make everything that came before seem like a mere prologue.

Certainly, like any company in the semiconductor industry, Intel was ready. In fact, it was almost too ready. In April 1999, it had introduced its next generation x86 microprocessor, the $950 32-bit 80486 (it was known publicly as the i486 or Intel 486 because of a recent court ruling that numbers could not be trademarked). It was a magnificent creation, the first x86 chip with more than one million transistors, an integrated floating-point (arithmetic) function, 50 megahertz clock speed, and the ability to execute more than forty million instructions per second. All of its advantages combined to give this new processor twice the performance of the 80386.

The 486, introduced at the spring Comdex computer show in Las Vegas in April 1989, stunned the crowd—and not just in a good way. Gordon Moore had long warned about the danger of moving faster than his law. The 486 hadn't quite done that; actually it tracked the curve of the law pretty accurately. Unfortunately, it also outpaced the market's *adoption* of new microprocessor technology. Even consumers, now that the general media were covering events like Comdex, got an early glimpse of the 486 . . . and wondered if they really needed a personal computer with that kind of power. Why not just save their money and stick with their old, perfectly adequate 386 machine?

Intel might have outrun its own abilities. The first production runs of most semiconductor chips were always a little buggy—not surprising, given their sheer complexity. But early reports suggested that the 80486 was suffering an unusually high number of bugs and software incompatibilities. That led a number of potential buyers to hold off into the new year. When the bugginess of the early production runs turned out to be true, Intel agreed to replace them. The market was understanding . . . *this time.*

This was rather disturbing to Intel. Here it had produced a supreme example of semiconductor technology, a more-than-worthy successor in the most important microprocessor family in the world, but computer manufacturers and consumers were questioning whether they actually needed or wanted it. There had been hints of such resistance when the 386 had replaced the 286, but now was it showing up everywhere, in every industry and every market.

Still, the momentum of the 386 and the PC boom was a huge wave carrying Intel forward. In October 1989, the company had its first billion-dollar quarter. And for 1990, it revenues crested just short of $4 billion.

The company was also restructuring itself. Flash memory, which would be a $500 million business for Intel by the midnineties, had proven itself so valuable that in April 1991, Intel announced that it was leaving the EPROM business. It was now a completely different business from what it had been when it had entered the 1970s. In Palo Alto, the state of California designated the spot where Bob Noyce invented the integrated circuit as the thousandth state historic landmark.

CHAPTER 49

Inside Intel Inside

<hr />

Intel had long had the reputation for being the most innovative semi-conductor company in the electronics industry. The reality was not quite that sterling—as already noted, on a number of occasions, Intel's best new designs were surpassed by superior offerings from Motorola, Zilog, and others. Nevertheless, the company's consistent innovation over the course of decades had no peer in the industry.

What Intel almost never got credit for, at which it was the best of its era and second only to Apple in the history of the electronics industry, was being an innovative *marketer*. And in the 1990s, Intel came close to being the best corporate marketer in any industry in the world.

This didn't happen overnight. Intel, after all, was a technology company, a component maker. That its products competed on their performance specifications and were sold almost exclusively to other electronics companies—or at the electronics departments of other manufacturing companies such as car makers—almost guaranteed that Intel would see little value to marketing beyond an up-to-date product catalog and price list and occasionally the usual trade press advertisement of the era—a blown-up photo of the chip against some factory floor, placed in a military photograph, or most flamboyant of all, held by a pretty girl next to her face. With few exceptions, semiconductor advertising hadn't changed much from the midsixties.

By the early 1980s, other, consumer-facing electronics companies—notably Apple, with its Macintosh 1984 commercial (perhaps the most famous single commercial in history), and IBM, with its Little Tramp PC ads—were experimenting with mainstream consumer advertising. Intel, influenced by Regis McKenna (who had been deeply involved in

Apple's advertising campaigns), tried its own series of innovative ads featuring monochromatic drawings of elegant men and women that looked like high fashion ads. They were so mismatched to the world of silicon wafers and binary code that Intel's audience didn't quite know what to make of them. One can easily imagine computer nerds and electronic engineers with pocket protectors staring at one of these ads, which looked like something from *Vogue*, in *IEEE Spectrum*—and scratching their heads in confusion.

Other than causing audiences to wonder if Intel had gone mad—and the curious kind of attention that followed—this ad campaign was largely a failure. Nevertheless, the "Vogue" campaign did shatter the prevailing marketing paradigm in the semiconductor world and served notice that Intel was willing to take great risks to get its message out to the world. And yet in 1989, when once again Intel embarked on a radically new advertising campaign, the whole sleepy world of tech marketing sat up in disbelief.

It was called the Red X campaign, and it was the product of growing frustration at Intel. As just noted, the introduction of the 486 had produced a considerable amount of backlash by potential customers who weren't sure they really needed the upgrade. But that backlash had already raised its head with the generation before and the introduction of the 386. Even then, some consumers and commercial customers, who were using their PCs for not much more than word processing and a little spreadsheeting, questioned the need for such a powerful CPU.

Intel responded with Red X. Its message was simple but unforgettable. In a full-page magazine ad, it consisted simply of the number 286 in block type with a big, spray-painted red X drawn over it. Only the Intel logo in the bottom right of the page gave any clue that this was actually an Intel ad, not an attack by one of its competitors.[1]

The campaign was led by Dennis Carter, vice president of Intel's Corporate Marketing Group, who had set out to break what he saw as a perceptual logjam in the marketplace. "The Intel 386 chip was a successful product, but it was mired in the high end of the market. The market was stagnant, and people perceived that the Intel 286 CPU was all that they would ever need. The Windows [3.0] operating system was coming, giving people a compelling reason to move to 32-bit processors, but that message wasn't coming across. We wanted a dramatic way to convey that the Intel 386 SX CPU was an affordable way to enter the 32-bit world."[2]

But however desperate Intel was to get its message out, the idea of undermining an existing and very profitable product line in order to promote its replacement was the biggest taboo in high-tech marketing. Just a few years before, Adam Osborne had been introducing one of his revolutionary new portable computers to the media at a press conference, and when he didn't get a strong reception, he proceeded to start talking about the new product's future replacement, still on the drawing boards. In the process, he killed not only the current product, but also his company.

Now Carter seemed to be doing exactly the same thing—and not with a niche product, but the most popular microprocessor to date and the source of at least half of Intel's nearly $3 billion in revenues. Running fashion ads for microprocessors had been crazy enough, but sabotaging your biggest product line in order to bolster a new and struggling one was more like corporate suicide.

None of this was lost on either Carter or Grove. Carter: "We were concerned that we could actually damage ourselves if people failed to understand the campaign. The red X slashing through the 286 was calculated to grab attention, to cut through the clutter of other advertising. But it could have killed the 286, without moving people to the Intel 386 CPU. Microcomputer Components Group president Dave House called it the 'Eating Our Own Baby' campaign."[3]

What's more, both the nature of the ad and its chosen placements, not just in the electronics trade press but the mainstream media, suggested that Intel wasn't trying to reach just its own customers, but its customers' customers, the consumers themselves. That had never been attempted before by an electronics component maker, and the response by consumers, as well as the manufacturers being circumvented, was unclear.

Carter: "We were speaking directly to PC consumers for the first time, rather than marketing only to OEMs. We weren't sure how our OEM customers would react."[4]

Intel was taking a huge chance, but that risk-taking went only so far. So before it rolled out the Red X campaign nationally, Intel first gave it a test run in Denver, Colorado (which happened to be near three big HP divisions filled with likely PC users). The results were so positive that Grove green-lighted the full campaign. As he did, he sent a memo to Gordon, saying, "The ad is imaginative—strikes me even as brilliant—but bold and aggressive. It has been blessed by all the top guns in marketing, including Gelbach, who was consulted as a disinterested bystander.

"I predict you'll hate it."[5]

Whether he did or not, Moore signed on. The Red X campaign proved a huge success—and the 386 was on its way. Before long, consumers were as likely to call their personal computers 386 machines as they were to use the brand names.

Intel, the rare big company that could actually learn from experience, took away from the Red X campaign two important lessons. First, the existing paradigm of information-heavy, drearily empirical semiconductor industry advertising was no longer valid. It was possible to craft a successful advertising campaign that was as sophisticated and clever as any offered by even the luxury consumer-products industry. Second, semiconductor suppliers no longer had to depend upon their commercial and industrial customers to do the marketing to end users. Rather, it was possible to circumvent those intermediaries and tell the company's story directly to consumers, to cultivate brand loyalty within them, and ultimately to teach them to look for Intel components in any electronic device they bought.

Now, in 1991, Intel was preparing to ship the replacement of the 386, and like that chip before it, the 486 was suffering pushback from those who didn't see the necessity of upgrading to it. Worse, at the end of 1990, AMD had finally managed to build its own 386 clone, and in the words of the *Microprocessor Report*, "The most valuable monopoly in the history of the semiconductor business is about to end." Within two years, AMD, with its cheaper and more powerful alternative, had captured more than 50 percent of the 386 market. Intel had to make the 486 work, or the consequences would be dire.

Carter hadn't forgotten the lesson of the Red X campaign, and now he set about crafting an even cleverer and more sophisticated version, one that might even make the Intel microprocessor CPU inside a computer more important than anything on the outside, including the computer manufacturer's logo. He searched for an ad agency that could understand what he wanted—and found it in the unlikeliest of places, Salt Lake City and the firm of Dahlin Smith White Advertising. Carter told the agency, "This is what we want to be. We want to make the processor more prominent in the computer. It's really important. It's invisible. People don't know about it. They don't know us. How do we do this?"[6]

Thus was born Intel Inside, one of the most successful branding campaigns in business history. When it began, Intel was still largely known only within the electronics industry or perhaps as the company once run by Bob Noyce. By the turn of the century nine years later, thanks to Intel Inside, surveys found that Intel was the second-best-known industrial brand (after Coca-Cola) in the world. That isn't just a success; that is a historic achievement, one not likely to be repeated for generations.

Though the common myth is that the phrase came from Intel Japan and its slogan: "Intel in It," in fact it arose out of an editing process between Carter and agency partner John White that began with "Intel: The Computer Inside." The now legendary logo, with its swoosh circle, lowercase lettering, and most of all, a break from Intel's original "drop e" design, was the creation of agency art director Steve Grigg under the direction of Andy Grove. It was specifically designed to look consumer friendly and casual and *not* corporate and formal.

Carter once again took the campaign to top management. "Everyone looks at it in disbelief. 'This is crazy. What are you doing? This is nuts.'" Some senior managers, despite the success of Red X, still wondered, 'Why in the world would our OEMs want to participate in such a program?'"

But once again, Andy Grove stepped up and announced that he thought the idea was "brilliant and that Carter should make it happen." Intel Inside was now Intel's marketing campaign for the rest of the decade, during which time it would spend $500 *million* on the initiative. This was a Procter and Gamble– or General Motors–type budget number, for which there was absolutely no precedent in the history of the electronics industry. It dwarfed even IBM's Little Tramp campaign.

In making the decision to back Intel Inside, though at the time he would have vociferously denied it, Andy Grove had behaved just like Bob Noyce. On a hunch, he had overruled his own lieutenants and made an enormous bet on a high-risk venture that might well have ended in serious damage to the company's finances and left the company open to both public ridicule and a customer mutiny. He had made this decision on only a limited amount of data and precedent, and he could have just as easily settled for a strong traditional campaign at a fraction of the budget.

But like Noyce with the microprocessor, Andy wasn't content just to gain a few more points in market share. He wanted to sweep the table, to gain a historic victory, and to create such a discontinuity in the industry that the competition wouldn't just be unable to come back; it wouldn't even know *how* to do so. That is what Noyce, and now Grove, had done. When the competition walked into the local electronics store and saw those rows of personal computers, all emblazoned with the Intel Inside sticker, and turned on prime-time network television and saw endless commercials featuring that logo, they knew the game was up. They might peel off a few points of market share here or there or enjoy a brief moment of technology leadership with a new chip design, but all in all, it was over. Now and for the foreseeable future, Intel had won the microprocessor wars.

Inside Intel, Intel Inside seemed so easy. All those years of struggling, sacrificing, and scrambling to stay at the head of the microprocessor pack, . . . and in the end, all it took was two words, a funky sticker, and a ton of money to put Intel over the top. Sure, none of this current success would have been possible without almost a quarter century of hard technological innovation and production, but what an odd turn of events. The company named after intelligence, founded by legendary scientists and famed for its hard-nosed engineering, was now best-known for staging the equivalent of Microprocessor! The Musical. And so while it was great for Intel employees that their kids' schoolmates finally understood what Billy's mom or dad did for a living, it was also a bit mortifying that this understanding revolved around prancing cleanroom technicians.

Still, Intel Inside worked brilliantly. And, compared with the cost of creating a whole new microprocessor generation, it had been a steal. By 1993, two years into the program, *Financial World* magazine would name Intel as the third-most-valuable brand in the world.

It would be another year after that before Intel would discover the *real* cost of the Intel Inside campaign.

CHAPTER 50

Wired World

---

ntel Inside may have been the most successful technology campaign of all time, but its impact was tiny compared to the next great tech wave that was already gathering throughout the electronics world: the Internet.

The Internet had been around for many years, of course. RAND scientist Paul Baran had made the technology possible in the early 1960s with his invention of *packet switching* as a way to break up and transmit messages over transmission lines. Along with the microprocessor, the Internet had been "invented" in tech's annus mirabilis of 1969—and the first message via this new technology was sent at about the same moment as Federico Faggin went into the lab to design the Intel 4004. But in those early years, it had been a Department of Defense project, led by the DoD's research wing, ARPA, the Advanced Research Projects Agency, and designed to enable military labs and universities with military/aerospace contracts to communicate easily and transfer large quantities of often classified data over secure high-speed Internet networks. As might be expected from a government operation, this early ARPANET was pretty rough—designed for efficiency, not comfort.

Over the subsequent decade, ARPANET, now called DARPANET (to reflect a Defense added to the beginning of the agency's name), continued to expand and become increasingly sophisticated. A critical breakthrough came in 1982, at about the same time as the IBM PC and the personal-computer boom, with the creation of the Internet protocol suite (TCP/IP), which standardized operations on ARPANET and opened it to interconnection with other networks around the world. With this, the modern Internet began, and by the end of the 1980s, with the help of the National Science Foundation, hundreds and then thousands of private Internet sites began to pop up. These sites were mostly chat rooms or bulletin boards, and their membership was limited by the fact that it was still hard to get on the Net and, once there, very difficult to navigate.

Those obstacles disappeared in 1989, when Tim Berners-Lee at the CERN Research Laboratory in Switzerland set out to find a way to allow computers throughout the world to become part of the Internet using common protocols. The result was the World Wide Web, and now millions of people began to devote their home and office computers to reaching out onto this new Web to find information and news that had never before been within their reach.

But the early Web was as much a frustration as a delight. In practice it was only half of a solution. Though it made access to the Internet much easier, once there, the user was in a great, busy city without addresses or street names. The only direction was often a long URL written on a scrap of paper, to be retyped every time. Thus in the early days of the World Wide Web, the Internet may have seemed infinite (though a tiny fraction of what it is today), but the experience of it was infuriatingly finite.

That changed in 1993, almost at the same moment that Intel was preparing to introduce its next-generation microprocessor to replace the 486. In 1993, a team at the University of Illinois at Urbana-Champaign, led by a brilliant graduate student named Marc Andreessen, developed a new and intuitive graphical interface for the Web—a browser—called Mosaic. Within a year, Mosaic had become Navigator, the product of the hottest new company in tech, Netscape, cofounded by Andreessen and located in Mountain View just around the corner from the old Fairchild plant.

It was Netscape Navigator that truly opened the Web to the average person. Now millions of sites could be found on the Web as easily as one might find a name and phone number in a phone book. Four years later, Google Inc. was founded; it became the dominant tool for searching the Web, which now could be "surfed" not only with a purpose, but merely following the whims of the imagination.

The most obvious impact on Intel was that the Internet finally convinced many to purchase personal computers and existing PC owners to upgrade. Less obvious was that the Internet, especially with the rise of applications such as eBay and Amazon later in the decade, now became appealing to the developing world. Early adopters there raced to the newly opened Internet shops and cafés to play in the global market, and hundreds of millions more people waited anxiously for Moore's Law to make computers cheap enough for them to afford.

Hidden behind all of this expansion was a new phenomenon that had even greater implications for the semiconductor business, and especially for Intel as the microprocessor standard setter. Until now, consumers had to be convinced, often without any evidence of real benefit, to upgrade to the next generation of processors. That's what Red X had been all about, and to some degree Intel Inside.

The advent of the Internet changed all that. The growing number of retail sites on the Web had a powerful motivation to make their sites ever more appealing in order to sell their wares, especially because customers couldn't actually touch and play with the products as they could at a bricks-and-mortar retailer. Thus in short order "e-commerce" sites that had been largely print based began to add images, then audio, then video—all to enhance the customer's experience. And customers responded by upgrading their computers in order to successfully obtain those enhanced experiences. And so, as the nineties went on, an astounding shift occurred. Now those same users who had resisted generational microprocessor upgrades suddenly began demanding that those upgrades be even more sweeping and arrive even faster. It was a metamorphosis of demand that could not have been better suited for Intel Corporation.

What happened next no one predicted. For the rest of the decade, the four-year business cycle that had defined the semiconductor (and most of the rest of the electronics) industry for a quarter century was effectively suspended. It wasn't that the cycle wasn't there. In fact, if you looked closely, it still had a downward effect upon the chip business in 1992 and 1996, as predicted. It was just that the historic growth of the Internet overwhelmed those up and down cycles, filling in the shallows and making the decade appear as one long upward curve.

In 1991, Intel had revenues of $4.8 billion. In 1992, when demand should have been slowing, those revenues climbed to $5.6 billion. That year, in which Intel became the world's largest semiconductor company (according to industry researcher Dataquest Inc.), the company had so much cash on hand that it broke with industry and Silicon Valley precedent and announced its first-ever dividend on common stock—10 cents—followed soon after by a stock split. At this point, Intel employees who had purchased stock at the founding of the company would have seen each of those shares, thanks to stock splits, turn into thirty-eight shares; each dollar invested, into more than $60.

That was just the start. Thanks to the Internet boom, beginning in 1992, the trajectory of Intel's growth, already impressive by the standard of any other industry, suddenly turned upward at a pace that seemed almost impossible for a company of its size. Without ever breaking its pace, the company grew from nearly $6 billion over the next four years to the nosebleed height of $16 billion, and by 2000 to $33.7 billion. This was one of the fastest run-ups by any big company ever recorded.

If, as noted, managing growth is one of the most underrated skills of CEOs, the ability to manage explosive growth in a billion-dollar company is one of the rarest. Each year, the company can jump in size and add more employees than would constitute an entire company at the bottom end of the Fortune 1000, and then do it again the next year, and the year after that. Single decisions can make or lose more money that the GDP of midsize countries and can affect the lives of thousands of employees and their families—and millions of consumers.

At the top of such a company, the CEO finds himself or herself dealing with a whole new set of challenges almost unknown when the company was younger and smaller. For example, instead of a handful of investors, the company may now have millions of shareholders. And thus, a single comment can not only have an impact on the value of the company's stock, but also affect the entire stock market, even the health of nations. The company that used to work in comparative isolation now finds itself the "barometer" of the nation's economic health. And that inevitably attracts the attention of national governments anxious to capture some of that success, rein in some of that power, and tap into some of those profits.

Most important and perhaps most insidious, this kind of success changes expectations of the company by employees, competitors, customers, and the government. These expectations can shift so quickly that the company can be caught flat-footed and still living with a past image that no longer obtains.

This shift in expectations toward Intel occurred in the first half of the 1990s. The company, including Andy Grove, still saw itself as the plucky little company that had bravely fought its way through a decade of mortal challenges (and had even seen one of its founders fall along the way) but had, through discipline, great engineering, incredible sacrifice, and a willingness to defend Moore's Law to the death, triumphed over its competitors. Now, as it stood alone on the battlefield, it was only right that it was rewarded for its victory. Especially so, since everyone in tech knew that victories were brief. Soon enough, new challengers would emerge, and there was no guarantee that next time the victory banner wouldn't be held by someone else. And whoever eventually supplanted Intel on top would never be as benevolent: Intel treated its customers like family, and rewarded them for their loyalty.

But Intel's customers, commercial and consumer, saw something very different. Both suppliers and business customers saw a juggernaut, a multibillion-dollar corporate giant growing at an astonishing rate and owning a near-monopoly on the core technology of the modern economy. With so much money and talent, Intel at any moment might decide to move up or down the supply and distribution chain—say, building its own fab equipment or its own personal computers—and crush them. Intel was as nakedly predatory as the other big winner of the decade, Microsoft, which was about to crush Netscape by bundling its own browser, Internet Explorer, into Windows. Intel knew how to play corporate hardball with the best of them. Intel liked to point to its new chip delivery list as an example of rewarding its most loyal, rather than just its biggest, customers, but in fact it was a cudgel. Without real competition, Intel could keep its chip prices high. And when supply failed to keep up with demand—as it had eventually with each generation of x86 chips—and a customer was reduced to buying clone chips just to stay in business, if Intel found out, that company would be plunged into the outer darkness, dropped several hundred places on the delivery list. Forever. That's why they secretly prayed but never verbalized the hope that AMD would win its lawsuits and the x86 family would continue to have a second source.

Meanwhile, for consumers, Intel was the happy electronics company with the dancing employees in bunny suits, whose friendly sticker on the front of their computer guaranteed that they could completely trust the performance of all of that complicated stuff inside and freed them to explore the mind-boggling new world of the Internet.

There is always a perceptual gap between a company and its strategic partners and customers. But in Intel's case, by the mid-1990s, this gap had grown to a chasm, to the point where it was impossible to reconcile the views of those three parties. Something had to break.

But in the meantime, Intel prepared its next-generation microprocessor, the one designed for the Internet age, designed to cement ownership of the industry for years to come. It would, in fact, be the name of a chip family that would still represent Intel twenty years later.

CHAPTER 51

Pentium

I n keeping with precedent, the new x86 processor should have been called the 80586 or i586. But the earlier court judgment that denied Intel the ability to copyright a number (numbers being deemed too indistinct), combined with Intel's desire to keep AMD and other potential cloners from anticipating the next Intel chip and using the number on their own version (such as had been the case with the AMD Am386), compelled the company to look for a new name.

The marketing firm Lexicon Branding came up with the solution, combining the Greek *pente*, five, with the chemical element suffix *-ium*, to suggest that this new chip would be a fundamental element of the computing world—Pentium.

Design work on the Pentium had begun in 1989 by the same design team that had created the 386 and 486. Ultimately, this team would include two hundred Intel employees, led by John Crawford (who had led the design of the 386) and Donald Alpert. General manager of the overall project was Vinod K. Dham, who would later say, "Speed was God for us when we designed Pentium."[1]

The Pentium was originally planned to be demonstrated at the June 1992 PC Expo trade show and formally introduced four months later, but it ran into some design snags, and was bumped to spring 1993.

By then, a number of new microprocessors—and even new companies—had jumped up to challenge Intel's hegemony. AMD, despite the lawsuit, went ahead and built its 80486 clone, the Am486. It was the one company that challenged Intel head-on.

Others took a different path, mostly pursuing a new vision of the microcode on microprocessors called RISC, for reduced instruction set computing (as opposed to what was characterized as Intel's complex instruction set computing, or CISC). The idea was that simpler instructions could be executed more quickly, thus improving processing speeds. Among the companies pursuing RISC processors were the high-end workstation makers Sun Microsystems with its SPARC chip and Digital Equipment Corp. with its Alpha VXP chip. A new chip company, MIPS, backed by money from a number of US and Japanese computer companies, also set out to deliver a high-performance RISC processor.

Motorola, of course, had its own competing processor, the 68040. As always, it was an impressive device, though its use was largely restricted to Apple Macintosh computers (that company still losing market share), later versions of Steve Jobs's failing NeXT computer, and the HP 9000 minicomputer.

But the most interesting competitor of all was a new consortium called AIM, formed in October 1991. The initials stood for Apple, IBM, and Motorola. That formidable trio decided, above and beyond their own companies' initiatives, to team up and try to break Intel's stranglehold on the personal computer CPU industry. AIM produced several doomed joint ventures, but its biggest achievement was the creation of the PowerPC microprocessor, introduced in 1992. It, too, had a RISC architecture and was a 32/64 bit processor, and it would prove to be as fast as contemporary Intel x86 designs. It also featured IBM's robust POWER instruction set architecture, which also positioned the chip for future generational jumps to 64-bit and 128-bit versions. Apple quickly adopted the PowerPC as its CPU, as did a number of video game console makers.

This was formidable competition. But the increasingly asynchronous nature of the two great microprocessor platforms—that is, Intel and everyone else—meant that most of these devices were introduced in 1991, giving Intel a two-year window to improve upon those designs, during which many customers were willing to wait—or scared not to.

It worked. When it was finally announced to the world, the Pentium was just about everything Intel and its customers dreamed it would be: a 32/64-bit microprocessor containing 3.1 million transistors capable of operating at speeds up to 300 million instructions per second (MIPS) at 100 MHz. It also had a floating-point unit that was up to fifteen times faster than the 486's, as well as extensive self-testing operations. But two features in particular set the Pentium apart from its predecessors and made it at least the equal of its new competitors.

First was a much wider data bus of 64 bits, which like a larger automobile carburetor, allowed data to be entered and retrieved from the processor at a much faster rate. Second and most important of all, the Pentium, following the lead of several of its competitors, adopted superscalar architecture. This meant that the chip had two data paths—"pipelines"—that enabled the chip to perform two operations at once. This was a major selling point of RISC, because it was believed that CISC processors could not run a dual pipeline. The Intel Pentium proved them wrong. Altogether, the Pentium was not only better than its new competitors, but *five* times as powerful as Intel's own 80486—for about the same price ($1,000).

In introducing the Pentium at this time and with this level of performance, Intel was fulfilling what risk analyst Robert Charette was formulating at the same time as the "spiral model." Here's the situation as described by a journalist at the time:

> This model resembles a spring, in which each turn corresponds to one generational advance in technology. The industry leader is at the very top of the spiral, setting the rate of spin; the other companies in the industry are arrayed at various points below. So, if a company does nothing, it stays in the same place, but loses its relative position to the other firms spiraling upwards. If it can move faster than the spiral spins, it will gain position against the leader—even overtake it and become the new leader, setting the new rate of spin. But that . . . is rare . . . as in Alice's experience with the Red Queen, you have to run as fast as you can just to stay where you are.[2]

With Pentium, Intel retained its position at the top of the spiral. And because it had set the rate of the spin at the pace of Moore's Law, its competitors had to maintain that breakneck pace *always* or risk falling back. And because (as we have seen) it was almost impossible to progress chip technology faster than Moore's Law, it was almost impossible for competitors to gain any ground on Intel *as long as it hewed to the law.* That's why Intel's new competitors tried the only option left open to them: change the paradigm and create a new spiral. Unfortunately, they hadn't changed it enough and—this time at least—Intel matched their innovations.

In terms of its adoption rate and magnitude, as well as its cultural impact, the Intel Pentium was probably the most successful microprocessor of all time. When *Fortune* magazine dedicated its cover to "The New Computer Revolution," those words were superimposed over a close-up of a Pentium chip. The Internet boom obviously helped, but the Pentium also fueled the Internet's expansion. It was Pentium-based personal computers that drove the arrival of e-commerce and the dot-com boom that followed. Millions of Pentiums inside servers and routers around the world had enabled Internet usage to grow during those years at a pace even greater than Moore's Law.

But other factors were at work as well. For example, among Intel's competitors, none had a new generation of chips to compete with the Pentium that was even close to ready. Even AMD, which tried to track Intel as closely as possible, had introduced its 486 clone just a couple years before and would need at least two more years to create a Pentium copy in quantity. By the same token, Intel Inside was also just two years old, and its real impact on consumer purchasing patterns was felt only after the Pentium introduction. And finally, the Internet was already beginning to morph from a print medium to an image medium and needed the Pentium's processing power to get there, and that need sold even more of the chips.

Whatever the resentments by partners and competitors and despite confused expectations by consumers, Intel during this period ran like a vast, well-oiled, supercharged machine. This was Andy, of course, and Craig Barrett (who was rewarded for his efforts by being named executive vice president and COO—and true heir apparent). But it was also company management, now numbering thousands of executives and managers in more than a dozen layers of leadership, and all 24,000 Intel employees.

During this period, Intel was opening, upgrading, or closing fabs all over the world, moving workers back and forth, all the while not allowing the company to lose a step, lest it fall behind those hungry competitors. This was possible only in a company that operated with a common purpose and tremendous mutual trust among its employees in one another's dedication and ability. Intel's culture and work environment was still too hard-nosed and spartan—and there was still way too much yelling—to ever put it in the ranks of the most progressive tech employers. On the other hand, there was probably no population of employees anywhere in any company in the world that could be called upon to produce such superhuman results with so much at stake—and do so consistently every few years for decade after decade. Intel's strategy was Moore's Law incarnate, but its secret weapon was its people.

With the Pentium successfully introduced and in full production, Intel essentially had a license to print money—mountains of money—and start preparing for microprocessor generation number six. Meanwhile, the company was hardly negligent about its other, smaller businesses. Grove had, after all, spent his early years running everything at Intel that *wasn't* microprocessors—and he wasn't going to let those products be second-rate now. More than that, he was acutely aware of the dangers of being a one-product company, even if that product was the most important of the age.

Thus in 1991, over the course of a single month, Intel announced *twenty-three* new networking products, the showpiece being a family of Ethernet adapter cards. That same year, its special big-computer processor, the i860, caught the computing world's imagination by being incorporated into a supercomputer and setting a new processing speed record of 32 gigaFLOPS (billion floating-point operations per second). With the EPROM business retired, the company was now cranking out flash memory chips as fast as it could make them. Intel even (and this only confirmed the fears of its suppliers) decided to stop buying silicon wafers and start growing and slicing wafers of its own.

It all couldn't have gone better, and this unprecedented level of success soon showed up in the bottom line. This was the moment when Intel's revenue curve suddenly turned upward, after tripling revenues during the previous six years, now growing sixfold in just four years.

Andy Grove knew enough not to let Intel become complacent during these good times. But he had never learned to deal with the implications of his own confidence. Now all of Intel shared that confidence. It had the best products and the best talent, and it knew the best way to do everything. If you doubted that, Intel invited you to look at its bottom line. With Steve Jobs (temporarily) out of Apple, Intel cornered the market on Silicon Valley hubris in the early 1990s.

CHAPTER 52

The Bug of All Bugs

E ven as the company steamed along, flags flying, that hubris was about to meet its nemesis. And it would come from the most unlikely source: a bespectacled college professor from Lynchburg, Virginia, with the unlikely name of Thomas Nicely.[1]

On June 13, 1994, in the middle of running several billion calculations on his new Pentium-based computer, Nicely was stunned to find an error in his results. Where 4,195,835 divided by 3,145,727 should have given the answer of 1.333820449, the computer instead read 1.333739068. *How could that be?* he asked himself. Computers don't make errors like that.

Being a good scientist, Nicely set out to discover what had happened. He spent the next four months reworking his calculations and concluded that the flaw was in the Pentium chip itself. So he called Intel. The company replied that he was the only person among Pentium's two million users who had detected such a problem. Intel took down the information and filed it away. The company wasn't particularly surprised by the phone call; as already noted, minor bugs were commonplace in each new generation of processor chips, and makers, once they had sufficient evidence of one, would notify their commercial customers and then fix the microcode in the next generation. So Intel didn't dismiss Nicely's claim outright; it was just one data point that would need to be confirmed by others.

What happened next caught Intel completely off guard and showed how much the world had changed just a couple years into the Internet revolution. Professor Nicely, surprised that no one else had seen this Pentium bug, e-mailed some of his friends and peers to ask if they'd encountered a similar problem. Those acquaintances, as surprised as Nicely, decided to post a message on CompuServe, the dominant electronic bulletin board of the time, asking if any of its twenty million users had ever heard of such a problem. "Then," recalled Nicely, "it went off like wildfire."[2]

Within days, the story—which just a few years before with the 486 would have been forgotten—took off across the Internet as one of the first great memes, then jumped to the mass media. It had all the right ingredients: little man David against a giant corporate Goliath, the fear of entrusting our lives to the incomprehensible complexities of electronics and computers, our misplaced trust in the wizards of high tech, etc. Soon all of the world's great newspapers and magazines carried the story, as did network news programs. The *Wall Street Journal*, recognizing that both the Pentium bug and Intel's stonewalling of the scandal might put the giant company in jeopardy, covered the story day after day, including a question-and-answer section for prospective buyers. The Associated Press raced a photographer to Lynchburg to get a color photograph of Nicely, which it then fired out on the wires around the world. *Newsweek*: "Overnight, it seemed, the computer world went nuts. If Intel did not disclose this flaw, techies speculated, how many others might there be? Buyers of Intel's chips, the nation's top computer makers, were furious. 'I am not pleased,' says a senior executive at one of the world's biggest companies, and called Intel's silence a 'cover-up.' Intel, he says, 'Should have come clean, right off the bat.' "[3]

Even before the blow-up, Intel had clues that something was going on. A few weeks earlier, some of its employees had seen the earliest impact of Nicely's queries: a number of comments posted in an Intel products chat room under headings like "Bug in the Pentium FPU" (floating point unit). On November 2, A. Uzzo & Company, a New York accounting firm, filed a lawsuit against Intel, alleging that the Pentium had been sold to them via false advertising, negligence, and unfair business practices. Before it was all over, there would be seven more similar lawsuits from other companies.

Contrary to the widespread belief that would soon emerge, Intel hadn't ignored Nicely's original query, especially not when a few other similar complaints arrived. Instead it had investigated the problem and found that it was real, but so rare as to not merit any immediate attention. Intel's investigators determined that it was a rounding error that only occurred once in every 9 *billion* division operations. Nicely had run into it because he was crunching such a vast number of divisions. The average spreadsheet user, Intel determined, might run into this same once every *27,000 years*. That made it so rare as to be immaterial—so Intel essentially shrugged and started a low-key search for a fix that could be sent out to scientists and other computation-heavy users. In other words, standard operating procedure . . . and if Intel didn't find the fix for Pentium, it would certainly do so in its next generation of chips.

But Intel didn't appreciate that the world had changed and so had its compact with its users. Those users were no longer just corporations, government agencies, universities, and other mainframe and mini-computer users, but millions of everyday folks—small-business owners, children, senior citizens, and college students—most of whom had bought these computers as their launch pads into the wonderful world of the World Wide Web. And though Intel didn't know it—and didn't really want to know—those Intel Inside stickers represented something special to those neophyte computer owners and Web surfers: trust. Through the welter of competing PC models, the Intel Inside sticker wasn't just a mark of quality, but a north star: if the Intel Inside sticker was on the case, then the purchaser would know in his or her heart that it was a good purchase, that the computer would do everything asked of it, and that the results it presented could be trusted.

In choosing to treat the Pentium bug as it did, Intel was treating an engineering problem with an engineering solution. But it was now more than that; the problem was now freighted with the expectations of millions of consumers who didn't care about mathematical odds but now worried that they could no longer trust their computers. They saw Intel's actions as a betrayal.

The anger was building, but as yet Intel couldn't see it coming. There were only a couple of clues: in his regular report of press coverage, Andy Grove noted a front-page story in one of the trade weeklies, and a week later a few smaller stories in other trade publications.

Then on Tuesday, November 22, 1994, two days before Thanksgiving, all hell broke loose. Over the next three weeks, Intel and Andy

Grove would undergo a trial so shocking that Andy would make it the opening of *Only the Paranoid Survive*. Grove: "I was about to excuse myself [from the class I was teaching at Stanford] to call my office when the phone rang. It was my office calling me. Our head of communications wanted to talk to me—urgently. She wanted to let me know that a CNN crew was coming to Intel. They had heard the story of the floating point flaw in the Pentium processor and the story was about to blow up."[4]

Andy rushed back to Santa Clara to meet an "aggressive and accusatory" CNN producer and crew. There was nothing Andy could do to divert the direction of the narrative and "CNN produced a very unpleasant piece," which aired the next day.

The media floodgates were now wide open. What began as straightforward coverage of the flaw, which was painful enough, soon morphed into something far worse: feature stories, editorials, and even guides that asked whether Intel could ever be trusted again and whether consumers should use the Intel Inside sticker as a signal to buy or to stay away. Grove: "Television reporters camped outside our headquarters. The Internet message traffic skyrocketed. It seemed that everyone in the United States keyed into this, followed shortly by countries around the world."

Almost overnight, the Pentium bug had become the locus of all the pent-up fears of average people toward the digital revolution, Moore's Law, and all of the unstoppable, ever accelerating changes taking place in their lives. And soon Intel's competitors, thrilled to see the industry's dominant company looking confused and crippled, piled on. So did many of Intel's biggest customers, who had grown tired of the company's high-handed tactics and delivery lists. It isn't often that a great company gets a glimpse of how its competitors and customers really see it. Intel now, to its bewilderment, got hit with both barrels.

It began on November 28, when high-end computer maker Sequent, which really did have a problem with the Pentium bug, announced that it would stop shipping its client-server machines that contained Pentiums.

"Then the biggest bomb of all landed: on December 12, IBM announced that it was suspending shipment of all of its Pentium-based computers. Some critics cried foul, calling Big Blue's move a sleazy publicity ploy to hurt Intel, while at the same time promoting PowerPC machines. But IBM replied that its own independent research on Pentium found that the flaw was much more serious than Intel claimed and that it was merely protecting its customers." IBM had finally gotten its revenge for Intel's shift of support to the PC clone makers.

Inside Intel, it was like a nightmare from which the company was unable to awake. Technical superiority and corporate pride had been the twin pillars on which Intel had been built and by which Andy Grove defined himself. Now both were under sudden and overwhelming assault. A few days before, Intel had been rolling along, the chip industry leviathan, seemingly unstoppable, owning its market and living up to its billing as the world's most important company. Now it was reeling from one body blow after another. Its image of technical superiority had been shattered, its failings decried by the world's biggest companies.

Now came the ridicule that shattered the company's pride. Perhaps the worst was the jokes:

Q: What's another name for the Intel Inside sticker they put on Pentium-based computers?
A: A warning label.

New label: "Intel Inside/Don't Divide"

"Intel apparently never recognized the fact [that] the company had entered not only a new market, but a new cultural dimension—faster, shallower and more fickle in its tastes. In this world, facts often took a backseat to emotion, and the mass panics—Alar-poisoned apples, toxic Tylenol, exploding [Ford] Pintos—regularly swept the landscape. All of the clues were there, but Intel, in the words of one anonymous Silicon Valley public relations executive, still seemed to think it was talking to *EE Times* or *Electronic News*.

As Thanksgiving turned into Christmas, Intel was still staggering. In terms of actual lost sales, the financial damage to the company was negligible. Nevertheless, the $150 million Intel Inside program, the greatest marketing campaign of the electronics age, had now been nullified. The trust the company had built over the previous twenty-five years was now in tatters and would take years to restore. And the company's competitors and some of its biggest customers were now racing to exploit the disaster.

And yet Intel still didn't seem to understand the nature or the magnitude of the crisis. All of its responses seemed sluggish and misdirected. The company agreed to replace *some* of the bad chips, but only for those customers whose applications were most likely to encounter the bug, i.e., those sophisticated users who were already well equipped to build their own work-arounds, not the everyday users who were most frightened by the bug.

Next, as the situation deteriorated, Grove decided to mobilize a thousand Intel employees to answer customer complaints. At the same time, the company announced that it was moving up the introduction date of the new, fixed, Pentium. It sounded good, and it involved an immense amount of work by company employees, but it was the right solution to the wrong problems. The markets—commercial, consumer, and not least, stock—wanted the problem to go away, and any wounds it created healed. But Intel still didn't get it, and other than these Band-Aids, the company plunged ahead with business as usual.

For example, it refused to pay the labor costs for those customers who did manage to score some good chips and undertook a replacement, the explanation being that the company had never done that in the past. Worse, Intel continued to resist the growing clamor for a full recall of all four million Pentiums already in use. On the contrary, Intel continued to manufacture the flawed chips as if nothing had changed.

Meanwhile, in the marketplace, a hysteria was beginning to form. The *Wall Street Journal* surveyed its corporate readers and found that most were still sticking with their orders for Pentiums but that they were also demanding replacement chips for the flawed ones they'd already received. The Associated Press reported:

A physics professor in Louisiana may have to scrap a semester's worth of work. An aggravated grad student in Massachusetts is reappraising communications research. A health consultant in Georgia frets that the computer software he created for clients can't calculate exactly right.

And at Los Alamos National Laboratory, birthplace of the atomic bomb and center of US weapons research, bosses are concerned enough to have set up a "Pentium hot line."

They've all been bitten by the Pentium bug.[5]

But perhaps the harshest appraisal came from the trade press within Intel's own industry. *MacWorld* magazine may have had its own agenda, but its words seemed to ring true in Silicon Valley: "This pattern of behavior shows that, like Microsoft, Intel is constitutionally incapable of tolerating rivals. Sadly, those who feel a need to crush their competitors often succeed only in creating real enemies. The Intel arrogance is now clear to the public, and that revelation has filled the company's rivals with new hope and determination."[6]

Then came the IBM announcement. Recalled Grove, "We were back on the defensive again in a major way." In fact, Intel, whether it recognized the fact or not, had never *not* been on the defensive since the scandal began.

Now the company itself was starting to buckle under the strain. Grove:

A lot of the people involved in handling [the Pentium bug] had only joined Intel in the last ten years or so, during which time our business had grown steadily. Their experience had been that working hard, putting one foot in front of the other, was what it took to get a good outcome. Now, all of a sudden, instead of predictable success, nothing was predictable. Our people, while they were busting their butts, were also perturbed and even scared.

And there was another dimension to this problem. It didn't stop at the doors of Intel. When they went home, our employees had to face their friends and their families, who gave them strange looks, sort of accusing, sort of wondering, sort of like, "What are you all doing? I saw such and such on TV and they said your company is greedy and arrogant." . . . And you couldn't get away from it. At every family dinner, at every holiday party, this was the subject of discussion.[7]

By this point, a number of Intel employees were starting to agree with the company's angry customers that it was time for the company to face the inevitable. And the prospect of spending the holidays manning phone lines from pissed-off Pentium users made that opinion even easier to hold.

But of course, it wasn't their choice. It could only be made by the CEO. And Andy wasn't yet ready to surrender. He knew that he was right; the Pentium bug was so rare as to be meaningless to 99.99 percent of Intel customers—and thus those customers, whatever their irrational response to this mess, didn't deserve a free replacement of their chips at the cost of millions of dollars to Intel. Andy Grove had always believed in rationalism; he was a scientist, and he believed that Intel had done nothing wrong. And he was too proud to cave in and make a mea culpa to what was nothing more than market hysteria.

"I had been around this industry for thirty years and at Intel since its inception, and I have survived some very difficult business situations, but this was different. This was much harsher than the others. In fact, it was unlike the others at every step. It was unfamiliar and rough territory. I worked hard during the day but when I headed home I got instantly depressed. I felt we were under siege—and under unrelenting bombardment. Why was this happening?!"[8]

Grove's biographer Richard Tedlow: "[Intel marketing manager] Karen Alter was there, saw it all, and couldn't get over the turmoil. Walter Mossberg of the *Wall Street Journal* told her that he'd seen Watergate, and Intel's cover-up was worse. As for Andy, according to Alter, 'he couldn't believe, on the one hand, that people were so stupid. And, at the same time, he was so furious at a couple of his general managers, who never really recovered in his eyes, that they had let the flaw get into production, that they hadn't found it earlier . . . he was just mad. He needed to be mad at somebody.'"

Alter thought the real hero of the story was Craig Barrett, "who has ice in his veins. He just stepped up and got us organized. . . . 'Okay, have to build a machine to deal with this.' We [soon] had a press machine and a Wall Street machine and an OEM communications machine." Barrett also ordered a quick redesign to address the flaw.

The bet in Silicon Valley was that Grove's ego wouldn't let him surrender to what he saw as irrational foolishness and that he would either let Intel die or get himself fired before he admitted that he had been wrong. Andy was a terrific CEO during good times and a great one during bad times, but few believed that he could cope with a mass demonstration of human weakness and irrationality. And as the holidays approached with no movement on Intel's part, that bet looked like a winner.

But once again, Andy Grove's reputation for being a take-no-prisoners competitor and an unsentimental boss camouflaged what was his greatest strength: his ability to learn and, when the evidence was clear, to change his mind, whatever the consequences.

As the Pentium crisis unfolded, Andy had set up a war room in a conference room just twenty feet from his cubicle. The room, with its oval table, was designed for twelve people, but as the scandal mounted, it was often packed with as many as thirty.

On Friday, December 16, the room was especially packed, when Andy entered and turned the conversation in a new direction. Grove: "The next Monday, December 19, we changed our policy completely. We decided to replace anybody's part who wanted it replaced, whether they were doing minor statistical analysis or playing computer games. This was no minor decision. We had shipped millions of these chips by now and none of us could even guess how many of them would come back—maybe just a few, or maybe all of them."[9]

The announcement, which raced across the Internet that evening and the media the next morning, brought few cheers but a tidal wave of relief. A month late but at last, Intel had reaffirmed the trust of its millions of users. Silicon Valley and the electronics industry—including even many of Intel's competitors—cheered the news. In those intervening weeks, the industry had been reminded just how important Intel had become and how great was the industry's dependence upon this ultimate keeper of Moore's Law. A cratered Intel, for which many of them had fervently prayed, now looked like the very last thing they wanted. Rather, what the industry needed was a humbler, fairer Intel, and with the Pentium bug scandal, that now seemed to be what they'd gotten.

On January 17, 1995, Intel announced that it expected to spend $475 million to replace an estimated 30 percent of the flawed Pentiums. That was a bigger write-off than the annual revenues of most companies in the electronics industry. But with thirty million Pentiums expected to be sold in 1995, that write-off would be quickly absorbed. And as a final touch to show that it had learned its lesson, Intel also announced that henceforth it would immediately disclose any and all defects in its future microprocessors.

In the end, though the Pentium bug would become an industry legend and a case study in what to do right and wrong during such a scandal, it had little financial effect on Intel. By spring, as the world turned to the exploding new world of electronic commerce and the rise of dotcoms, the Pentium bug had become an old story. Somehow through it all, Intel's stock price had managed to hold steady—which took the ground out from under most of the lawsuits against the company over the bug. By March, the company had settled the last of these for just $6 million, less than one day of Intel's revenues.

Even Intel's competitors, who had rubbed their hands at the prospect of a stricken and distracted Intel, didn't make out as well as they hoped. In this, Intel had been very lucky that the Pentium crisis hadn't taken any longer to emerge, because when it did, the other microprocessor manufacturers didn't yet have their next-generation chips ready to serve as true Pentium competitors. They could only sell what they already had, and that wasn't enough to strip Intel of anything more than a few temporary points of market share.

As for Intel, the whole episode seemed like a nightmare from which it had finally awakened. And the company had come away changed from the experience. It now understood that it wasn't enough to stay innovative and compete hard. Intel had tried to be like Andy Grove, but now even Andy was seeing the electronics world anew. As a boy, he had learned that the world could be deeply, dangerously, and unpredictably irrational. But science and then the semiconductor industry had served as a sanctuary from this darker world. Sure, it had its subjective, improvisational, and crazy side—it was Bob Noyce's ability to inhabit that side that he despised and envied most in the great man—but it was nothing that Andy couldn't handle.

Now, after the Pentium bug, Andy Grove realized that good

management, clever marketing, and ruthless competition weren't enough. Even in the precise engineering world of semiconductors, where the equivalent of entire cities were now being etched on slivers of silicon the size of a fingernail and the lines of those etches were now measured in the diameters of atoms, unexpected and even fatal threats could come from anywhere at any time—even from children and even at the zenith of one's success. And that meant eternal vigilance before those threats arrived and humility after they left their mark. It was no surprise, then, that a year later, the Pentium bug story would open *Paranoid*.

Intel was done with the Pentium bug scandal. But Andy Grove was not. He would continue to learn from it for years to come. And a good thing too, because there was one other party that wanted to pursue this story as well: the United States government.

CHAPTER 53

The Endless Lawsuit

L awsuits—over products, recruitment, contracts, and claims—have been a part of the electronics industry, Silicon Valley, and especially the semiconductor business almost from the beginning. Bob Noyce may have resisted the desire to sue the hell out of the various teams of employees who walked out of Fairchild (sometimes with valuable trade secrets and manufacturing recipes) and started their own companies, but he hardly set a precedent.

After Fairchild blew up and scattered new chip start-ups all over the Valley, these suits took on a personal edge as former friends and workmates now found themselves at each other's throats competing for the same markets. Meanwhile, as the more successful companies grew, went public, and became rich and their products reached down to consumers, those companies became ever more attractive targets for shareholder and class-action suits, especially when they indulged in their usual promotional bravado, or tipped over into the quadrennial industry recession.

But nothing in that long history of litigation ever matched *Intel v. Advanced Micro Devices* in all of its incarnations. It remains the longest, bitterest, and most famous series of suits in high-tech history. It is the *Jarndyce v. Jarndyce* of the electronics age.[1]

The genesis of the case has already been discussed. Basically, the relationship between Intel and rival AMD began for three reasons: Bob Noyce's desire to help his protégé Jerry Sanders get AMD off the ground; the demand by Intel's major industrial customers—especially IBM with its PC—to have a strong second source for Intel's microprocessors; and finally, once all of this was under way, AMD's desire to maintain this hugely lucrative contract for as long as possible—in the face of new CEO Andy Grove's decision to end it.

The two companies cemented that relationship in 1982 when they signed a technology-exchange agreement making AMD the second-source supplier of the 80286. A key phrase said that AMD may "copy micro-codes contained in Intel microcomputers and peripheral products sold by Intel." The two companies interpret this phrase differently, and each to its own advantage. For Intel, it means only the 80286 and nothing more. AMD reads *microcomputers* as any and all Intel microprocessors that contain the x86 microcode—the 386, the 486, and even the Pentium.

Andy Grove, who had hated and envied the Noyce-Sanders relationship from the beginning, was not going to let this arrangement stand— for a couple very good reasons. First, it had been a shotgun marriage that Intel had been forced into by an overly conservative IBM that had nothing to lose in making such a demand. Intel had never wanted it, and once the PC clone makers began to outpace IBM, Intel—by now a giant corporation with many fabs and capable of meeting whatever demand arose—saw no reason why it needed to support a competitor. Second, AMD had grown fat on what Grove considered to be the theft of Intel's hard work and massive investment. He knew that by the time each x86 generation ended, AMD had taken at least 15 percent of the total revenues. That was money that would have helped Intel through the tough times and financed future processor generations.

That's why in 1984 Grove decided to cancel the AMD license, thus disallowing AMD use of its new 80386 design. AMD originally thought Intel was bluffing and trying to negotiate a better deal. It went ahead with its 386 clone and filed suit in 1987 when that chip was ready for market and Intel still refused to budge. In its filing, AMD accused Intel of employing illegal tactics—notably using its priority delivery list as a cudgel against customers who considered buying AMD chips—in a "secret plan" to create a monopoly on personal-computer microprocessors. AMD said that it would have sold many more chips if not for Intel's underhanded tactics.

When this suit didn't have its desired impact, AMD took it up a notch, filing an antitrust complaint in Northern California in 1991 accusing Intel of engaging in monopolistic actions. Meanwhile, Jerry

Sanders, the cleverest figure in Valley history, waged his own campaign in the media, portraying Intel as a predatory giant and making fun of Andy Grove as being stiff and soulless. But against the antitrust suit, Grove decided to fight back. In a statement (prepared with Intel PR director Jim Jarrett) on the filing, Grove announced, referring to the pop duo that won a Grammy with dubbed recordings: "What we have here is the Milli Vanilli of semiconductors. Their last original idea was to copy Intel."[2] It was a line so nasty and funny that it not only topped everything Sanders had said to date, but turned the ridicule back onto AMD. It was the best one-liner of Andy Grove's career.

The semiconductor industry held its breath waiting on the verdict of the suit. An Intel win would give it de facto ownership of PC processors. An AMD win would set off a land rush of chip companies hungry to dive back into the business they had lost years before. In the end, the verdict went to . . . Intel. Or was it AMD? In arbitration, a judge ruled that AMD had no right to Intel's 80386 under the old second-source contract, but that the company could still sell its 386 clone, and it was awarded $10 million from Intel and a royalty-free license to any Intel patents it used in its own 386 clone. Exactly what all this meant was anyone's guess. One thing was certain: it wasn't over.

At the beginning of the case in 1987, AMD's counsel Richard Lovgren predicted that the trial would take twelve weeks.

Eight years later, after warehouses had been filled with legal documents, and after the two companies had spent an estimated $200 million in legal fees (in 1994 AMD spent $30 million on lawyers alone), the two parties ended 1994 apparently as disputatious as ever. The year had been as inclusive as those that preceded it: in March, AMD won a federal retrial over the right to clone the 486, then in October, Intel won back a portion of the 486 case. Finally, on December 30th, the California Supreme Court upheld AMD's right to make 386 clones.

Intel and AMD, along with the rest of the semiconductor industry, were now five generations past the originalchip, which was now sold for pennies by scrap dealers. The leading processors of 1995 were a thousand times more powerful. One writer compared it to Boeing and Lockheed fighting over the technology of the B-17 bomber in the age of the B-52. And the dispute became even more absurd if you knew that after the 1987 suit, AMD had been forced to write its own microcode, so strictly speaking, AMD wasn't even cloning Intel chips anymore.

But this was personal, less *Intel v. AMD* than *Andy v. Jerry*. And if left to those two tech titans, it would never be resolved. The solution would have to come from elsewhere.

In mid-1994, the two counsels, Lovgren of AMD and Thomas Dunlap of Intel, who had begun the summer joking with each other about a settlement, began to grow more serious about the idea. Finally, after one of the endless court sessions, the two attorneys decided to take a walk together down a side corridor. They were surprised to find themselves in considerable agreement—a good thing to know when, a few days later, the federal magistrate on the case, Patricia Trumbull, told them in no uncertain terms to resolve the case.

The one thing both sides immediately concurred upon was that it would be a terrible mistake to bring Grove and Sanders together in the same room. The two men had been at war, in one form or another, for thirty years. They avoided each other at industry events. They made a point of never crossing each other's path. Sanders would say: "I'm so out of touch with Andy Grove that I don't know what sort of person he has become. I must admit that as long as Andy was at Intel, they would never settle forever, because I would never surrender."

The two counsels decided instead to bring in the chief operating officers from each company, Craig Barrett and Rich Previte, to negotiate. They met, symbolically, at a law office unaffiliated with the case, on the Sunnyvale–Santa Clara border, halfway between the two firms. The December 30 ruling accelerated the negotiations; so did a mediator brought in in early January to deal with one last sticking point. The agreement was reached.

"When it started, I was 32 years old," said Lovgren. "I'm now 40. You get pretty tired."[3] Said Dunlap, "The market changed. It wasn't a change of heart or personalities. I think the companies started looking to the future."[4]

At the conclusion of the negotiations, the two attorneys stood and shook hands. Then Barrett and Previte shook hands. Everyone joked that the final step would be for Sanders and Grove to shake hands . . . but they never did.

On Wednesday, January 11, 1995, at 2:00 p.m., Intel and AMD officially signed the settlement. Intel agreed to pay AMD $18 million. AMD agreed to pay Intel $58 million. "After hundreds of millions of dollars in legal fees, costs passed on to customers, the greatest feud of the microprocessor era concluded on a minor key."[5]

Industry consultant Tim Bajarin called it "the end of one of Silicon Valley's soap operas." Jerry Sanders called it "global peace." Andy Grove had no comment.[6]

Was it really all over? Of course not. In 2000 AMD filed a complaint with the European Commission claiming that Intel had violated European antitrust laws through "abusive" marketing programs. Then, in 2004, after both Sanders and Grove had fully retired from their respective companies, the Japanese Fair Trade Commission raided Intel's offices for documents, then ruled that Intel Japan had stifled competition by offering rebates to big companies—Fujitsu, Hitachi, Sony, and others—for not buying AMD chips. A year later, AMD filed an antitrust suit against Intel in the US District Court in Delaware charging that Intel had maintained its monopoly in the x86 microprocessor market by coercing customers not to buy AMD chips.

And so on and on it went, *Intel v. AMD*, unlikely to end before one of the companies dies . . . if then.

CHAPTER 54

Mea Culpa

The AMD feud was a nuisance for Intel. So was being forever bolted to Microsoft in the personal computer business. But in the late 1990s, the presence of both threatened to become much worse for Intel.

As the dot-com bubble began to swell and the electronics industry grew increasingly frenzied—thanks to the Internet boom, it had been in nearly continuous growth now for a decade—the US Justice Department opened a case against Microsoft for antitrust practices. Thirty state attorneys general quickly joined in.

The feds had a pretty good case. Microsoft held 90 percent of the world's personal-computer operating-system business and roughly that much of the office productivity software and word processing (Excel and Word) business, but after being blindsided by Netscape Navigator, it had also come back and crushed that company with Internet Explorer, thus gaining a near-monopoly in that market as well.

That Microsoft recovered so quickly was a credit to the competitiveness of Bill Gates and the resourcefulness of the company in that era. But it was also the product of Microsoft "bundling" Internet Explorer with its already industry-dominant operating system—if you bought one, you got the other. And you quickly found that Internet Explorer was easier to install and worked better with Microsoft Windows.

If this kind of competition-crushing behavior wasn't enough to get a federal investigation, then Gates's obvious contempt for Washington certainly sealed the deal. The technology world cheered at the news; computer users were a little more ambivalent.

One of the dangers of being so closely tied to another company, as Intel was with Microsoft, was that if one company went down, it might pull the other down with it. And it wasn't long before the feds were knocking at Intel's door. Soon leaks appeared that suggested that senior Intel executives had been deposed about a particular meeting between them and Bill Gates. According to *U.S. News & World Report*, these leaks described "heretofore secret testimony that paints Bill Gates as a big bully, this time leveling 'vague threats' against giant chip maker Intel Corp. in a 1995 meeting with the company's chairman, Andy Grove."[1]

According to the leaks, Gates was furious with Intel's plans to compete with Microsoft in new multimedia software. Said *U.S. News & World Report*: "One executive said in an internal memo . . . that Gates 'made vague threats about support from other [non-Intel chip] platforms' if Intel didn't give in. Microsoft's chief was 'livid' about an Intel lab's investments in Internet technology 'and wanted them stopped.' These details are especially shocking because Gates, well known for his merciless attitude toward the competition, seems to be threatening Intel, a close—if wary—Microsoft partner."[2]

Andy Grove wasn't afraid of Microsoft, but he was worried about the Justice Department nosing around Intel's business. He knew that Gates, with his pompous attitude, was at great risk of being hammered by the Justice Department and Microsoft eviscerated. And if Microsoft could be put through this meat grinder, why not Intel? After all, it owned an even greater market share of an even more valuable industry, and as the court records of the AMD suit showed, Intel certainly hadn't been above using its own muscle—the delivery list—to threaten its own customers. As Andy privately admitted to at least one former Intel executive, "We may have gone a little too far."

As it turned out, Andy Grove had good reason to be concerned, though the threat came from a different direction: the Federal Trade Commission. In June 1998, The FTC alleged that Intel had used "unfair, monopolistic business practices to the detriment of computer vendors, processor manufacturers, and graphics chipmakers."[3] The corporate "victims" of Intel's shady business practices were listed as Intergraph (a software and services company), Digital Equipment, and personal computer/server maker Compaq. The trial was slated to begin in January 1999 but, because of the complexities of discovery, ultimately was moved to early March and was expected to last up to two months, thanks to a witness list that topped fifty—including executives from the three companies, plus AMD and most of the top management of Intel, including Grove, Barrett, Otellini, Alter, and microprocessor division senior VP Albert Yu. Intel was also expected to present witnesses or statements from competitors, including National Semiconductor, Sun Microsystems, and Integrated Device Technology.

Intel's strategy was to fight the case with three arguments. The first was that the company had every right to refuse to share its company secrets. "Did we withhold our intellectual property from Intergraph? Yes," answered Peter Detkin, general counsel for Intel, and said Intel was within its legal rights to do so. The second argument was that the case was just the federal government, flush from winning against Microsoft, trying to get another high-tech notch in its gun. "You won't find the evidence that you have in the Microsoft trial," he warned the government. The company got support from one of the Senate's most senior members (and chairman of the Judiciary Committee) Orrin Hatch, who warned the government about overreaching on the case. And third, Intel argued that in the volatile world of semiconductors, Intel, whatever its market share, was anything but a monopoly. "What monopoly?" Detkin asked CNET, noting that Intel had been losing market share to AMD and others all year.

The delay in the trial date may have won the case for Intel. The nine months between the FTC's charges and the trial had been one of those intervals, at the tail end of one x86 generation (in this case, the Pentium Pro) and the next (the Pentium III), when the cloners temporarily caught up with Intel. Moreover, a new graphics chip announced by the company at the beginning of 1999 had suffered a slow start. And as noted, the impending Pentium III was not as appealing to the low-end market as its x86 predecessors. As a result, by the time of the trial, Intel's overall market share had temporarily fallen from just under 90 percent, according to market watcher IDC, to just over 75 percent, in the retail market to a little more than 50 percent, and in the budget market to only 30 percent. These numbers didn't help the government's case.

The case was Intel's to win. The FTC had a weak case. Intel's timing was perfect. It had a team of top-notch lawyers. It had the support of one of the Senate's most powerful figures. And Intel had some of the biggest companies in tech behind it.

Intel might have won. But it would have been a Pyrrhic victory. Intel had seen what had just happened to Microsoft when it decided to go nose-to-nose with the feds. Bill Gates, acting like the Andy Grove of old, set Microsoft up for years of litigation and misery. Intel might win the case—now—but the FTC would just come back at it from a different direction (as in fact it did anyway in 2009).

What would the new post–Pentium bug Andy Grove do? Would he dive into the fight with the feds with a sense of holy justice, or would he put the health of Intel first?

In the end, literally on the eve of the trial, Andy caved.[4] On March 8, 1999, Intel settled the case with the FTC, without admitting any guilt, by agreeing to change several of its business practices, notably to stop changing the operation of its microprocessors to make them incompatible with clones and to stop punishing current customers for purchasing similar chips from competitors. Andy Grove, having learned the lessons of the last few years—the most important of them being humility—had humbled himself before the feds, . . . and likely saved Intel.

CHAPTER 55

Bunny Hop

I n November 1995, even as Intel was finally replacing the buggy Pentium chips, it announced that processor's successor, the Pentium Pro. It was the most powerful microprocessor yet, containing 5.5 million transistors (more than 2 million more than the basic Pentium) and capable of running at 200 MHz. Despite its name and the fact that, like its predecessors, it could still run the original x86 instruction set, the Pentium Pro was a radically new device. For one thing, its microarchitecture was a complete break from that of the Pentium, rather than just being an upgrade of it. It also had a 36-bit address bus, which meant that it could access as much as 64 gigabytes of memory, and in many important ways, it behaved more like a RISC than a CISC device, showing the impact of Intel's competition.

But perhaps the most obvious difference between the Pentium Pro and its predecessors was actually visible to the naked eye: instead of just one processor per chip, the Pentium Pro was available in dual and even quadruple processor configurations, that is, two or four processor engines sharing the same chip. Needless to say, this made the Pentium Pro physically a whole lot bigger than earlier x86 models, and at $1,325, it was also a lot more expensive. Still, it was hard to argue with performance that was as much as 35 percent greater than the original Pentium and slightly exceeded that of its RISC competitors.

The Pentium Pro met with strong response, at least from the makers of big computers, workstations, and servers. In fact, it was quickly incorporated into the design of the ASCI Red supercomputer, built by Intel for the government's Sandia Labs, which in late 1997 became the first computer to exceed one teraFLOPS (trillion floating-point operations per second), making it the fastest supercomputer in the world.

But the Pentium Pro wasn't the same hit with Intel's bread-and-butter customers, the personal-computer companies. For one thing, it was just too powerful: once again, the power of microprocessors was outstripping the needs of everyday users. It was also too big and expensive. When competition was making personal computers smaller (laptops were now coming on the scene) and cheaper, a bigger and more costly CPU wasn't particularly welcome.

The reality, which Intel and the other semiconductor companies were beginning to appreciate, was that the quarter-century run of single-product microprocessor companies was coming to an end. The computer industry was now spread across a vast spectrum, from cheap desktops up to high-powered servers and supercomputers. For a while, the challenge of top-end machines could be solved with arrays of standard microprocessors, but that was just a stopgap. Intel and the others now had to offer multiple product lines of processors—low-power chips for consumer products and budget PCs, midrange processors for professional and small-business computers, and powerful processors (usually with multiple processor cores) for high-level scientific and business applications, as well as the growing army of servers powering the Internet.

In Intel's case, three product families would define the company in the early years of the new century: the high-end 1.4 billion–transistor Xeon line for powerful computers; Pentium models II, III, and 4 for the midrange; and the Intel Celeron for lower-powered, budget applications. Even in later Intel generations of these three lines—generation number 9 appeared in 2011—they still remained x86 microprocessors, still backwardly compatible with and capable of running the original 8086 code.

Intel was back, and in a big way. In 1998, it had revenues of $26.3 billion; a year later, $29.4 billion. It was quickly becoming one of the largest companies in electronics. And it wasn't making that money just from microprocessors either. It continued to press forward on innovation in its other businesses as well. In 1995, the company focused on its networking product line, adding hubs, switches, routers, and all of the other accoutrements of the Internet's infrastructure. In 1997, the company announced a major innovation to flash memory—StrataFlash, which multiplied the density of those devices by letting more than one bit of data be stored in each cell.

In 1995, the company also began to pay attention to its public image. The first clue was Andy Grove's appearance at Telecom '95 in Geneva, the largest conference in the telecommunications industry. There, in a memorable image that Intel promulgated everywhere, he gave the keynote address, sharing the stage with Nelson Mandela. It signaled that Andy was now a global figure—hopscotching Bob Noyce and even Dave Packard.

The following year, Intel went at community service with a vengeance. It created the Intel Involved program to free its employees to give time in public service.

It also began negotiating with photographer Rick Smolan to create a big photo volume called *One Digital Day* on the microprocessor's impact on modern life. It would also be a special edition of *Fortune* magazine. In an unlikely move that testified to Intel's power in the US economy, the company also sponsored a giant traveling show of the Smithsonian Institution's greatest artifacts. In addition, Ted Hoff, Stan Mazor, and Federico Faggin were inducted into the National Inventors Hall of Fame—icing on the cake.

In 1997, Intel took over as sponsor of the International Science and Engineering Fair high school competition. This, the famous GE Science Fair of the fifties and sixties, had declined considerably in the intervening years. Intel's involvement, combined with the serendipitously timed hit movie *October Sky* two years later, gave the program new life, and in the decades to come, it may prove to be one of Intel's biggest contributions to science and engineering.

Also in 1997, having lain low for a couple of years after the Pentium Bug, Intel came roaring back into the marketing and advertising game at the biggest venue of all. That year, the company bought time during the Super Bowl to debut the Bunny People—dancers in colored metallic clean-room suits who pranced and gyrated, promoting Intel Inside and two new processors, the Pentium with MMX technology (an added instruction set for the growing number of multimedia applications), and the 7.5-million-transistor Pentium II. The ads were silly and campy, but despite that—or more likely because of that—the ads were a huge hit, further softening Intel's image after it had looked like a hard-hearted megacorporation during the Pentium bug scandal.

If this was Andy Grove's plan—to invert Intel's, and his own, image from tough, uncompromising, and unsentimental (which had worked perfectly well in the early days) to the warm, family-oriented service provider you were happy to welcome into your home, even into your car, your kids' game console, and all of your appliances—it was brilliant for both parties. The Intel Bunny People, the science fairs, and all the rest made Intel the most consumer friendly company in tech (and laid the groundwork for Apple's historic advertising under Steve Jobs in the new century).

But it did even more for Andy Grove. For most of the decade, Silicon Valley had recognized that Andy was doing something remarkable at Intel. That he had not only turned the company into an international giant, but made it, against all odds, even more dynamic and innovative than ever. After the schadenfreude of the Pentium scandal faded away, there was a new respect for Andy and the way he had resolved the crisis by swallowing his pride and *changing*. Most Valley executives knew that Andy was astoundingly smart and capable, but few believed he could actually admit to error and *evolve*. When he did, many of them looked at themselves in the mirror and asked if they could have done the same in similar circumstances.

Others were watching too. Ironically, the Pentium bug had drawn public attention to Intel like never before. Add to that the company's rapid growth, the publication of *Paranoid*, with its almost unique candor from the CEO of a giant corporation, and now the beginnings of the dot-com bubble pulling the country upward into unprecedented prosperity, and the nation's mass media began to turn their attention to Silicon Valley and Intel in particular. The networks and major newspapers and magazines began to regularly send reporters to the Valley, even set up news bureaus. Suddenly, with thousands of new companies being created in a matter of months, the products and people of Silicon Valley weren't just the stuff of the business pages, but of features, even front-page news.

But some prescient journalists had already been looking at Intel for a long time. Peter Jennings, the anchor at ABC, was one of them. Another was Walter Isaacson at *Time*. At only forty-four, after a stint as editor covering new media (when he had learned the impact of the digital revolution), he had just been named managing editor of the magazine. Isaacson, who would show his deep understanding of the tech world a dozen years in the future when he authored a best-selling biography of Steve Jobs, now set out to wake up *Time*'s readers to the incredible changes taking place around them. And to Isaacson and the magazine's editors, no one seemed to embody this new digital world more than Andrew S. Grove.

Intel and Grove were told only that the magazine was working on a feature about Andy, by all indications just another of the many pieces that *Time* and other magazines and newspapers were writing about him. Then on December 23, 1997, timed for the fiftieth anniversary of the invention of the transistor, America awoke to find that Andy Grove had been named *Time*'s Person of the Year, an honor he now shared with some of the greatest figures of the century, including popes, presidents, national leaders, war heroes, and even two men who at different times had tried to kill him: Adolf Hitler and Joseph Stalin.[1]

Writing the lead story himself, Isaacson noted that in a year of big stories, the biggest was "the new economy." At the center of that was the "microchip," and at the very center of the microchip industry was Andy Grove. Andy, Isaacson wrote, combined a paranoia bred in his childhood, the entrepreneurial optimism of the immigrant, and "a sharpness tinged with arrogance" from having a "brilliant mind on the front line of a revolution." He combined, Isaacson wrote, "a courageous passion" with "an engineer's analytic coldness."[2]

The reaction in Silicon Valley and the rest of the electronics industry was shock at first. Why hadn't Noyce gotten the recognition? Or Packard? Or Brattain, Bardeen, and Shockley? Steve Jobs? Faggin and the microprocessor team? Or Gordon Moore?

But that shock soon turned to acceptance. Why not Andy? Who in the high-tech world more deserved the honor? Then it turned to pride. The world finally appreciates what we've done here. We now belong to history.

For Andy Grove, the announcement came as satisfaction, vindication—but also, being Andy, it was yet another data point in his planning for the future. He was rich, famous, arguably the most powerful and admired business executive in the world. Intel provided more than 80 percent of the world's microprocessors and was growing as fast as any big company ever. He was a best-selling author, and now he had been named *Time* magazine's Person of the Year, an honor reserved for the likes of Churchill, Martin Luther King Jr., and Gandhi. He had reached the zenith of his career and his life. How would he ever top this? Yet one more x86 processor? Another couple points of market share? One more medal from the IEEE?

No, the time had come. The man who had wanted to be a scientist, and before that a journalist, had instead become the best businessman in the world. On March 26, 1998, sixty-one-year-old Andy Grove announced that he was stepping down as CEO of Intel Corporation. He would move up to become the company's chairman of the board (Gordon Moore would become chairman emeritus). Craig Barrett would become Intel's fourth CEO, and the first nonfounding employee in the job.

The Noyce-Moore-Grove era—the reign of the Trinity that had been at the helm of Intel Corporation for thirty years—was over.

CHAPTER 56

Running the Asylum

I ntel Corporation began the 1990s in mourning . . . and ended them in madness. The company began the decade with just under $4 billion in annual revenues; it ended it with more than $30 billion. It started the decade with 24,000 employees and ended with 70,000. In 1990 it had eight wafer-fabrication factories costing $800 million each in two countries; in 2000 that had grown to twelve fabs at $5 billion each in three countries. At the beginning of the decade, Intel was making its chips on six-inch-diameter silicon wafers, and the surface details of those chips were as small as 1 micrometer; by the decade's end, the company was working with eight-inch wafers, with details as small as 0.25 microns. And its flagship microprocessor, which contained 1.2 million transistors in 1990, had grown to contain nearly 50 million transistors . . . at a lower price. Each of those later chips was fabricated in laboratories with more complexity than the Manhattan Project, using far-ultraviolet light (because the wavelengths of visible light were now a thousand times too long), washed with water purer than that used in human surgery, and ultimately would perform more calculations in a second than the number of human heartbeats in a thousand lifetimes.

Few companies in history had ever matched such an achievement over the course of a single decade, but those statistics, as incredible as they were, still didn't capture the true nature of Intel's experiences over the course of those ten years. Intel had begun the decade as a midsize electronics company surrounded by a number of comparable-size competitors but having a few important advantages. It still had its two founders, as well as its top operating executive. The company was celebrating its first million-transistor microprocessor and preparing to experiment with mass marketing. The market was strong, and the economy was finally recovering from the latest recession.

Ten years later, one of those founders had been dead so long that most Intel employees had never seen him; for them he existed only as the name of a building and as the subject of Carolyn Caddes's famous photograph mounted in the lobby of that building. The second founder was preparing to retire, which he would do in 2001. And the man who led the day-to-day operations of the company for thirty years, who had gone from angry anonymity to less-angry global fame, had left the daily operations of the company to become chairman.

In those intervening years, there had been no real downturn; it had been a decade of almost continuous nationwide, industrywide, and companywide expansion. Now, at the end of 1990s, it had reached fever pitch, as even the rules of economics seemed suspended; everyone was growing rich, and companies like Intel seemed ready to burst at the seams. Now for the first time in its history, Intel, which had always been a magnet for job applicants, found itself bleeding talent as many an ambitious employee left a high-paying job to join one of the thousands of new dot-com start-ups in the Valley that seemed to be making their founders into overnight billionaires.

High tech, and especially Silicon Valley, seemed to be heading into chaos, where notions like sales, profits, and market share no longer mattered as long as the stock price kept climbing and everyone kept getting richer. Even a company like Intel, which had always prided itself on being rational and strategic, couldn't help but be caught up in the whirlwind, not when orders kept pouring in and you couldn't drive out of the Intel parking lot without hearing endless ads for dot-coms on the radio, seeing planes towing ads overhead, and being passed on the freeway by flatbed trucks carrying billboards because there wasn't a square inch of commercial space in the Valley that hadn't already been taken by dot-com advertisers.

Interviewed by *Forbes ASAP* magazine in the last days of 1999, Gordon Moore, the most rational man in high tech, could only say that he didn't understand what was happening, that every rule he had ever learned about business was now out the window, and that it had gone on so long that he wondered if perhaps it was he, not the market, who was out of touch with reality. Asked to make a prediction about the months to come, Moore demurred, saying that he didn't even understand how the boom—it was now increasingly being called a bubble—had lasted this long.[1]

The irony was that Intel, without planning to do so, was playing at least two crucial roles in keeping this bubble inflated. First, it continued to track Moore's Law, producing one generation of more powerful processors and flash memory chips after another, making the Internet ever more powerful, an increasingly rich experience, and more pervasive around the world. More users spending more time on the Web and doing more things (like spending money to buy things) buoyed up the boom and spared the dot-coms from facing reality. Second, the stock market, recognizing the central role Intel played in the infrastructure of the Internet, making the basic building block of all of its components, rewarded the company by sending its stock soaring. In 1993, the company's stock had sold for about $3.40 per share, with about fifty million shares sold each day. By mid-1997, at the beginning of the dot-com boom, it had jumped to almost $19 per share, with ninety million shares being traded daily.

But that was just the beginning. Two years later, at the apogee of the market boom and the ultimate limit of the dot-com bubble, Intel's stock crossed $33. As exciting as this stock price was, and as impressive as it was to be sitting at the very apex of the business world, Intel's management knew it had to end. The whole US economy, especially the electronics industry, was becoming distorted by these crazy numbers and insane valuations. Talent and money were flowing in the wrong direction, toward companies that couldn't possibly survive. A reckoning was coming, and one of the earliest clues was that even as Intel's stock price reached record highs, the number of shares being traded had fallen back to just over thirty million shares per day. The stock market, losing momentum, was straining to keep from stalling.

In early January 2000, Cisco CEO John Chambers returned from his holiday vacation to a shattering piece of news. He had left two weeks before, content in the knowledge that the company's back orders were still strong enough to keep the company healthy through the first quarter of the new year. He returned to find that all of those back orders had *disappeared*. Cisco, which had been as high-flying as any big company in the boom, now faced not just hard times, but potential bankruptcy. Before the morning was over, Chambers had already put the company in crash mode and was preparing lists of cutbacks in every office. Within days, Yahoo, which had helped set off the dot-com boom, was in similarly critical condition.

The bubble had burst, . . . but not yet for Intel. As the new dot-com companies died and even the big tech companies saw their stock prices collapse (the stocks of Amazon, Cisco, and Yahoo lost as much as 80 percent of their value seemingly overnight), Intel's stock continued to climb—not least because it seemed the last safe harbor for investors in tech.

On August 31, 2000, long after most of the rest of the tech industry had collapsed, Intel's stock crested at more than $78 per share. With that nosebleed price, the company had a market capitalization of nearly $500 billion—one-half *trillion* dollars—which made it greater than the combined value of the *entire* US automobile industry. It was at that moment that Intel became the most valuable manufacturing company in the world. It was the first time ever in tech, and it wouldn't happen again for another decade, when Apple made its own historic run.

It was the perfect coda to the Andy Grove era at Intel.

PART VI Aftermath

CHAPTER 57

The Full Weight of the Law

The men who followed Andy Grove into the CEO position at Intel had to contend with a series of challenges in the twenty-first century that would have been almost unimaginable in the twentieth. On top of that, they also had to deal with a void inside Intel left behind by the departure of its three great leaders—most of all, Andy Grove. Just as crushing, they also had to deal with the living Andy Grove looking over their shoulders, second-guessing their every move.

Craig Barrett had inherited the chief executive position at Intel just as the dot-com chaos began to overwhelm rational thinking. It was not an easy transition. For one thing, there was the obvious problem of dealing with thus runaway growth. When would the bubble burst? How many of the orders that Intel was booking were real? How do we retain employees being lured away by the gold rush taking place just miles away? And how can we prepare for the inevitable bust that could arrive at any moment?

Another important question, though one rarely faced except by big companies that have experienced stratospheric stock prices, is *what to do with all of that cash*? One answer, which Intel executed on the same day that Grove became chairman and Barrett CEO, was to reduce the available stock pool to make the remaining shares more valuable to holders—a hundred-million-share buyback.

A second possibility was acquisition. While still CEO, Andy had seriously pondered purchasing Cisco, then available for less than $1 billion. It would have instantly made Intel a major player in the networking world. But in the end, he decided against it: "We considered half-heartedly—buying Cisco . . . [but] we didn't have the distribution channel for a Cisco acquisition. But maybe we could have made it work."[1] A bad decision? After all, Cisco became a $100 billion company. On the other hand, it was so far out of Intel's core business that Cisco might have been neglected and eventually jettisoned, never becoming the corporate giant (and major Intel customer) it was destined to be.

Soon after Barrett became CEO, he, too, considered a major acquisition, Fore Technologies. But in a reversal of its usual acquiescence, and with Andy abstaining, the Intel board of directors voted Barrett down. It was, Andy would later say, "the only time the board rejected management's recommendations." He would also say that he agreed with the board's decision.

This was a stunning rebuke to Barrett, one that had never happened to Noyce, Moore, or Grove. Clearly the balance of power had shifted between the board of directors and company management. It would be a source of great frustration to Craig Barrett during his tenure as CEO, and yet he would retain that same power when he became chairman seven years hence.

Why did Intel ultimately pass on Cisco, Fore, and the numerous other companies it looked at over the years, especially in the late nineties, when it had more cash than it knew what to with? The roots went back twenty years. In the late seventies and early eighties, chip companies were being snapped up by large electronics firms and even nontech manufacturing companies (for example, Fairchild by French-American oil giant Schlumberger). These acquisitions inevitably failed, largely because their new owners simply didn't understand the unique dynamics and culture of the chip industry. The surviving independent chip companies—desperately installing shareholder "poison pills" and other rules to spare themselves the same fate—promulgated these stories of failure. It worked, and the acquisition binge ended.

For good or bad, chip companies believed their own story. And that belief was underscored by comments from the likes of former Intel executive–turned–dean of Valley venture capitalists John Doerr, who said that Intel, as the leader of the microprocessor world, had no margin of error in its work and thus should never be distracted by other activities.

Even Andy Grove seemed to believe the myth. His biographer Richard Tedlow, who attended a class Andy taught at Stanford in 2005, heard him tell his MBA students that Intel had always been "shitty" at diversifying, whether through intrapreneurship inside the company (he apparently forgot the microprocessor) or through acquisition. The problem, he said, was a lack of "strategic recognition" and "strategic will." Tedlow concluded that because of this attitude Intel, when faced with an investment opportunity, simply gave up too soon.

The third strategy for dealing with a mountain of extra cash was to use it as a war chest: invest some of it to supercharge company operations and save the rest as a buffer during hard times. After the rebuff on the Fore Technologies acquisition, this was the path that Barrett took, and it would prove to be one of the most controversial but ultimately the most successful of the decisions he made as an Intel's CEO.

It wasn't easy being in Craig Barrett's shoes. On the one hand, he was in charge of the most valuable manufacturing company in the world, the linchpin of the global Internet economy and one of the key pillars of US economic health (in 1999 Intel was named one of the thirty Dow Jones Industrial stocks). On the other hand, as he told the *Wall Street Journal* a decade later on the occasion of his own retirement, he was acutely aware that his career would be spent "in the shadow of the guys who preceded me." And the biggest shadow of all was that of his lifelong boss, Andy Grove. "Was it hard to follow a legend? What do you think?" he snapped at the reporter.

The new relationship began well. Andy had only good things to say about his successor. "Craig is more purposeful, more organized—and tougher—than I am. . . . He transformed Intel from a medium manufacturer to an outstanding one. He is now ready to exercise those qualities on business matters as well, hence his promotion to president."[2] Gordon Moore also extolled Barrett's qualities; and Intel as a whole, though exhibiting the usual fear about a change in command after all of these years, seemed happy that the job had gone to one of its own, and it welcomed what looked to be a kinder, gentler Intel.

Industry analysts and outside media were not so welcoming. The same journalists who had decried Andy's heavy-handed tactics now wondered if Intel could continue to thrive without them. Was Barrett tough enough? Could he be Mr. Outside for Intel the way he had been Mr. Inside, especially after a global business superstar like Grove? Most of all, Andy Grove's genius had not been in avoiding mistakes but in admitting them and keeping the company moving forward. Did Barrett have that talent? The only way to find out would be during Intel's next great crisis—and then it might be too late.

In the end, when the crises came, Craig Barrett did pass muster. He held Intel together during both the dizzyingly good times and the desperately sad ones. He would serve as Intel's CEO for seven years, beginning in 1998. Then in May 2005, he advanced to become Intel's chairman, as Andy Grove formally retired from the company. When he left the CEO's job, Intel's annual sales, at $38 billion, were bigger than when he had inherited the company despite four of the worst economic years in American history. On the other hand, the company's stock at the end of his tenure was at the same price as at the beginning. And it is that last fact that neatly encapsulates how history has treated Craig Barrett's tenure as Intel CEO: he didn't wreck the company, but neither did he make it any greater. He did some brilliant things, but also some wrongheaded and expensive ones. And unlike Grove—and certainly unlike Bob Noyce—he left less with a bang than a whimper.

If Barrett's tenure as Intel's CEO has become something of a footnote in Intel's history, it wasn't for a lack of effort on his part. The man who had to drive less than ten miles to go to college became one of the most peripatetic CEOs of his era, the very embodiment of the chief executive as road warrior: "Mr. Barrett is legendary for a working week that begins at the Intel facility near Phoenix, where he has lived for a quarter-century ('I haven't been home for five days in a row in 20 years,' he says) and moves Tuesday mornings to Santa Clara, Calif., and Intel headquarters—with stops around the world (30 nations per year on average) in between. It's back to Phoenix on Thursday night and then off to Montana to hunt and fish at his celebrated Three Rivers resort. He returns to Phoenix late Sunday night."[3]

In person, Craig Barrett was almost a perfect antithesis of Andy Grove. Andy was small and nattily dressed, a quick-motioned, fast-talking bantam rooster with a face that, when he was angry, would darken, the veins standing out on his temples. Barrett was huge and rambling, a rancher with callused hands and a sunburned neck who looked uncomfortable in a tie, and had a hard-edged voice with a slight drawl. Both men had tempers, but Andy was fire—loud, obscene, and emotional; Craig was ice—hard, cold, and unforgiving. Many Intel employees who had suffered under Andy's "Attila the Hungarian" rule and welcomed the new CEO soon found themselves nostalgic for the old order. Andy liked debate because to him it was a game that brought out his passion and his love of conflict. For Craig Barrett, the big-game hunter, who cared most for solutions, you *were* the game. Andy might scream at you and hold a grudge against you for a decade—but then, out of the blue, forgive you and be delighted to see you. When Barrett turned his face from you, you were confined forever to the outer darkness. Craig Barrett, perhaps better than anyone in the tech world, could create a setting for you to do great things. But Andy, in large part because of the Noyce in him that he so despised, could convince you to die for Intel.

Part of what Barrett was up against was apparent from the start of his career as CEO of Intel: he wasn't Andy Grove. Nobody was. Intel had borne the stamp of Grove's oversize personality for more than thirty years; it was embedded in the company's DNA, in its culture, even in the way it looked at the world. The employees told "Andy stories," repeated his one-liners, and thrilled at his appearances on the global stage. What's more, Andy was still there, still looking over Craig's shoulder, the court of last resort for anyone not thrilled with Barrett's leadership.

Barrett, for his part, wasn't a performer or public figure. He didn't speak in clever phrases, nor was he much more than a competent public speaker. And though he raced around the world, he wasn't going to share a stage with Nelson Mandela or write best-selling books or compare AMD to Milli Vanilli. Rather, Barrett still saw himself as a problem solver, as someone who took on a compromised or deteriorating situation, looked at it honestly and without prejudice, and then built (to use Karen Alter's term) a "machine" to solve it. Arguably, Craig Barrett had been the best COO in Silicon Valley history, far better even than Andy Grove.

Barrett: "I remember being in a meeting at one of our plants with 21 of our manufacturing managers. We started talking about changing our factory model and one of the veteran managers—one of those guys who kind of ran his own little kingdom like a prince, said, 'Are we going to discuss this? Are we going to get a vote on this?' And I said, 'Yeah, we'll vote—the only problem for you is that I get 22 votes.' In other words, there's a time to let everyone twist the knobs and a time to make a decision."[4]

But being a great operations guy didn't always translate into being a great chief executive. The former requires superior competence; the latter, great vision. Had this been the early 1980s, when Intel's task was pushing forward the x86 technology and holding off the competition, Barrett, with his focus on making Intel the most tightly run ship in the chip business, might well have been a hugely successful and famous CEO. But this was the beginning of the new century—and Craig Barrett had the bad luck of having his tenure as CEO occur during a period when Intel was buffeted by huge historical forces that were beyond his control. First there was the dot-com bubble, a massive investment by the venture capital industry into the emerging Internet economy that created thousands of companies, most of which were never competitive. Then the bursting of that bubble and the subsequent electronic-industry crash. Next came 9/11 and a global economic paralysis. And finally, coming off the Internet bubble, industry scandals at Enron and WorldCom (among others) led to new regulations—notably Sarbanes-Oxley and options expensing—that crushed the underlying dynamic of places like Silicon Valley.

Thus Barrett was doomed to spend his entire tenure as Intel CEO in crisis mode, perpetually defending the company from these giant external threats. The only way to actually take the offensive during this period was on the path being cut by Steve Jobs, newly returned to Apple Computer, as a charismatic myth maker with a cult following, reinventing the company as he went. That wasn't Craig Barrett, . . . and that wasn't Intel, or at least it hadn't been Intel for twenty years. Andy Grove had kept Intel exciting, interesting, and more than a little dangerous. Now, for the first time in its history, Intel—at least compared to Apple and newcomers like Google, Facebook, and Twitter—looked old and boring.

Craig Barrett didn't help his case. During the bubble, his decision to hoard Intel's cash proved to be one of the most consequential in the company's history. After the bubble burst, Intel's stock managed to defy gravity for a while (thanks to that shift in investment that made it worth a half-trillion dollars) as other high-flying tech companies crashed. But eventually—roughly ten months later—it fell as well. The difference was that most of those other top tech companies went broke, or nearly so, and suffered devastating layoffs, cancelation of future product lines, and even restructuring. Intel, by comparison, remained largely awash in cash—and though he was widely attacked for it at the time ("I took a lot of shit for that," he told a reporter for the *Wall Street Journal*), Barrett decided to use that money to maintain Intel's usual high level of R&D and employment during the hard times of the postbubble, post-9/11 economy.[5] Time has shown Barrett's decision to have been a brilliant one. Andy Grove would call it the finest performance of Barrett's career. And once again, as the economy picked up, Intel had more momentum than any of its rivals.

But then Barrett made a decision that would haunt his legacy. He decided to take Intel, largely through acquisitions, into the communications business—out of Intel's core competence and almost a decade too early. Asked about it later, Barrett shrugged. "I bought high and sold low. But at least money was cheap in those days."[6]

By 2005, when Barrett stepped up to chairman and Grove officially retired (he remained as a "special adviser"), the market was all but screaming that the future was mobile—MP3 players, cell phones, and light portable computers. Yet even as Intel employees listened to their iPods and lined up (and even camped out) at local electronics shops for bragging rights to be among the first to own an Apple iPhone, Intel still seemed to see mobile electronics as a sideline, a niche market, and continued its focus upon the personal computing industry. But now a new generation of chip companies saw in Intel's distraction an enormous opportunity, . . . and they pounced.

This emerging challenge now belonged to Paul Otellini, Intel's new CEO. Otellini had at this point spent more than thirty years at Intel, his only employer as an adult. Smart, well liked, low-key, and confident, Otellini, who looked like a former athlete–turned–business executive, was cut from the Gordon Moore mold—perhaps not surprisingly, since both had grown up in the Bay Area and attended Cal (a quarter century apart). Indeed, Otellini was as pure a product of San Francisco and the

Peninsula as anyone in Valley history. He had always lived there, his brother was now the priest in charge of a large parish in Menlo Park, and Otellini was deeply immersed not only in the new Valley, but also in the old Santa Clara Valley. For example, he was a member of the area's Bohemian Club counterpart, the Family Farm, and heavily involved with Santa Clara University.

Otellini had risen through the ranks of Intel largely through solid competence, hard work, and the fact that he was almost superhumanly, consistently excellent at whatever assignment he had been given. He had been technical assistant to Andy Grove, had managed Intel's relationship with IBM, had served as executive vice president for sales and marketing, had run the company's microprocessor and chip-set business, and in 2002 had been named Intel's chief operating officer and elected to the board of directors.

At this point, other than Barrett, nobody at Intel had more wide-ranging executive experience than Paul Otellini. It seemed obvious that he would be Barrett's successor, and yet within the company, especially at the board level, this promotion was not so obvious. Why? Because Otellini was not a technologist.

When he talked about it years later to Santa Clara MBA students, he had developed a workable answer to that question. He would begin by noting that by the time you reach senior management at a company like Intel, most of your technical training is irrelevant anyway. Then he'd add, "In terms of a company like Intel and not having an engineering degree—I've managed this stuff for nigh on four decades now. I have found that sometimes the ability to ask the obvious question—people come in to me and say—I've got this great feature, and we've got this chip, it can do this—and I say great. So why would anyone want to buy it? What can they do with it? And sometimes, that grounding of reality has been the most important thing."[7]

It sounded good, but the reality was that in 2005, when Barrett prepared to retire to the chairmanship and the board began its search for his replacement, Otellini's lack of technical degrees weighed heavily against him. Most important, the person who resisted Otellini's promotion most on these grounds was Craig Barrett, the man to whom he had reported for the previous three years, and to whom he would have to report if he got the job.

This was not a happy position to be in. And until the phone call came one evening, Paul Otellini was uncertain which way the board vote would go—and whether, all things considered, it was a job worth taking. In the end, the call came, and Otellini accepted the job. In the years to come, with both Craig Barrett *and* Andy Grove second-guessing his every move—even in public—Paul Otellini sometimes wondered if he had made a terrible mistake.

In May 2005, when Otellini stepped up to CEO, Intel "was in a funk like never before." The situation was so bad that Intel couldn't even decide if it wanted to celebrate its fortieth anniversary. He had no choice but to swing the ax. Nearly twenty thousand employees were lost to layoffs, attrition, and the sale of business units. Managers (more than a thousand of them) took the hit first.

"It was miserable," Otellini would later recall. After all, he had worked with most of these managers for decades. "There were a lot of nights when I didn't sleep. I remember saying that this was not what I'd planned for my first year on the job."[8]

The marketplace reality that Otellini and Intel faced in 2005 was that, despite propelling the technology forward in all of its businesses, from microprocessors to flash memory, Intel had been caught nodding on almost every front. Had the company grown too old, too rich, too distracted by Barrett's other projects? Perhaps all of the above. And now new competitors were challenging in all of those markets.

The most surprising and scariest challenge came in flash memory. As already noted, Intel had built the first commercial flash chip in 1988 and had owned the market ever since—it would grow to $25 billion by 2012—using it as a reliable revenue source year after year.

Then in 2006, seemingly out of nowhere, the South Korean giant Samsung announced a new type of flash memory chip (NAND) so good that it instantly tore a large chunk out of Intel's (NOR-type) flash business. Intel, stunned, raced to respond . . . and quickly discovered that Samsung was a uniquely formidable competitor. For one thing, the company was huge and in numerous businesses; thus it was its own best customer. Moreover, being based in the Far East, it had unique access to the various game and consumer-products companies located there. Finally, and this made Intel executives visibly shake when discussing Samsung with outsiders, Samsung had a commitment to winning the market that almost seemed otherworldly in its ferocity, the kind of intensity that Intel hadn't known in a long time. Samsung even publicly announced its goal of surpassing Intel to become number one in chips. Intel quickly set up new operations in South Korea to build the relationships it had failed to do in the past, but as late as 2013, Samsung was still challenging Intel for the title of the world's biggest chip maker.

In microprocessors, the new threat was ARM, a Cambridge, United Kingdom, company that didn't even build chips but merely licensed the intellectual property it had developed for chips to be used in smartphones. ARM had spotted the opportunity in mobile early and created superb designs specifically for use in laptop computers, palmtops, and the new generation of cell phones. That was impressive enough, but even more important, one of ARM's founding investors was Apple, and that gave the company an inside track to be maker of the processor of choice for the greatest run of innovation in tech history. The iPod, the iPhone, and the iPad—each created new multibillion-dollar market categories, in which Apple (at least at first) had a monopoly.

Watching this, Intel could only kick itself. It had spent the last few years focused on WiMax wireless networking, only to see it killed by G4 wireless technology. It had developed a new, very-low-power processor family called Atom but had designed it for laptops and netbooks, not tablets and smartphones. Intel had blown it, and for a company that had driven the industry standard for forty years, to see it now held by other, younger, smaller companies was devastating.

"I don't feel badly about missing tablets," Otellini would say in defense of the company, "because everybody else did, too."[9] He would remind audiences that by 2012 there were fifteen tablets in the world shipping with Intel processors in them, but that it didn't matter because Apple still owned more than 90 percent of the market; by the end of 2010, 1.8 billion ARM processors were in use.

On the other hand, Otellini would admit, "On phones, I feel badly about it—because we had an early position. Fact, the first RIM [Blackberry] device had a 386 in it. Who knew? We didn't really do anything with phones."[10]

He told the *Atlantic Monthly*, "The lesson I took away from that was, while we like to speak with data around here, so many times in my career I've ended up making decisions with my gut, and I should have followed my gut. . . . My gut told me to say yes."[11] Eventually Otellini took to responding to questions about Intel's failure in the mobile business with "We're going to stay at it till we win it."

Whether this would prove a decisive strategy or just whistling past the graveyard, Otellini was playing a very long game, the results of which might not be known for a decade, long after he'd be gone from the company. In the near term, Intel would be fighting against a strong headwind. The Great Recession of 2008, and its attenuated aftermath, guaranteed that Paul Otellini's tenure, like Barrett's before it, would (financially at least) be a holding pattern, with Intel stuck for most of a decade with annual revenues just short of $40 billion. Only in the last three years of Otellini's leadership did Intel finally start growing again, in 2011 finally kissing $50 billion in revenues, only to slip back slightly the next year.

Barrett, who had joined Intel during a recession, retired in 2009 in an even bigger recession. He went back to his hunting, running his resort lodge, funding an honors college at Arizona State, and accompanying his wife, Barbara (a former US ambassador to Finland), in her quest to become an astronaut.

On May 2, 2013, a tired and relieved Paul Otellini retired as CEO of Intel after thirty-nine years with the company. The announcement, made at the end of 2012, had come as a unwelcome surprise, as he was still three years short of his official retirement date. His only official comment was that "it's time to move on and transfer Intel's helm to a new generation of leadership."

While he would remain a director of Google, Otellini would hold no more positions at his lifelong employer. In a final grace note to the leitmotif of his time as Intel's CEO, on the day he retired, Intel, which still held less than 1 percent of the mobile market, announced a new family of processors designed specifically for mobile applications. Brian Krzanich, who had been with Intel since 1982 and was currently the company's COO, was appointed Intel's new CEO.

Whatever the legacy of Craig Barrett and Paul Otellini, one fact would endure above and beyond everything else: both men preserved Moore's Law throughout their tenures as CEO. Thus they propelled the digital revolution forward one more human generation and eight more digital ones into the world of mobile computing, social networks, and three billion people around the world using the Internet to transform their lives. This was not inevitable, nor was it easy. Barrett in the *Wall Street Journal* at his retirement:

> "[Moore's Law] is something to hang in front of the bright, bushy-tailed new young graduates and tell them: 'We've kept this thing going for 40 years now, so don't screw it up'—and by God, they don't."
>
> Inevitably, Mr. Barrett says, every few years "some company will say, 'What's with the pell-mell rush to improve our technology every two years? Let's slow down to say, four years, and only have to invest half as much capital.' It always sounds like a cool idea, and it always ends up with that company losing market share."
>
> Mr. Barrett has personal experience. Early in this decade, Intel hung on to the Pentium IV microprocessor too long and watched smaller competitor Advanced Micro Devices (AMD) gobble up half the market. Mr. Barrett sent out a blistering, all-hands memo that still makes employees shudder. "We won't let that happen again," he says, with finality.[12]

Otellini, also in the pages of the *Wall Street Journal*, in 2008: "There is one goal Mr. Otellini is adamant Intel will reach during the next five years: the continuation of Moore's Law. 'I guarantee you that Moore's Law will not end on my watch,' he says with a shudder. 'Nobody in tech wants to be the guy who goes down in history for killing Moore's Law—and it sure won't be me.' "[13]

To Intel's perpetual chagrin, there is really only one photograph of the Trinity together, so the company regularly crops it or reverses it in a poignant attempt to find something novel. Other than that, there is also a grainy outdoor photograph, taken just months before Noyce's death, that shows a gathering of the six original Intel employees who still remained at the firm after twenty years. Noyce, looking shockingly old, stands in the back with Jean Jones, Tom Innes, Ted Jenkins, and Nobi Clark. Grove and Moore kneel in the grass in front with Les Vadász and George Chiu. All are squinting in the brilliant sunlight.

And so it comes down to the one professional photograph of the three men who led Intel to greatness. It was taken in the late 1970s, and it shows its age. The three men are posed behind a table in what appears to be a conference room, with classic wood-paneled aluminum-trimmed closet doors behind them. On the table before them lies a poster-size transparency of a chip design—perhaps one of the early microprocessors. In small reproductions, the transparency looks merely like a tile or marble design on the tabletop.

The three men are arrayed behind the table in a kind of triangle. Moore is on the right, name badge hanging from his shirt pocket, knuckles resting on the table. Noyce is in the back, hands hooked in pockets, wearing a loud tie that is too wide and too short for modern tastes. Both Noyce and Moore have their characteristic smiles.

Grove is on the left, visually and symbolically the odd man out. He isn't the indistinguishable Andy of the original Intel group photo, but neither is he yet the iconic Andy, business titan. The glasses are gone, but now he sports a big Mario Brothers mustache. Unlike the other two men in their dress shirts and glasses, Grove sports a turtleneck and a pair of rust-colored corduroy jeans.

But most telling is Andy's pose. Where the other two men stand straight and comparatively stiffly, Andy has flung his right leg up on the table, so that he half-sits while the others stand. The pose not only creates a low wall in front of Noyce, visually pushing him back to the rear of the shot, but Andy's knee points right at the groin of his boss. Andy appears to be laughing.

That is the photo with which Intel has had to work for all of these years. There are, in fact, more photos of the Traitorous Eight, even of Bob Noyce and Steve Jobs. And of course, there are numerous photos of Noyce and Moore, including some of the most iconic in Valley history, and there are endless shots of Moore and Grove taken over the course of decades. Unless they are buried in the Grove vacation photo albums, there are no shots of Noyce and Grove alone.

And so it comes down to this solitary photograph of the three men who built Intel. Thousands of Intel employees over the course of generations have looked at this photo during good times and bad and asked themselves what Noyce, Moore, and Grove would have done. The only answer this photograph, taken during one of Intel's darkest eras, gives is *be confident and fearless.*

As Intel approached its half-century mark, both of those traits seemed in short supply. It was still one of the most valuable companies on earth, a corporate giant employing tens of thousands of people in scores of countries, its overall dominance of the semiconductor industry intact, its role still central to the fate of the electronics industry. Still, the malaise that had almost paralyzed the company in 2005 remained and even deepened despite the efforts of Barrett and Otellini to counter it. There was a sense, especially in light of the inroads made by ARM and Samsung, that Intel had become too complacent, too careful, too *old.*

But companies, unlike people, don't have to age. Steve Jobs came back to a dying Apple, in far worse condition than Intel, and made it the nimblest company on earth. Hewlett and Packard, both old men, went back into HP—"the Great Return"—and restored the giant company, bigger than Intel, to its old energy. And Lou Gerstner Jr. took the oldest company in tech, IBM, and completely rewrote its business strategy on the fly.

As has been noted over and over in this book, Intel's greatest strength has never been its vaunted technical prowess, formidable as it is, but its willingness to take huge risks, even betting the company. Sometimes it has won those bets, crushed the competition, and changed the world. More often it has failed but then clawed itself back into the game through superhuman effort and will, . . . and then immediately gone on take yet more risks.

Intel had become the most important company in the world because of not just the technological genius of Gordon Moore, but also the vision and jaw-dropping risk taking of Robert Noyce and the business wizardry, intellectual flexibility, and superhuman energy of Andy Grove. The company now seemed to have forgotten the influence of the last two men, content to just keep moving the technology forward.

Bob Noyce had quit Fairchild because he couldn't bear bureaucracies, slow decision making, and most of all, safe and conservative practices. For him, business was the Great Game, to be played with every bit of artfulness and skill he could bring to it. For his part, Andy Grove fired or demoted people for *not* taking enough risks and for devoting themselves not to winning but to preserving the status quo. For Andy, business was war—brutal, unforgiving, and the ultimate way to assert one's identity and place in the world.

Most of all for the three men of Intel's Trinity, even in the worst times, working at Intel was *exciting*. It was the most engrossing thing in their already remarkable lives. And that excitement—of competition, of advancing technology, of transforming the world, and most of all of being part of the Intel family—made it worthwhile to come to work each day even though they were already living legends and billionaires.

As it approached its golden anniversary, Intel seemed to have forgotten this excitement or come to believe that it belonged to the company's younger, wilder days, when giants walked its hallways. Now the task was to protect the company's assets, to guard the markets already captured, to not move too quickly after new business opportunities, and most of all, not to risk and perhaps lose the legacy of the great men who had built the company.

And yet walking those same hallways and sitting in one of those thousands of cubicles and offices was the man or woman—or perhaps a duo or trio with the right mix of skills—who could take Intel not only to its next level of success after a decade of near stasis, but to its old level of excitement and glory. But that would never happen until the leadership of Intel, not least its board of directors, was willing to take the risk, as Bob Noyce and Gordon Moore did every day, of letting an Andy Grove run the company.

Intel was a company built for glory. And even now, the greatest part of that glory still remained beating at the heart of the enterprise, as it had during the time of the two founders, as it had when being carried on with unmatched skill by Andy Grove, and as it had while being actualized by tens of thousands of committed and tireless Intel employees over all of those years, in good times and bad: Moore's Law.

At its founding, Intel, the world's most important company, had made a tacit promise to the world to uphold and preserve Moore's Law into the indefinite future. Intel had never faltered, and as long as there was the tiniest hope at the furthest limits of human effort and imagination, it never would.

Through it all, the few good times and the many bad, the employees of Intel, from the founders to the newest interns, had carried the flame of the law forward. This ferocious, uncompromising, *heroic* dedication to the law made Intel Corporation the most singular of organizations, a company of greatness. More than anything on the balance sheet or in the company product catalog, this was Intel's greatest achievement. This was the mark Intel would leave on history.

Now the company faced its own reckoning. Could it find, in the flame of Moore's Law, in the courage and vision of its founders, and in its remarkable history of brilliant successes and brilliant failures, a new Intel worthy of standing with the old, a new leadership unafraid to step out of the long shadow of the Trinity?

Epilogue: Roaring at the Night

A ndy Grove, age seventy-six, sits in his office, a scarred old lion in his lair.[1]

Despite the fact that his name is still often in the news—especially now that the promotion has begun for the forthcoming premiere of the PBS documentary—he is a difficult man to find. The man who once led the most important company in the world has traded his legendary cubicle for an equally tiny office, with an even tinier reception area for his longtime secretary, Terri Murphy, up one flight of the atrium stairs of a small two-story, six-office professional building among the precious boutiques and antique shops in downtown Los Altos.

On the sidewalk, few people recognize him, though his is one of the most famous names in Silicon Valley history. One reason is that the Valley is so intensely focused on the future that it rarely looks back and almost never celebrates its founders.

Bob Noyce, for example, the man who made it all possible, the archetype of all of those thousands of entrepreneurs in the Valley and millions around the world, is almost forgotten just twenty years after his death. His widow, Ann Bowers, waited too long to allow his biography to be written. And though she found the perfect biographer in Stanford's Leslie Berlin, an entire generation had grown up knowing only that Noyce was a mentor to their hero, Steve Jobs. When Jack Kilby received the Nobel Prize in Physics in 2000 for co-inventing the integrated circuit, he publicly mourned that Bob Noyce couldn't be there to share it—and the media, including the *San Jose Mercury-News*, had to explain who Bob Noyce was and why he was important.

But Andy Grove is also less recognized now, even among old Valleyites, because he no longer looks like Andrew S. Grove, master of the semiconductor universe, the confident, hard-eyed figure in a black turtleneck who stares out at the reader of *Paranoid*. Parkinson's disease, now in its late stages, has twisted Andy's face, pulling his jaw off to one side, canting his head, and as it destroys his large muscles, it has reduced his gestures to sudden, oversize, theatrical jerks.

But his eyes haven't changed. They are as ferocious, and even terrifying, as ever—not pitiless, like the stuffed cobra among all of the Intel-era tchotchkes that fill his office wall alongside the photos of his grandkids, but those of a man spoiling for the excitement of a good intellectual fight. With his bantamweight size, Andy was never a physically threatening figure, though the shouting and swearing and table pounding were meant to convince you otherwise.

But at verbal fisticuffs, which he loves as much as anyone ever, he is unbeatable. He would rather die than lose an argument, and unless you are feeling the same, you'd better not challenge him. So you go into any conversation with Andy Grove knowing that it will become a debate and that he will win. He will win because he is smarter than you. And if that doesn't work, he will win because he knows more than you. Or if not, because he's cleverer than you. Or because he will work harder than you. Or because he is willing to make decisions you are afraid to make. Or he will reset the terms of the debate. Or, if all else fails, he will make it personal and just emotionally run over you, shouting and browbeating you into submission. Whatever it takes, he will win, because that is what it means to be Andy Grove. He will do all of these things to win because he believes he is right and because he believes his being right is the natural order of things.

If later you prove him wrong, he will do the most amazing thing. He will accept your argument, and he will apologize. But that will be no advantage to you in your next encounter. Meanwhile, for now, all you can do is fend off as many punches as possible and hope you are still standing when the bell rings.

It is shocking at first to see Andy's physical transformation, especially if the image in your mind is that of a fast-moving, crisp, and precise figure, forever going with a purpose down Intel's hallways, barking orders, laughing explosively, throwing off wry comments. But that shock quickly disappears the moment the conversation begins. Clearly, growing older and sicker has not made him softer. "I choose to face reality," he says, with that Hungarian accent that still hasn't faded after more than a half century. "I don't think I'm growing more compromising on what I think."

One of the opinions he won't compromise on is his opinion of Intel's current state. He makes it abundantly clear that he isn't happy that the company isn't what it was or what it could be. One can only imagine what it must have been like to have been Craig Barrett or Paul Otellini on the receiving end of Andy Grove's opinions.

One of the remarkable aspects of Andy's illness is how long he has had to endure it. In 1993, the Grove family doctor retired, and when the family found a new one, he ordered physicals for all of them. That's when Andy's elevated PSA count was first discovered. Subsequent tests found prostate cancer. The doctor gave Grove three choices: radiation, surgery, or nothing and rely upon the disease's slow growth. Being a scientist, Andy did his own research, made himself an expert—and an impatient one at that—and opted for high-dose radiation.

Keep in mind that all of this—prognosis, treatment, recovery, and aftermath—occurred from 1993 to 1997, during one of the most intense periods of Andy's career. The Pentium introduction, the beginnings of the dot-com boom, Intel Inside, globe-trotting, standing onstage with Nelson Mandela, and most overwhelming of all, the Pentium bug crisis all took place during this period, at a time when many patients radically cut back on their commitments and become near invalids. At Intel, only his administrative assistant, Karen Thorpe, knew of the diagnosis. And incredibly, Andy only lost three days of work.

In fact, though eventually the people around him saw the real cost of the illness and its treatment, the announcement of Andy Grove's cancer passed by quickly and uneventfully. It was a subject of considerable discussion for a few days, but it quickly passed from memory. Two years later, when he announced his retirement, only a single publication even mentioned the possibility of Grove's illness playing a role in his departure. But the reality, as he told Thorpe, was that he could not both run Intel and become a leader in the fight against prostate cancer.

As soon as he retired as CEO in May 1996, Andy threw himself into this new challenge. Already, in a stunning act of self-exposure, he penned a cover story about his cancer for *Fortune* magazine. Here is how it began:

> My secretary's face appeared in the conference-room window. I could see from her look that it was the call I was expecting. I excused myself and bolted out of the room. When I stepped outside, she confirmed that my urologist was on the phone. I ran back to my office.
>
> He came to the point immediately: "Andy, you have a tumor. It's mainly on the right side; there's a tiny bit on the left. It's a moderately aggressive one." Then, a bit of good news: "There are only slim odds that it has spread." The whole conversation was matter-of-fact, not a whole lot different than if we had been discussing lab results determining whether I had strep throat.
>
> But what we were talking about was not strep throat. We were talking about prostate cancer.
>
> Let me start at the beginning.

That single piece likely had more impact than all of the articles Andy would have written if he had stuck to a career in journalism. It was, as always with Andy Grove, brutally honest, opinionated, and in his assessment of the current state of medical care, controversial. Intel PR, fearful of damage to company stock, didn't want him to do it, but Andy plunged ahead. The effect was electric. Soon his office was flooded with calls from friends offering their condolences—and even more from other men suffering the same disease or fearful that they would be. The message to the callers was the same: get a PSA test, or if you have the disease, learn about it and understand your options. In the end, Andy's public confession likely saved uncounted lives.

By the turn of the century, Andy's PSA count was back to normal. Though the cancer could reappear at any time, a decade later it was still gone. Once again, Andy Grove had met a challenge and through a smart strategy and unrelenting commitment to success—and a willingness to challenge the status quo no matter whose pride was humbled—had triumphed. He had beaten cancer.

But now even as he defeated one mortal threat, he was attacked by another. In 2000, while on a walk with an old friend who happened to be a psychiatrist, the doctor noticed a slight tremor in Andy's right hand and suggested he get it checked.

It was Parkinson's disease. No cure, only a steady decline that might be slowed using some new drugs and other methodologies. It could not have come at a worse time. Parkinson's, with its devastating diminishment of physical powers, often brings with it deep depression. And bad news and difficult changes seemed to come from every direction during those years. A doting father, he was now an empty nester with his two daughters grown and gone. Then Eva's mother died, leaving her in deep mourning. Finally, in 2002, Andy's mother, Maria, the heroic and sacrificing woman who had been his savior and his rock during the most terrible years of the twentieth century, died. The growing numbers of his grandchildren were a solace, but he would have to face this terrible disease with less support than he had ever known. He had gone public with his prostate cancer, but now, with Parkinson's, Andy went private, in part because revelations of his illness could seriously reduce the value of Intel's stock.

Still, he refused to buckle but once again turned to attack his biggest challenger, if only to beat it to a temporary draw. He was a wealthy man, though his estimated $300 million net worth paled next to Gordon's (the difference between being a founder and an employee), and now he and Eva put it to work to find a treatment for the disease. They created the Grove Foundation and used it fund the Kinetics Foundation (to keep their name out of the news), which supports Parkinson's research around the world. Meanwhile, in his own life, he used his powerful mind and will to cope with the changes taking place in his body. He put the glasses he hadn't worn in twenty-five years back on, took speech training, used speech recognition software instead of typing, and restructured his life and workday to work around his growing limitations. He also participated in tests of experimental drugs, such as L-dopa, some of which dramatically, though temporarily, improved his symptoms.

All of these efforts culminated in Andy Grove's speech on retiring from Intel in 2005. Everything was carefully choreographed to diminish the appearance of Andy's disease. Compatriots old and new were there, as was Eva. It was part of the company's annual sales and marketing conference, an internal gathering with more than 4,500 employees in attendance—most of them from field sales, a group that had been among Andy's biggest fans. They cheered almost continuously. Craig Barrett gave a memorably passionate introduction, choking up at the end. And then Andy rose to make a few remarks. He thanked Intel's employees for the chance to be mentored by them, he extolled the virtues of Intel's new CEO, Paul Otellini—and for those few minutes it was the Andy Grove of old. It was an unforgettable performance . . . but one Andy knew he would never again be able to duplicate.

While Andy Grove fought his battles, Gordon Moore essentially retired to the honorary title of chairman emeritus, a job with no real duties other than to continue being Gordon Moore. And Moore was ready to leave. Betty's health had deteriorated, and she now lived almost full-time in Hawaii. Gordon, always inseparable from her, found the shuttling back and forth wearying. Retirement allowed him to spend more time with her.

And yet if Moore was less a physical presence in Silicon Valley, spiritually he was more present than ever. For one thing, there was the law. By the beginning of the twenty-first century, historians had begun to appreciate that Moore's Law wasn't just a measurement of innovation in the semiconductor business or even the entire electronics industry, but it was also a kind of metronome of the modern world. Moreover, with such a remarkable yardstick, they could even obtain a glimpse of the future—e.g., Ray Kurzweil's theory of the impending singularity, a merger of human beings with their machines.

Spoken or unspoken, the recognition of the power of Moore's Law underpinned everything from government planning to military weapons development to education and stock-market investing. It was implicit in every TED talk, Davos gathering, and science-fiction movie. And from there it echoed across the culture: twenty-first-century Americans now expected a world of perpetual, rapid change in which their tools and toys would become obsolete every couple years. Changes that in the past would have seemed miracles—mapping the human genome, 3-D television, entire libraries on handheld tablets, automobiles driving themselves, earning a college degree online, storing the memory of every minute of one's life—now were accepted with a shrug, even frustration that it all wasn't progressing even faster. More even than Intel, this was Gordon Moore's greatest legacy, maybe the greatest legacy of anyone of his generation, and he knew that it would outlive him.

Now that he was retired, there was one more mark that he and Betty wanted to leave on the world. He was immensely rich—in 2000, *Forbes* put his wealth at nearly $12 billion—as only the founder of a Fortune 50 company could be. In fact, after Oracle's Larry Ellison, Moore was the richest man in California and one of the fifty wealthiest in the world. Four years later, *Business Week* named Gordon and Betty the most generous philanthropists in the nation, ahead even of Bill and Melinda Gates, the world's wealthiest couple. In between, the Moores had created a foundation in their name, headquartered in Palo Alto (so Gordon could travel easily back and forth to Hawaii) and endowed it with a jaw-dropping $5.8 billion. This amount, combined with other giving (such as a total of $600 million to Gordon's alma mater, Caltech, part of it to build the world's largest optical telescope), totaling more than $7 billion, represented more than three-quarters of the couple's net worth—a percentage of giving that dwarfed that of any of the other billionaire philanthropists on the list. *Business Week*:

"The Moores, like a growing number of big givers, take a businesslike approach to philanthropy. Rather than throwing money at problems, they try to ensure the most productive use of their dollars by funding projects they believe can produce 'significant and measurable' results."[2]

These endeavors have included work in environmental conservation (by such as the Center for Ocean Solutions), science (the Caltech programs), patient care (the Betty Irene Moore School of Nursing at UC Davis), and close to the heart of the Pescadero boy, the Bay Area's quality of life. Gordon, who was deeply involved in the selection of grants, brought the same talents that he had used all of his life: the ability to discern, as with integrated circuits and microprocessors, precise changes that could scale up and possibly change the world—and of course, projects with precisely measurable effects. In his own way, he was bringing his law to the world.

As the years passed, Gordon never seemed to age. His hair grew grayer, but his soft voice and gentle manner, combined with his natural reserve, had always made him seem a bit of a wise old man. Even when he became a little more stooped and began wearing hearing aids in both ears, he still seemed the Gordon Moore of Shockley and Fairchild and Intel. But he was now in his eighties, and because he had always lived so quietly—Noyce got his biography, Andy got his, too, and wrote his own memoir just in case; but Gordon refused all such overtures, preferring his last document to be a monograph for the Chemical Society—the public, even Intel employees, never knew of his declining health. Perhaps he felt no need to defend any of his actions. After all, he was the richest man in the world without enemies.

In the spring of 2010, a runaway infection put Gordon in the hospital and nearly killed him. His family asked Intel not to announce the news, but to be prepared; the company prepared an obituary. By summer's end, Moore had rallied, and by autumn 2010, he was back attending meetings at the foundation. Still, it was a reminder that, unlike his law, Gordon Moore was mortal and that Intel's remaining time with him was precious.

The state of Andy Grove's health, though less precarious, was also more obvious. In the late spring of 2012, as he prepared for his interview for the PBS documentary, he knew that his appearance would shock not only viewers around the country, but even more his old employees, who had seen little of him over the last decade. But he refused to let the world's response stop him. Parkinson's disease was just another bug, like all of the others, to be met head-on and without fear. "You have to face reality," he will later say.

The filming threatened to be exhausting. The camera crew had taken over a conference room in Intel's Robert Noyce Building, just around the corner from the lobby and Bob's photo. The darkened room, with its tiny limbo set in one corner, was one giant obstacle course of cables and equipment, and there was some concern that Gordon, and even more Andy, scheduled just after, would have trouble negotiating the passage.

Each interview lasted almost three hours, exhausting work for both men, and more than once the memories provoked strong and long-buried emotions. In between the two shoots, the two old men—boss and employee, mentor and student, partners and friends for more than fifty years—had a warm reunion. Andy would later say that the meeting—the two of them hadn't seen each other in months—was especially resonant because he feared (and likely Gordon did too) that it would be the last time they would ever meet.

Happily, that wasn't the case. And now, six months later, as he sits in his little office, Andy says he is looking forward to seeing Gordon again at the upcoming premiere. For now, the matter at hand is jousting with an author, winning the conversation—at one point he glares and announces, "I thought you were a real asshole back in the old days when you were writing about Intel"—and trying to control what will be the key chapters in this book. He's already prepared his own list of what those chapters should be.

It is also important to Andy that the world understand that his reputation as a tough guy is vastly overrated and that, for example, Craig Barrett, whom he calls "the most underrated of CEOs," was far scarier. "I didn't intimidate people that much. I was just loud."

He has also thought a lot about the culture of Intel and what he did right and wrong in cultivating it under his leadership. "There were certain 'soft spots' that I just couldn't accept," he says, "and I punished anyone I caught: politicking, exhibiting sloppy thinking—Gordon was even harder on that—lying or [shading the truth] to be clever, giving up, or talking to people in a way they wouldn't want those people to speak to them." After a half century in corporate leadership, it has been distilled to this.

As the conversation continues, Andy softens. Conducting a long conversation now has become an exhausting experience. He wants to talk about the past. He reminisces about the keynote with Nelson Mandela in 1995, clearly one of the high points of his life. And he talks about the Pentium bug and how in the end, over his kitchen counter, Dennis Carter ("I adored him because he always spoke his mind") finally convinced him to accept the inevitable. "I yelled, cried, cajoled, . . . and in the end, apologized."

Behind all of the bravado and self-confidence, was he ever scared? "Yes," he replies, "but I learned to use that energy to pump myself up."

Finally, his thoughts go back to the beginning, to Fairchild, the birth of Intel, his enduring relationship with Gordon, and his battles with Bob Noyce.

"Gordon did impress me at the beginning," Andy says, "but more than that, he always stood by me. Bob impressed me, too, but there was so much I didn't like about him. His charisma put me off. His management style put me off. His inability to make decisions put me off. So did his unwillingness to actually learn the business. I didn't like those things in him that the world most admired him for.

"But I can remember, in 1971, when we were at the Stanford Chalet with the Noyces on a skiing trip. It was time to go home, . . . and it started snowing. We didn't know what to do. And Bob just crawled under my car and put the chains on, while me and my wife and our daughters just stood there, watching helplessly. That was the best of him. So was his risk taking, his impressive physical courage, his intellectual clarity. That was the part of him I loved, not all of the famous stuff."

His rigid face can no longer betray his emotions. But in his eyes, tears begin to well. "After all these years, I miss Bob the most."

Appendix: A Tutorial on Technology

Digital electronics can be confusing not only to people outside the industry, but even to folks inside it. This has become increasingly the case as the years pass and ever more new layers of products and services are piled atop the existing core.

Ask someone who works in online gaming or social networking how an integrated circuit works—a component upon which that person's job, employer, and industry depends for survival—and you are likely to get a blank stare and perhaps something mumbled about "silicon" and "semiconductor." This isn't surprising: after all, the worlds of social networks and code writing are six or seven levels removed from the largely chemical business of making computer chips. It would be like asking someone preparing a Big Mac at a McDonald's in Prague about cattle feed in Tulsa.

This is also true in the media, even in the trade press. The reporter who writes incisively on, say, Apple and its next generation iPhone may have little knowledge about the chips inside that device, other than perhaps the name of the manufacturer of its central processor and perhaps its memory chips. That's why you read very little these days, other than financial news, about the semiconductor industry. The men and women who once covered the business have mostly retired, and the new generation of technology and business reporters are much more comfortable writing about Twitter or Facebook.

That's a pity. It is precisely because those companies (and the thousands more like them) depend for their existence upon the Internet or cellular telephony, both of which rest upon semiconductor components, that the semiconductor industry is more important to the modern global economy than ever before. Unlike their predecessors, these journalists aren't following the twists and turns of the semiconductor book-to-bill (orders-to-deliveries) index, so they are consistently caught by surprise by technology industry booms and busts.

For that reason, let's take a brief lesson or refresher on the computer chip, the integrated circuit: how it is made, how it works, its many types, and how they are used.

The Silicon Family Tree

Electronics, as the name suggests, revolves around the management of individual electrons (as compared to the older *electricity*, which deals with the flow of very large numbers of electrons).

At the heart of all electronics—indeed, the reason for the science of electronics—is the *switch*, and the story of electronics is ultimately the story of the evolution of the switch. The most primitive switches are *mechanical*—think of a swinging door or the big switching device on train tracks or even the clutch on a transmission. All essentially turn things on and off.

An *electrical switch*, like the light switch on a wall, turns an electrical current on and off. So far, we are at the state of the art circa 1840. The next great breakthrough was the *electromechanical switch*—think of a telegraph key—which uses an electrical current to create a magnetic field that pulls a mechanical switch back and forth far more quickly. Still, you need to physically move something, and moving parts are slow, get hot, and wear out quickly.

Hence the invention in the early twentieth century of the first true electronic device, the Audion vacuum tube. This device was essentially a light bulb with its internal vacuum, but instead of running electricity through a filament wire glowing with current, the wire was replaced with an *emitter*, which spewed electrons across a space to be collected by a *receptor*.

The crucial thing about the vacuum tube was what you could do with that spray of electrons. For example, you can put a second emitter in the middle from another source to add electronics to the spray (an amplifier), dampen the spray in a controlled way (with a resistor), or change its frequency (remember: electrons are both particles and waves—the oscillator). Or you can take out the target collector and instead shoot the electrons the length of the tube, where they will strike and excite a light-sensitive material painted on a flat end, . . . and you get television and radar.

By the end of the 1920s, vacuum tubes were everywhere, most famously in radios. But with time and use, their limitations became clear, and those limitations were manifold. For one thing, they were fragile. That was bad enough on a table in the family living room, where the radio might get bumped or knocked to the floor. It was infinitely worse in the harsh, rattling conditions of a fighter plane, tank, or ship. They also ran hot—put them too close together or in an unventilated area, and they could burn themselves out. And, since that heat is just lost electrical energy, that meant that tubes were also comparatively expensive to run. Most tellingly, the very success of tubes only fueled the demand for even faster performance that tubes could not achieve. The result, epitomized by the gigantic ENIAC computer, the size of a building, hot as an oven and sucking power from the grid, burning out a tube every few seconds, underscored the fact that the tube era was coming to an end.

Thus the epiphany of John Bardeen and Walter Brattain when they saw the demonstration of semiconductor materials. *Here* was a way to reproduce the performance of the vacuum tube in a solid material—hence "solid-state electronics." Semiconductors, with their property of conducting a main current that could be turned on and off by a smaller, perpendicular side current, were just perfect for switches.

Transistors had no moving parts, nor were they fragile, hot power wasters. Largely because of those features, transistors could also be made much smaller than tubes and could even run on batteries. They offered the potential for a whole new world of portable consumer electronics—its emblematic achievement the hugely popular transistor radio.

The transistor, in its ultimate incarnation as a tiny metal pot on top of a tripod of "leg" contacts, was revolutionary. All the electronic instruments that had been previously developed could be upgraded by replacing their tubes with transistors, and the devices themselves were more rugged, for applications like avionics and rocketry.

Moreover, the existence of these tiny, low-power switches—and later amplifiers and other types of circuits—opened the door to all sorts of new inventions that had been all but impossible with tubes. Those new inventions would spark huge new industries, each of them bigger than the entire electronics industry just a few years before. Home audio systems, reliable car radios, affordable TVs, the ubiquitous transistor radio—and most important, computers.

Computers had already been around for a decade, but they were largely electromechanical, were limited in speed, and broke down often. The big leap came in the United States with ENIAC and its vacuum tubes. Though the first transistor-based computers didn't appear until the midfifties, it was obvious even at the beginning of the decade that solid-state would be the future of computing. Meanwhile, both the Cold War and—beginning with Sputnik in 1957—the space race all but guaranteed that government demand for transistors would be almost limitless for the indefinite future.

That's why the transistor took off so quickly and was such a landmark invention. It was an order of magnitude faster than a vacuum tube, an order of magnitude smaller (and getting much smaller than that), it would soon be cheaper, and you could drop it off a building (or soon, fire it into outer space) and it would still work. With the transistor, the recipe for the modern integrated circuit was in place. And though the pathway was hardly obvious and it took the contribution of a few geniuses and thousands of smart engineers along the way, it was essentially a straight line over the next twenty years from the transistor to the IC to the microprocessor.

From Analog to Digital

Now we need to follow a second historic thread. Don't worry, the two will soon intertwine.

The natural world is continuous because time and space are continuous. Everything is curved; straight lines almost never occur. Objects are essentially edgeless. Time never stops. We experience the universe as a continuum.

But what if you want to *measure* the natural world, then manipulate the information you gain from those measurements? It proves to be very difficult. Because the world is continuous, events and things appear as waves of different amplitudes, so the information you capture is also wavelike; it is *analog*. But analog data are hard to manipulate, and the useful signal is hard to distinguish from the surrounding noise.

However, if you measure the natural world not by its behavior but by whether something is there or not, events are easy to spot. Then if you make a whole lot of individual measurements really fast, you can also get a sense of the shape and magnitude of the event. This is looking at the world from a *digital* point of view. Digital data have the advantage of being easier both to capture and to manipulate. The disadvantage is that they will always be—like calculus—an approximation of reality. But like calculus, if you take enough samples you can make that approximation pretty darn close.

That's why there were few applications for digital data until the twentieth century. It was taking the temperature twice a week and coming up with an average; it wasn't that useful. But once those vacuum-tubes and transistor switches came along, it was possible to make those measurements ten times, a hundred times, and today, almost a trillion times per second. It is still an approximation of the natural world, but so close that the difference doesn't matter (and soon, when the measurement speed crosses the threshold of the briefest events in the universe, it *really* won't matter).

Hence the rise of digital electronics: here was a way to measure and manipulate discrete data so fast and in such volume that the result could be applied without fear to any human activity (even spaceflight) and any natural event. That reality is the underlying truth of Moore's Law: that is, every couple of years digital technology becomes twice as good at capturing and replicating reality. And each time that happens, more new products, businesses, and industries become possible. That's why every smart businessperson and scientist over the last thirty years has done everything he or she can to get aboard the Moore's Law express, even if, as in the case of the Human Genome Project, only part of the industry (in this case, the empirical half called genomics) can climb on.

Still, even with the transistor, the process of capturing data about the natural world was not yet complete. It wasn't enough to simply reduce the natural world to arithmetic. You can't just tell a machine to add two numbers, even if you explain to it how addition works.

Happily, the solution was already at hand. Boolean algebra, devised in 1854, was designed to be a mathematics of values, in which true corresponded with one and false with zero. It turned out to work just as well to make 1 and 0 correspond to a switch—in the chip era, Federico Faggin's transistor gate—being on or off, open or closed. In Boolean algebra, any number, letter, or symbol could be converted to a series of 1 or 0 *bits*, strung into *bytes* of anywhere from 4 to 128 bits for ever greater precision (hence 8-bit to 128-bit processors, the latter capable of much greater functionality).

The combination of transistor technology, especially in the form of integrated circuits, with Boolean algebra kicked off the digital age, of which we are still a part.

The process of taking all of the applications found for vacuum tubes and applying them to transistors (along with a whole bunch of new applications) was only partially completed when the integrated circuit came along and started the transferral process all over again. What Jack Kilby's idea, Robert Noyce's design, and Jean Hoerni's planar process accomplished was to make the transistor simpler, easier to produce, and most of all, *scalable,* i.e., replicable in large volumes—and better yet, on a single chip. Now the race was on to apply this new integrated-circuit technology, or at least its underlying process, to a panoply of electronic applications. Pretty quickly, the semiconductor industry broke into pieces in pursuit of these different opportunities.

There were basically three directions they could go. *Discrete* devices were continuations of the stand-alone transistor lines they were already building, such as light-emitting diodes for control panels. *Linear* devices used semiconductor technology to build analog chips, such as amplifiers for high-end audio systems. This was where Bob Widlar showed his genius. Finally, there were *integrated* devices—the integrated circuits that, by the billions, changed the world.

There were also several paths to take in terms of what to build these chips from and how to fabricate them. Most of the early transistors were made from germanium as the doped insulator. Germanium is especially resistant to shock, radiation, and heat, which made it particularly appealing in aerospace and military applications. Unfortunately, germanium crystals proved difficult to grow without impurities and to a sufficient diameter to be cost effective to slice up into wafers and then chips. That's why the commercial chip industry settled on silicon, which today is grown in cylindrical crystals up to fourteen inches in diameter.

Just because silicon has dominated the chip industry for the last thirty years—and defeated other contenders, such as artificial sapphire—doesn't mean that it will retain that throne forever. Nanotechnology could make questions of purity moot and even eliminate the need to grow giant crystals.

As for how you build integrated circuits, as we've seen, there are two primary methods, defined by the order and form in which the silicon, nonsilicon (epitaxial), and metal layers are laid down. The *bipolar* method is intrinsically faster, and the product is more resistant to heat and radiation. The metal-oxide semiconductor (MOS) method produces a chip that is more fragile but allows for greater integration, is cheaper, and is easier to fabricate in multiple layers. MOS won that race (and drove a number of competitors, like Bob Noyce's old employer Philco, out of the business in the process), but bipolar has never really gone away because it fills a necessary niche. As you'll remember, at its founding, Intel was expected to be a bipolar company, because Fairchild was, but then fooled everyone by becoming a major pioneer in MOS, leaving most competitors behind.

The next question facing new chip companies, in particular the Fairchildren, was: what kind of digital chip should we build?

Again, there were several choices. *Logic* chips perform operations on the incoming data as determined by the *software* instructions running the computer or other system. Together they constitute the *central processing unit* (CPU). A classic logic chip was the TTL (transistor-to-transistor logic) chip.

Memory chips come in two forms: *RAM* (random-access memory) chips hold large quantities of data for an extended period of time. They are the semiconductor equivalent of disk memory, which in fact they have largely supplanted in all but the largest memory-storage applications ever since the second-generation Apple iPod, which removed the tiny magnetic disk and replaced it with flash-memory chips. RAM chips have not only undergone a mind-boggling expansion in their capacity over the last fifty years (they were the basis of Gordon Moore's original graph), largely because they are the easiest chips to build, but they have also evolved architecturally, from *static* RAMs (SRAMs), which require as many as six transistors per chip and can retain residual memory even after erased, to *dynamic* RAMs (DRAMs), which have unmatched density but lose their memory once turned off, to *flash* memory, which is actually a read/write form of read-only memory that finds universal application because of its ability to be easily erased and rewritten and to retain memory without requiring a power refresh. That's they are popular in memory sticks, aka thumb drives.

ROM (read-only memory) chips are traditionally those that can be read from but not easily recorded on. Typically, ROM chips hold the operating memory of the system, including the set of instructions by which it does its various tasks, including rules for operating on the data coming from the RAM and external memory devices. The challenge facing ROM chip makers has long been: how do you change those instructions once the chips are in the machine—without taking them out and replacing them? That led to the invention of ROM chips that could be erased (using such techniques as the application of ultraviolet light) and reprogrammed on the fly.

In the minicomputers of the 1960s and 1970s, all of these chips could be found in abundance on the main printed-circuit *motherboard* of the computer. In even greater numbers, other chips helped manage power and the movement of data into and out of the CPU.

Then the engineers began asking new questions. As these chips become smaller and more powerful, thanks to Moore's Law, why do there have to be so many? More important, why do they all have to remain *monolithic*, each one performing only one function? If we can integrate one type of circuit, why not different kinds?

Thus in the late 1960s, work began on taking the functions of almost all of the chips on a computer's motherboard and moving them to different regions on the surface of a single chip, with the metal channels on the surface replacing the interconnected wiring on the printed circuit board.

The result was the Hoff-Faggin-Shima-Mazor *microprocessor*, of which Intel was the sponsor and more than forty years later still is the protector and greatest developer. Happily for the sake of this narrative, the story of Intel—after a decade in the complicated world of memory and even logic chips—largely revolves around this microprocessor, the supreme invention of the semiconductor industry (and arguably of any modern industry). Because of that, the story shifts early from advancing across the ever broader front of the entire spectrum of semiconductor devices to a series of regular and predictable generational advances of new microprocessor models: the 286, 386, 486, Pentium, etc.

At least that was true until the middle years of the 2000s. Then two forces emerged that caused even Intel's line of microprocessors to itself divide and move forward on several fronts. One of these was cost: it simply became too costly—and the number of customers too few—to justify continuing to build monolithic (in this case meaning a single microprocessor per chip) processors. The other was market schism: at the high end, the Internet infrastructure companies, such as Cisco, and supercomputer makers, still wanted the most powerful processors available. The powerful chips that Intel (it dominated this world) built for these companies, such as its Itanium line, continued to track with Moore's Law.

On the other end lies the mobile market, which is willing to sacrifice some performance for the sake of size, price, and most of all, low power consumption. Here, ARM has stolen the lead, with Intel struggling to catch up.

Because microprocessors and flash memory get all of the attention these days, it's easy to forget that all of those other integrated, discrete, and linear devices—many of them with roots dating back to the nineteenth century—are still being fabricated in abundance, in most cases more than ever before. Chips are still there at the heart of every electronic device. We've just stopped looking for them. And waiting in the wings, likely to appear sometime over the next couple decades are radically new forms of switches—single transistor gates, molecular switches, quantum dots—that may usher in the era of postdigital electronics.

A Note on Sources

Anyone who attempts the history of a giant corporation that is a half century old faces the inevitable problem of weighting eras and subjects. Which is more important: the fabled beginning of the firm, when it had great ambitions and few successes? The scrappy middle years, when it is clawing its way to the top? The golden age, when the company becomes a less interesting but more influential colossus? Or the mature years, when the company's true character shows through?

It becomes even more complex when you try to tell that story through not one or two founders, but three—all of them very different in personality and not even necessarily liking one another. Finally, there is the challenge of writing about a technology company. How deep do you go into the arcana of bits and bytes, silicon and software, transistors and teraFLOPS, without losing the average reader or insulting the tech-savvy reader?

Telling the story of Intel and its trio of founders presents all of those challenges and more. Luckily, the company and its founders have been blessed with the attention of some very good biographers and historians, not to mention the literary skills of one of its founders.

The two works that everyone turns to—and will for generations to come—are Leslie Berlin's *The Man Behind the Microchip,* her definitive biography of Bob Noyce, and Richard Tedlow's equally exhaustive *Andy Grove: The Life and Times of an American*. No one can write about those two individuals without paying considerable homage to, and drawing heavily upon, those two works. It is impossible to give these authors enough credit for their achievements, though I have tried throughout this book.

Tedlow's achievement is particularly impressive because he managed to deal with such a powerful personality as Grove's and still emerge with a strong, independent narrative.

But Berlin's achievement is even greater because, despite never having met Noyce and finally getting the chance to write the book only after much of the world had forgotten Bob, she managed the herculean achievement of finding and interviewing the key figures in Noyce's life—a project she began with her dissertation and continues even now at Stanford University—and constructed a narrative so complete, so accurate, and so compelling about this elusive figure that she almost single-handedly restored Bob Noyce to his proper preeminence in the story of Silicon Valley and the high-tech revolution.

There is a third author whose work is inescapable when writing about the history of Intel: Andrew Grove. Andy is one of those rare famous individuals, like Churchill, who adopt the dictum that the only way to establish your place in history is to write that history yourself. It goes without saying that his autobiography, *Swimming Across*, provides vital source material both for Tedlow's book and this, but so do his other books. His classic *Only the Paranoid Survive* remains one of the best narratives about the nearly catastrophic Pentium bug scandal. What makes these books remarkable—beyond the fact that they were written by a man who was running a giant global corporation at the same time—is their honesty. Andy is a man with strong opinions about other people, many of them wrong, but he is brutally honest about himself and his failings—more so than any other cultural titan I've ever met. It is his greatest attribute.

I was in the unique position, as a twenty-four-year-old cub reporter at the *San Jose Mercury-News*, of being the first mainstream journalist to cover Intel Corporation. Even then, the company was almost a decade old. In the years before that, while working at Hewlett-Packard, I had tracked the company's story on the pages of Don Hoefler's scandalous newsletter. Regis McKenna, the legendary Intel and Apple marketing consultant, was also my childhood neighbor, and regular conversations with him over the last forty years have given me a rare glimpse into Intel available to few outsiders.

Because I began my career as a journalist in Silicon Valley so early and have continued it for so long, I am now in the surprising position of being perhaps the last journalist to have known well not just Gordon Moore and Andy Grove, but also Bob Noyce. Probably 90 percent of Intel's current employees cannot say that. I spent considerable time interviewing Noyce for my first book, for the PBS miniseries *Silicon Valley*, and on my own public-television interview program. Indeed, I probably interviewed Bob more than any other reporter; his last major interview was with me, and I wrote his obituary. Because of that unique access, I think I have a better understanding of what Noyce was like in person—including that incredible charisma, so unlike Steve Jobs (with whom I also spent considerable time), to whom he is often compared. You can't fully appreciate the presence of Bob Noyce and the depth of the relationship between him and Moore if you didn't see the tears in Gordon's eyes when he described his lost business partner.

My relationship with Andy Grove has been much more complicated. I interviewed him numerous times as a reporter—until he decided I wasn't sufficiently appreciative of Intel. At that point, he refused to speak to me for almost a decade. It was Bill Davidow, with whom I was writing a book (*The Virtual Corporation*), who started the rapprochement. Since then, we have been friendly—to the point that Andy was willing to meet with me, despite his ill health, in what became the epilogue for this book. As always, for the first hour of that meeting, he told me, chapter and verse, what he hadn't liked about my reporting over the years. But I knew that was coming, and I'll always cherish that meeting.

Gordon Moore is a far more difficult figure to pin down. My first interview with him was in the late 1970s, and there have been numerous casual and formal conversations and encounters since. In a way, Gordon created Silicon Valley, and he is its most extraordinary creation. But unlike the other two members of the Trinity, there may never be a major biography of Gordon Moore. Before I began this book, I approached him about writing one—and I know I'm not alone. Characteristically, Gordon replied that he didn't want such a book written, that his legacy would be a monograph on Moore's Law that he was preparing to author with his son for the Chemical Heritage Foundation. It was a classic Moore moment. Happily, between my public-television series *Betting It All* and numerous conversations and interviews over the years, I was able to piece together a biography of Gordon comparable to that of his two counterparts. Perhaps it will convince him to let me write the book he deserves.

Over the last thirty-five years, I have written several hundred articles, long and short, about Intel and its Trinity, visited the company or attended its events scores of times, and known hundreds of Intel employees—many of them my neighbors and friends. Many of these experiences (for example, the early Intel employee who had a heart attack at a Sunday morning meeting was the husband of my secretary at Hewlett-Packard) have made their way into my books—*The Big Score*, *The Microprocessor: A Biography*, and *Betting It All*—that in turn have served as source material for the likes of Tedlow and Berlin . . . and now for *The Intel Trinity*.

Notes

Introduction: Artifacts

1. Conversation with the author.

Chapter 1: The Traitorous Eight

1. Jillian Goodman, J. J. McCorvey, Margaret Rhodes, and Linda Tischler, "From Facebook to Pixar: 10 Conversations That Changed Our World," *Fast Company*, Jan. 15, 2013.

Chapter 2: The Greatest Company That Never Was

1. Michael S. Malone, *The Big Score: The Billion Dollar Story of Silicon Valley* (New York: Doubleday, 1985), 89.
2. Ibid.
3. Goodman et al., "From Facebook to Pixar."
4. "The Founding Documents," special insert, *Forbes ASAP*, May 29, 2000, after p. 144.
5. Ibid.
6. Ibid.
7. Malone, *Big Score*, 92.
8. Ibid., 91.
9. Ibid., 95–96.
10. Ibid., 97.
11. "Silicon Valley," *The American Experience*. PBS, Feb. 19, 2013.
12. Malone, *Big Score*, 150.
13. Leslie Berlin, *The Man Behind the Microchip: Robert Noyce and the Invention of Silicon Valley* (New York: Oxford University Press, 2006), 139.
14. Ibid.
15. Tom Wolfe, "The Tinkerings of Robert Noyce," *Esquire*, Dec. 1983, pp. 346–74, www.stanford.edu/class/e140/e140a/content/noyce.html (accessed Oct. 25, 2013).
16. Malone, *Big Score*, 105.

Chapter 3: Digital Diaspora

1. Malone, *Big Score.*
2. Ibid., 108.
3. Ibid., 106.
4. *The Machine That Changed the World*, documentary miniseries, WGBH/BBC 1992.
5. "Silicon Valley," PBS.
6. Malone, *Big Score*, 109.
7. Ibid., 110.
8. Berlin, *Man Behind the Microchip*, 152.
9. Malone, *Big Score*, 85.

Chapter 4: The Ambivalent Recruit

1. "Resignations Shake Up Fairchild," *San Jose Mercury-News*, July 4, 1968.
2. "Interview with Don Valentine," Apr. 21, 2004, Silicon Genesis: An Oral History of Semiconductor Technology, Stanford University, http://silicongenesis.stanford.edu/transcripts/valentine.htm.
3. "Industry Leaders Join in Kennedy Tributes," *Electronic News*, June 10, 1968.
4. Richard S. Tedlow, *Andy Grove: The Life and Times of an American* (New York: Portfolio, 2006), 111.
5. Berlin, *Man Behind the Microchip*, 158.
6. Andy Grove interview by Arnold Thackray and David C. Brock, July 14, 2004, in Tedlow, *Andy Grove*, 95.
7. Peter Botticelli, David Collis, and Gary Pisano, "Intel Corporation: 1986–1997," Harvard Business School Publishing Case No. 9-797-137, rev. Oct. 21, 1998 (Boston: HBS Publishing), 2.
8. Tedlow, *Andy Grove*, 98.
9. Ibid.

Chapter 5: Intelligent Electronics

1. "Making MOS Work," *Defining Intel: 25 Years/25 Events* (Santa Clara, CA: Intel, 1993), www.intel.com/Assets/PDF/General/25yrs.pdf (accessed Nov. 9, 2013).
2. Ibid.
3. Ibid.
4. Ibid.
5. Berlin, *Man Behind the Microchip*, 166.
6. Gupta Udayan, *Done Deals: Venture Capitalists Tell Their Stories* (Boston: Harvard Business School Press, 2000), 144.
7. Leslie Berlin interview with Art Rock, *Man Behind the Microchip*.
8. Ibid., 182.
9. Ibid., 183.

Chapter 7: The Demon of Grinnell

1. As it happened, the author's father and father-in-law were part of this brief national sensation. His father-in-law, who flew in Detroit, never forgot the experience. His father, whose family, in Portland, Oregon, was too poor to pay for a ticket, was heartbroken. He became a pilot—and just before his death in 1988 was given a ride on one of the last Tri-Motors, in Morgan Hill, California.
2. Malone, *Big Score*, 75.
3. Berlin, *Man Behind the Microchip*, 16.
4. Ibid., 17.
5. Malone, *Big Score*, 75.

Chapter 8 : The Pig Thief

1. Berlin, *Man Behind the Microchip*, 22.
2. Ibid., 22.
3. Malone, *Big Score*, 77.
4. Ibid.
5. Ibid.
6. Ibid., 78.
7. Berlin, *Man Behind the Microchip*, 31.
8. Ibid., 35.
9. Ibid., 37.

Chapter 9: A Man on the Move

1. Berlin, *Man Behind the Microchip*.
2. Malone, *Big Score*, 79.
3. Berlin, *Man Behind the Microchip*, 50.

Chapter 10: Gordon Moore: Dr. Precision

1. Berlin, *Man Behind the Microchip*.
2. Ibid., 141.
3. Ibid., 142.
4. Intel Museum.
5. Ibid.
6. Ibid.
7. Michael S. Malone, *Betting It All: The Entrepreneurs of Technology* (New York: Wiley, 2002), 152.

Chapter 11: A Singular Start-Up

1. "Journey Through Decades of Innovation," Intel Museum, www.intel.com /content/www/us/en/company-overview/intel-museum.html.
2. Malone, *Big Score*, 147.

Chapter 12: The Wild West

1. Don C. Hoefler, *MicroElectronics News*, July 3, 1976, http://smithsonianchips.si.edu /schreiner/1976/images/h76711.jpg (accessed Nov. 9, 2013).
2. The author was offered the editorship of *MicroElectronics News* by Mr. Hoefler in 1980. He refused, but not without regret.
3. Tom Junod, "Tom Wolfe's Last (and Best) Magazine Story," *Esquire* blogs, Feb. 21, 2013, www.esquire.com/blogs/culture/tom-wolfe-robert-noyce-15127164 (accessed Nov. 9, 2013).
4. Malone, *Big Score*, 150.
5. Ibid.
6. Malone, *Betting It All*, 151.

Chapter 13: Bittersweet Memories

1. Elkan Blout, ed., *The Power of Boldness: Ten Master Builders of American Industry Tell Their Success Stories* (Washington, DC: Joseph Henry Press, 1996), 84.
2. Berlin, *Man Behind the Microchip*, 200–201.

Chapter 14: Miracle in Miniature

1. Michael S. Malone, *The Microprocessor: A Biography* (New York: Springer, 1995), 3.
2. Berlin, *Man Behind the Microchip*, 183.
3. Ibid., 184.
4. Ibid.
5. "Least Mean Squares Filter," Wikipedia, http://en.wikipedia.org/wiki/Least _mean_squares_filter.
6. Malone, *Microprocessor*, 7.
7. Berlin, *Man Behind the Microchip*, 184.
8. Ibid., 185.
9. Malone, *Microprocessor*, 7–8.
10. Berlin, *Man Behind the Microchip*, 185.
11. Ibid., 186.
12. Ibid., 187.
13. Ibid.
14. Malone, *Microprocessor*, 8.
15. Berlin, *Man Behind the Microchip*, 188.

Chapter 15: The Inventor

1. Malone, *Microprocessor*, 10.
2. Ibid.
3. Ibid., 11.

Chapter 17: Dealing Down

1. Tedlow, *Andy Grove*, 138.
2. Berlin, *Man Behind the Microchip*, 189.
3. Ibid.
4. Ibid., 190.
5. Author interview, 1985.

Chapter 18: The Philosopher's Chip

1. Malone, *Microprocessor*, 11.
2. Ibid., 14.
3. Ibid., 15.
4. Berlin, *Man Behind the Microchip*, 198.
5. Ibid.
6. "The Chip Insider's Obituary for Bob Graham," The Chip History Center, www.chiphistory.org/legends/bobgraham/bob_obituary.htm (accessed Nov. 9, 2013).

Chapter 19: Product of the Century

1. Malone, *Microprocessor*, 17.
2. "Intel 8008 (i8008) Microprocessor Family," *CPU World*, www.cpu-world.com/CPUs/8008.
3. Malone, *Microprocessor*, 17.
4. Ibid.
5. Ibid., 18.
6. "Intel 8080," Wikipedia, http://en.wikipedia.org/wiki/Intel_8080 (accessed Nov. 9, 2013).
7. Berlin, *Man Behind the Microchip*, 200.
8. Ibid.
9. Malone, *Microprocessor*, 16.
10. "About," Regis McKenna, www.regis.com/about.
11. As a reporter for the *San Jose Mercury-News*, the author had this experience.
12. Malone, *Microprocessor*, 130.
13. Ibid.
14. Berlin, *Man Behind the Microchip*, 203.
15. *Defining Intel: 25 Years/25 Events.*
16. Ibid.
17. Malone, *Microprocessor*, 18.

Chapter 20: Crush

1. This chapter is derived from Malone, *The Microprocessor: A Biography*; and William H. Davidow's *Marketing High Technology: An Insider's View* (New York: Free Press, 1984).
2. Davidow, *Marketing High Technology*, 3–4.
3. Ibid.
4. Ibid., 4.
5. Ibid., 6.
6. Ibid., italics added.
7. Scott Anthony, *The Little Black Book of Innovation* (Boston: Harvard Business Review Press, 2012), 61.
8. Davidow, *Marketing High Technology*, 6.
9. Malone, *Microprocessor*, 158.
10. Davidow, *Marketing High Technology*, 8.
11. Malone, *Microprocessor*, 160.
12. Davidow, *Marketing High Technology*, 7.
13. *Defining Intel: 25 Years/25 Events*, 14.
14. Regis McKenna, *Relationship Marketing: Successful Strategies for the Age of the Customer* (Reading, MA: Addison-Wesley, 1991), 4.

Chapter 21: Silicon Valley Aristocracy

1. First heard by the author c. 1988.

Chapter 22: Public Affairs

1. Jerry Sanders quote part of interview by author for *Betting It All*, public television series, 2001.
2. Berlin, *Man behind the Microchip*, 205.
3. Ibid., 201–2.
4. Ibid., 216.

Chapter 23: Consumer Fantasies

1. Tedlow, *Andy Grove*, 145.
2. Ibid., 167.
3. Ibid., 146.
4. Ibid., 167.
5. Author interview for public television series *Silicon Valley Report*, c. 1986.

Chapter 24: A Thousand Fathers

1. Malone, *Microprocessor*, 19.
2. Ibid.
3. Ibid., 131.

4. The author was in attendance.
5. Malone, *Microprocessor*, 132.
6. Author interview with Dr. Federico Faggin, Feb. 3, 2014.
7. Malone, *Microprocessor*, 152.
8. Larry Waller, "Motorola Seeks to End Skid," *Electronics*, Nov. 13, 1975, 96–98.
9. George Rostky, "The 30th Anniversary of the Integrated Circuit," *Electronic Engineering Times*, Sept. 1988.
10. Owen W. Linzmayer, *Apple Confidential 2.0: The Definitive History of the World's Most Colorful Company* (San Francisco: No Search Press, 2004), 4.
11. Michael S. Malone, *Infinite Loop: How the World's Most Insanely Great Computer Company Went Insane* (New York: Doubleday, 1999), 61.
12. Ibid., 49.
13. Berlin, *Man Behind the Microchip*, 223.
14. Ibid., 224.
15. Ibid., 225.
16. Mimi Real and Glynnis Thompson Kaye, *A Revolution in Progress: A History of Intel to Date* (Santa Clara, CA: Intel, 1984), 14.

Chapter 25: The Knights of Moore's Law

1. Quoted in Alexis C. Madrigal, "Paul Otellini's Intel: Can the Company That Built the Future Survive It?" *Atlantic Monthly*, May 16, 2013.
2. Michael S. Malone, "From Moore's Law to Barrett's Rules," Weekend Interview, *Wall Street Journal*, May 16, 2009, http://online.wsj.com/article/SB124242845507325429.html (accessed Nov. 9, 2013).
3. Ibid.
4. Michael S. Malone, "Intel Reboots for the 21st Century," Weekend Interview, *Wall Street Journal*, Sept. 27, 2008.
5. Ibid.
6. The author was in attendance.

Chapter 26: (Over) Ambitions

1. Tedlow, *Andy Grove*, 163.
2. Tom Foremski, "Interview with Intel Employee #22—Surviving 30 Years," *Silicon Valley Watcher*, Dec. 3, 2012, www.siliconvalleywatcher.com/mt/archives/2012/12/interview_with_9.php (accessed Nov. 9, 2013).
3. Tedlow, *Andy Grove*, 165.
4. Author conversation with Gordon Moore, 1999.
5. Malone, *Big Score*, 319–20.
6. Ibid.
7. *Defining Intel: 25 Years/25 Events.*
8. Ibid.
9. Ibid.
10. Ibid.

Chapter 27: Beatification

1. *Defining Intel: 25 Years/25 Events.*
2. Steve Hamm, Ira Sager, and Peter Burrows, "Ben Rosen: The Lion in Winter," *Bloomberg Businessweek*, July 26, 1999, www.businessweek.com/1999/99_30/b3639001.htm (accessed Nov. 9, 2013).
3. Berlin, *Man Behind the Microchip*, 248.
4. Herb Caen column, *San Francisco Chronicle*, Feb. 5, 1980, 1B.
5. Author conversation with Tom Wolfe, 2000.
6. Malone, *Microprocessor*, 186–87.
7. David Manners, "When the US IC Industry Was Rocked on Its Heels," ElectronicsWeekly.com, Aug. 31, 2012.
8. Malone, *Big Score*, 248.
9. Ibid., 249.
10. "'Disbelief' Blamed in Computer Sting," Associated Press, June 24, 1982, http://news.google.com/newspapers?nid=1314&dat=19820624&id=NvlLAAAAIBA J&sjid=hu4DAAAAIBAJ&pg=4786,4368515 (accessed Nov. 9, 2013).
11. Ibid.

Chapter 29: Mother and Child

1. *Andy Grove*, 1–2.
2. Ibid, 19.
3. Andrew S Grove, *Swimming Across: A Memoir* (New York: Hachette Books, 2001), 40.
4. Ibid.
5. Ibid.
6. Ibid.
7. Tedlow, *Andy Grove*, 27.

Chapter 30: Father and Child Reunion

1. Grove, *Swimming Across*.
2. Ibid.
3. Ibid.
4. Tedlow, *Andy Grove*, 38.
5. Grove, *Swimming Across*.

Chapter 31: Andy in Exile

1. Grove, *Swimming Across*.
2. Ibid.
3. Ibid.
4. Ibid., 142–43.
5. Ibid., 290.
6. Ibid., 156.
7. Ibid., 45.

8. Ibid., 170–71.
9. Ibid., 214.
10. Tedlow, *Andy Grove*, 55.
11. Ibid., 57.
12. Ibid., 59.

Chapter 32: Freedom Fighter

1. Grove, *Swimming Across*, 249.
2. Ibid., 262.

Chapter 33: A New Life, a New Name

1. Kate Bonamici, "Grove of Academe," *Fortune*, Dec. 12, 2005, 135.
2. Tedlow, *Andy Grove*, 73.
3. Ibid., 77.
4. Ibid., 76.
5. Ibid., 81.
6. Ibid., 86.
7. Conversation with Regis McKenna, Sept. 2012.
8. Tedlow, *Andy Grove*, 96.
9. Leslie Berlin, "Entrepreneurship and the Rise of Silicon Valley: The Career of Robert Noyce, 1956–1990," PhD dissertation, Stanford University, 2001, 150.

Chapter 35: Riding a Rocket

1. Malone, *Microprocessor*, 171–72.
2. Ibid.
3. The author was in attendance.
4. The author was that reporter.
5. Malone, *Microprocessor*, 172.

Chapter 36: For the Cause

1. Malone, *Microprocessor*, 172–73.

Chapter 37: East of Eden

1. Michael J. Lennon, *Drafting Technology Patent License Agreements* (New York: Aspen Publishers, 2008), appendix 4B-28.

Chapter 38: Turn at the Tiller

1. *Defining Intel: 25 Years/25 Events.*
2. Ibid.
3. Malone, "From Moore's Law to Barrett's Rules."
4. Ibid.
5. "Turning On to Quality," *Defining Intel: 25 Years/25 Events.*
6. Ibid.

Chapter 40: Crash

1. Author conversation with Jim Morgan, 2007.
2. Semiconductor Industry Association, board of directors minutes, June 16, 1977.
3. Berlin, *Man Behind the Microchip.*
4. Ibid., 264.
5. Malone, *Betting It All*, 151.
6. "Atari Democrat," Wikipedia, http://en.wikipedia.org/wiki/Atari_Democrat (accessed Nov. 9, 2013).
7. Berlin, *Man Behind the Microchip*, 268.
8. Clyde Prestowitz, *Trading Places: How We Allowed Japan to Take the Lead* (New York: Basic Books, 1988), 149.
9. *The State of Strategy* (Boston: Harvard Business School Press, 1991), 57.
10. Berlin, *Man Behind the Microchip*, 269.

Chapter 41: Memory Loss

1. Andrew S. Grove, *Only the Paranoid Survive: How to Identify and Exploit the Crisis Points That Challenge Every Business* (New York: Doubleday, 1996), 88.
2. Malone, *Betting It All*, 152.
3. Grove, *Only the Paranoid Survive*, 89.
4. Ibid., 91.
5. Ibid., 89.
6. Malone, *Betting It All*, 152.
7. "Downsizing Intel," *Defining Intel: 25 Years/25 Events.*

Chapter 42: Andy Agonistes

1. "Going It Alone with the Intel 386 Chip," *Defining Intel: 25 Years/25 Events.*
2. Ibid.

Chapter 43: Mentoring a Legend

1. Berlin, *Man Behind the Microchip*, 252.

Chapter 44: The Man of the Hour

1. Berlin, *Man Behind the Microchip*, 285.
2. Rodgers quote from public television series *Malone*, hosted by the author, 1987.

Chapter 46: Empyrean

1. Andrew Pollack, "Sematech's Weary Hunt for a Chief," *New York Times*, Apr. 1, 1988.
2. Katie Hafner, "Does Industrial Policy Work? Lessons from Sematech," *New York Times*, Nov. 7, 1993.
3. Ibid.
4. Robert D. Hof, "Lessons from Sematech," *Technology Review*, July 25, 2011.
5. Ibid.
6. Noyce interview with author, *Malone* show, 1988.

Chapter 47: The Swimmer

1. Michael S. Malone, "Robert Noyce," *Upside* magazine, July 1990.

Chapter 49: Inside Intel Inside

1. "The Red X Ad Campaign," *Defining Intel: 25 Years/25 Events*.
2. Ibid.
3. Ibid.
4. Ibid.
5. Tedlow, *Andy Grove*, 255.
6. Ibid., 256.

Chapter 51: Pentium

1. Harsimran Julka, "Speed Was God When We Created Pentium: Vinod Dham," *Economic Times*, Nov. 16, 2010.
2. Malone, *Microprocessor*, 167–68.

Chapter 52: The Bug of All Bugs

1. Malone, *Microprocessor*, 236–43.
2. Dean Takahashi, "The Pentium Bypass," *San Jose Mercury-News*, Jan. 16, 1995, D1.
3. Michael Meyer, "A 'Lesson' for Intel," *Newsweek*, Dec. 12, 1994, 58.
4. Grove, *Only the Paranoid Survive*, 11.

5. "Heavy Duty Users Reassess Work," Associated Press, *San Jose Mercury-News*, Dec. 24, 1994, D8.
6. Adrian Mello, "Divide and Flounder," *MacWorld*, Mar. 1995, 20.
7. Grove, *Only the Paranoid Survive*, 14.
8. Ibid., 15.
9. Ibid.

Chapter 53: The Endless Lawsuit

1. Michael Singer, "Intel and AMD: A Long History in Court," CNET News, June 28, 2005, http://news.cnet.com/Intel-and-AMD-A-long-history-in-court/2100-1014_3-5767146.html?tag=nw.20 (accessed Nov. 9, 2013).
2. Andrew Pollack, "Rival Files Antitrust Suit against Intel," *New York Times*, Aug. 30, 1991, www.nytimes.com/1991/08/30/business/company-news-rival-files-antitrust-suit-against-intel.html (accessed Nov. 9, 2013).
3. Malone, *Microprocessor*, 245.
4. Ibid.
5. Ibid., 244.
6. Ibid.

Chapter 54: Mea Culpa

1. Russ Mitchell, "Microsoft Picked on Someone Its Own Size: Leaks Detail a Past Spat with Intel," *U.S. News & World Report*, Aug. 30, 1998.
2. Ibid.
3. Michael Kanellos, "Intel Antitrust Trial Date Set," *CNET News*, July 10, 1998, http://news.cnet.com/Intel-antitrust-trial-date-set/2100-1023_3-213195.html (accessed Nov. 9, 2013).
4. "FTC Antitrust Action Against Intel," *Tech Law Journal*, www.techlawjournal.com/agencies/ftc-intc/Default.htm (accessed Nov. 9, 2013).

Chapter 55: Bunny Hop

1. Tedlow, *Andrew Grove*, 389.
2. Walter Isaacson, "Andy Grove: Man of the Year," *Time*, Dec. 29, 1997.

Chapter 56: Running the Asylum

1. Gordon Moore interview with the author, *Forbes ASAP*, Dec. 1999.

Chapter 57: The Full Weight of the Law

1. Tedlow, *Andy Grove*, 406.
2. Ibid., 403.

3. Malone, "From Moore's Law to Barrett's Rules."
4. Ibid.
5. Ibid.
6. Ibid.
7. Paul Otellini, "Mission Matters," *Santa Clara* magazine, Spring 2012, www.scu.edu/scm/spring2012/otellini-talk.cfm (accessed Nov. 9, 2013).
8. Alexis Madrigal, "Paul Otellini's Intel: Can the Company That Built the Future Survive It?" *Atlantic Monthly,* May 16, 2013.
9. Otellini, "Mission Matters."
10. Ibid.
11. Madrigal, "Paul Otellini's Intel."
12. Malone, "From Moore's Law to Barrett's Rules."
13. Malone, "Intel Reboots for the 21st Century."

Epilogue: Roaring at the Night

1. This section is based upon an interview with Andrew Grove by the author, Feb. 2013.
2. "Intel Co-Founder Gordon E. Moore Takes Over No. 1 Spot on *Business-Week*'s Annual Ranking of 'America's Top Philanthropists,'" PR Newswire, www.prnewswire.com/news-releases/intel-co-founder-gordon-e-moore-takes-over-no1-spot-on-businessweeks-annual-ranking-of-americas-top-philanthropists-55688447.html.

Index